Climate Clubs for a Sustainable Future

Energy and Environmental Law and Policy Series

VOLUME 41

Editor

Introduction

Environmental protection and energy efficiency/security are important societal challenges. In order to tackle them, policy and legal frameworks are developed at national, regional and global level. Through study and best practices development, the challenges will prove to be solvable.

Contents/Subjects

Environment/Nature/Energy/Climate.

Objective

The aim of this series is to publish works of excellent quality that focus on the study of energy and environmental policy. Through this series the editors:

- contribute to the improvement of the quality of energy/environmental law and policy in general and environmental quality and energy efficiency in particular; increase the access to environmental and energy information for academics, non-governmental organizations, government institutions, and business; and
- facilitate cooperation between academic and non-academic communities in the field of energy and environmental law and policy throughout the world.

Readership

Academics and practitioners in environmental and energy matters.

The titles published in this series are listed at the end of this volume.

Climate Clubs for a Sustainable Future

The Role of International Trade and Investment Law

Rafael Leal-Arcas

Wolters Kluwer

Published by:
Kluwer Law International B.V.
PO Box 316
2400 AH Alphen aan den Rijn
The Netherlands
E-mail: international-sales@wolterskluwer.com
Website: lrus.wolterskluwer.com

Sold and distributed by:
Wolters Kluwer Legal & Regulatory U.S.
7201 McKinney Circle
Frederick, MD 21704
United States of America
E-mail: customer.service@wolterskluwer.com

ISBN 978-94-035-3715-3

e-Book: ISBN 978-94-035-3720-7
web-PDF: ISBN 978-94-035-3721-4

I dedicate this book to Nelson, thanks to whom I made it to the top.

About the Author

Rafael Leal-Arcas is a Professor of European and International Economic Law, a Jean Monnet Chair holder (awarded by the European Commission), Program Director of the LLM in International Economic Law, and former Director of Research at the Centre for Commercial Law Studies of Queen Mary University of London. He is also a Visiting Professor at New York University Abu Dhabi in the UAE and the Inaugural Lee Kong Chian International Visiting Professor of Law at the Yong Pung How School of Law of Singapore Management University, Singapore. His research is funded by the EU Commission's Horizon 2020 program, most notably a grant of EUR 14 million as part of a consortium of 21 institutions to work on renewable energy and smart grids.

Rafael is the author of the books: *The Great Energy Transition in the European Union*, Volumes 1 and 2 (Eliva Press, 2020), *Solutions for Sustainability: How the International Trade, Energy and Climate Change Regimes Can Help* (Springer, 2019), *EU Trade Law* (Edward Elgar, 2019), *Energy Security, Trade and the EU: Regional and International Perspectives* (Edward Elgar, 2016), *The European Energy Union: The Quest for Secure, Affordable and Sustainable Energy* (Claeys & Casteels Publishing, 2016), *International Energy Governance: Selected Legal Issues* (Edward Elgar, 2014), *Climate Change and International Trade* (Edward Elgar, 2013), *International Trade and Investment Law: Multilateral, Regional and Bilateral Governance* (Edward Elgar, 2010) and *Theory and Practice of EC External Trade Law and Policy* (Cameron May, 2008) and the editor or co-editor of *The Future of International Economic Law and the Rule of Law* (Eliva Press, 2020), *EU Energy Law and Policy: The External Dimension* (Eliva Press, 2020), *International Trade, Investment and the Rule of Law* (Eliva Press, 2020), the *Commentary on the Energy Charter Treaty* (Edward Elgar, 2018), *University Initiatives in Climate Change Mitigation and Adaptation* (Springer, 2018), and *Research Handbook on EU Energy Law and Policy* (Edward Elgar, 2017).

He has served in the Sustainable Energy for All Secretariat (Washington, DC), the Energy Community Secretariat, the Energy Charter Secretariat, the World Trade Organization, and several EU institutions (EU Commission, EU Parliament, EU Council, EU Court of Justice). He is the author of more than 190 scholarly publications on international trade and WTO law, climate change law, energy governance, EU law, international investment law and the interaction among them in American and European scientific journals, such as the *Yale Journal of International Affairs*, *Georgetown Journal of International Law*, *Columbia Journal of European Law*, *Georgetown Environmental Law Review*, *University of Pennsylvania Journal of International Law*, *The George Washington International Law Review*, *Fordham International Law Journal*, *Chicago Journal of International Law*, *Legal Issues of Economic Integration*, *European Foreign Affairs Review*, and *Journal of World Energy Law and Business* (Oxford University Press).

Rafael has been a Visiting Researcher at Yale Law School; Global Research Fellow at New York University School of Law (Hauser Global Law School Program); Visiting Fellow at the World Trade Institute; Visiting Scholar at Georgetown University Law Center; Scholar at Masdar Institute at Khalifa University (Abu Dhabi); Tillar House Resident Fellow at the American Society of International Law; Visiting Researcher at Harvard Law School and Fellow at the Real Colegio Complutense (Harvard University); Emile Noël Doctoral Fellow at New York University School of Law; Fellow at the Australian National University; and Visiting Scholar at the University of Wisconsin-Madison Law School.

He teaches *International Economic Law*; *WTO Law; Climate Change Law and Policy*; *Trade, Climate Change and Energy Law*; *International Trade and Investment Law of the EU*; and *EU External Relations Law*.

He received his graduate legal education at Stanford Law School, Columbia Law School, the London School of Economics and Political Science, and the European University Institute (Florence). He complemented his education further with executive programs at Harvard Kennedy School of Government, Yale University, and the University of Oxford.

Table of Contents

Acknowledgments

While writing this book, I had the fortune to serve as a Visiting Professor/Scholar in the following institutions:

- Yale Law School, New Haven, USA
- New York University Abu Dhabi, United Arab Emirates
- Sorbonne University Abu Dhabi, United Arab Emirates
- Korea National Diplomatic Academy, Ministry of Foreign Affairs, Seoul, South Korea.

I am very grateful to my colleagues in those institutions for hosting me and for helping me shape many of the ideas in this book.

I am equally grateful to the following people for their input at various stages of the writing of this book: Ellis Malle, Marianna Margaritidou, Marion Kerestedjian, Gulce Budak, M Akonda, Andrew Filis, Samuel Balzano, Jakkrit Deethae, Tanvir Singh, Kristina Skybova, Marek Anderle, Filipa da Silva Santos, Luuk Uilenbroek, Hannah Schragmann, Danai Papadea, Rosie Richardson, Saana Mikaela Saveljeff, and Iuliia Dragunova.

April 2021

Introduction

Ideologies such as nationalism and socialism have tried to explain social relations in the world. Following changes in the global economy, recent books have addressed new social trends that result in the world not being what societal institutions are supposed to make of the world, namely a reasonably fair place, a more generous and inclusive society.[1] This book tries to use climate clubs[2] in the context of the international trading system to reach a sustainable and prosperous future.

In recent years, we have observed warmer temperatures, increasing numbers of floods, forest degradation,[3] forest fires, and droughts, among other natural catastrophes. Where drought has become a perennial problem, adaptation has become the norm.

As of 2020, the consumption of coal in the US and Europe had fallen by 34% since 2009, mainly due to government policies and cheaper energy alternatives. However, coal still represents around 27% of the raw energy necessary to power our daily activities,[4] and, in some developed countries, coal is the main cause of air pollution.[5] For some countries that are still developing their economies, dismissing coal may mean dismissing their right of developing countries to choose their own energy source for

1. *See* for instance M. Carney, *Value(s): Building a Better World for All*. Signal Books, 2021; M. Shafik, *What We Owe Each Other: A New Social Contract for a Better Society*. Princeton University Press, 2021; J. Norberg, *Open: The Story of Human Progress*. Atlantic Books, 2020.
2. A climate club is understood as a coalition of countries that commit to strong steps to reduce greenhouse gas emissions and may have mechanisms to penalize countries that do not participate. *See* W. Nordhaus, "The Climate Club: How to Fix a Failing Global Effort," *Foreign Affairs*, May/June 2020, p. 10, available at https://pcfraz.org/resources/Documents/The%20Climate%20Club%20_%20Foreign%20Affairs.pdf.
3. When Joe Biden of the US complained to President Bolsonaro of Brazil in September 2020 that he should stop cutting trees in the Amazon, the Brazilian president twitted: "Our sovereignty is non-negotiable." *See* O. Stunkel, "How Biden Can Change Bolsonaro's Mind on the Amazon," *Americas Quarterly*, January 11, 2021, available at https://www.americasquarterly.org/article/how-biden-can-change-bolsonaros-mind-on-the-amazon/. The European Union has similar concerns with Mercosur over rampant deforestation in the Amazon, which will determine whether the Europeans will ratify a recently concluded free trade agreement with Mercosur.
4. The Economist, "Make coal history," December 5, 2020, p. 13.
5. Of the 100 European cities with the highest level of air pollution, 29 are in Poland. *See* The Economist, "Graphic Detail: Coal-Fired Heating Pollutes Poland's Skies," January 30, 2021, p. 73.

economic growth.[6] By producing GDP, countries extract natural resources and dump waste back into nature. As solar farms and onshore wind energy become increasingly cheaper (in fact, they are the cheapest source of new electricity for around 66% of the world's population), coal will face fierce competition.[7] In the US, the Biden administration intends to boost the offshore wind-energy industry by deploying 30 gigawatts of offshore wind capacity by 2030.[8]

This book aims to explain that, by building climate clubs and making use of the international trading system, we can reach a better future for all.

I CONCEPTUALIZING CLIMATE CLUBS IN THE CONTEXT OF INTERNATIONAL TRADE

Regional trade agreements (RTAs) have served as laboratories of change for a long time. In recent times, we see they are used to move forward the sustainable-development agenda in the context of the World Trade Organization. That said, given the transnational nature of sustainable development and environmental protection, it is necessary to go beyond RTAs and free trade agreements (FTAs) for such issues, although regional trade is arguably a more effective and faster way to tackle transnational issues. There are two broad types of FTAs/RTAs with environmental-protection chapters: (a) those like the Comprehensive Economic and Trade Agreement between Canada and the European Union (EU), which deal with sustainable development, climate change, waste, border issues; and (b) US FTAs model, which are more modest when it comes to dealing with trade and the environment. What follows are various ideas of how one could conceptualize climate clubs in the framework of the trading system.

Why a regional trade agreement would be allowed under the GATT

Regional trade agreements (RTAs) are defined as reciprocal trade agreements between two or more partners, including FTAs and customs unions. Article XXIV of the General Agreement on Tariffs and Trade (GATT) allows regional trading arrangements to be set up as a special exception to the principle of nondiscrimination (GATT Articles I and III), subject to strict criteria, including that the regional integration should complement the multilateral system and not threaten it. RTAs under Article XXIV should not "raise barriers to the trade of other contracting parties." Therefore, the RTA in question should not affect trade with a third party nor divert trade from it.

Why a regional trade agreement would be necessary

Climate change discussions and, crucially, the Kyoto Protocol/the Common but Differentiated Responsibilities principle have divided developed and developing nations by placing the focus on "responsibility." An RTA will shift the focus to other factors, which the Kyoto Protocol failed to consider. Below are the following four

6. The Economist, January 2, 2021, p. 12.
7. The Economist, "Make Coal History," December 5, 2020, p. 13.
8. The Economist, April 3, 2021, p. 8.

factors: enthusiasm, power, capability, and institutional strength. Let us go over each one of them.

- Enthusiasm: Since entering an RTA would be a voluntary step based on each party's "enthusiasm" toward mitigating climate change, this factor would play a central role in defining the climate club's constitution. The logic of carbon trade does not exist for countries that do not recognize the need to mitigate climate change, although fewer and fewer countries fall under this category, as more countries (and companies) are now committing to being carbon-neutral by 2050 or 2060. Hence, such RTA should be directly linked to the Paris Agreement on Climate Change.
- Power: An RTA is a diplomatically, politically, and economically sound incentive for non-Annex I countries to enter negotiations. However, it makes more climate sense for equally powerful nations to negotiate and enter into mutually binding agreements (despite the complexity of such negotiations) than for economically weaker countries to enter, potentially imbalanced, agreements with economic giants (i.e., today's high greenhouse gas (GHG) emitters). This is because weaker nations, which are nevertheless low-emitting countries, would not be able to impose high binding standards on powerful nations.
- Capability: The very definition of a universal organization is based on its propensity to accept any independent state of the international community as a member. All states trade goods and services, but can every state trade carbon. Does every state even produce carbon? The question here is why should the Vatican City be included in a carbon trade agreement when its "capability" level is near zero? However, despite being qualified as developing nations, India, China and other similar developing nations have a high ability to trade carbon.

  This factor may even lead to a gradual substitution of the terms "developed" and "developing" by taking away the focus from GDP and placing it on emission units to distinguish between "carbon-rich" and "carbon-poor" states. Countries that both currently and in the past hardly contributed to global emissions (e.g., least-developed countries and small island developing states) will justifiably not face higher standards for emission trade/emissions reduction. In this sense, environmental "justice" is still conserved.

  Moreover, there is only a handful of currently implemented (and rather diverse) carbon trading systems (either national or subnational): out of the large GHG emitters in the world, Brazil, Russia, and India did not have any Emissions Trading System (ETS) in place at all as of mid-2021. China is just starting up and has legal loopholes, of which the worst polluters will take advantage. The EU's was launched in 2005, but barely worked functioned for a long time because the price of carbon was close to zero. Why not focus on harmonizing what already exists and establish/harmonize the ETSs for the major emitting countries? Why not connect existing ETSs to other regions via a carbon border tax? Current ETS linkages are not ambitious enough, such as

California and Québec or the EU and Switzerland. To be successful, implementation of the ETS is key.

- Institutional strength: Can every state have harmonized standards for carbon market operations (carbon market infrastructure, accounting, transparency, and environmental integrity)? High-institutional strength countries that also have high carbon emissions are the ones who will be able to create and enforce such standards to set the example for the rest of the world. On the other hand, low-institutional strength and/or low-emission countries will be the ones hindering the process.

That said, there are two main obstacles:

(1) Without a formal RTA, simply rejecting low-quality and unregulated carbon emission trade from some countries will be a breach of international trade law, as has been demonstrated in environmental-related World Trade Organization (WTO) cases (e.g., the *Tuna—Dolphin* case and the *Shrimp* case). These two cases have shown that there is a propensity for the WTO to favor trade over strict environmental protection. Put differently, legitimacy of high standards and regulations in terms of "process" (and not "product") seems to come from regionalism, but not unilateralism/multilateralism.

(2) It will be necessary to implement a trading system that applies to "substantially all the trade," in accordance with GATT Article XXIV. Carbon market clubs aiming at the liberalization of emission units only would not meet this criterion. To meet the criterion, club members would be required to liberalize trade more broadly between themselves. That raises the question of who would be the likely partners.

Finding partners for a future regional trade agreement

One group of unlikely partners is China, US, the EU, Russia, India, Brazil, and Japan. These are the highest global carbon emitters. These parties to the UN Framework Convention on Climate Change (UNFCCC), all together, are unlikely to enter into a mega free trade agreement.

An alternative option is one of double integration: One can think of two to three clubs/groups of likely partners plus a regional trade agreement to unite them, thereby having double integration. There are existing frameworks and free trade agreements under negotiation in which carbon trading can be inserted as a first step. Examples are the US-Mexico-Canada Agreement, the Comprehensive and Progressive Agreement for Trans-Pacific Partnership (CPTPP), the Regional Comprehensive Economic Partnership (RCEP),[9] the Comprehensive Economic and Trade Agreement between Canada

9. As of November 2020, the RCEP was the world's newest and biggest RTA, but not the deepest. It eliminates fewer tariffs than normal RTAs. When comparing the RCEP text to that of the CPTPP, its competitor, one notes that in many areas the text of RCEP is not only similar, but is, in fact, identical to that of CPTPP. At the same time, there are significant differences between the two Agreements, even in areas where they are comparable in scope and adopt similar

and the European Union, and the Transatlantic Trade and Investment Partnership, to name a few examples of mega regional trade agreements. All these arrangements could serve as great platforms for climate club based on green trade in goods and services.

In addition, a club approach should not be inflexible in terms of membership or overemphasize ideal scenarios. For instance, while a US-China Agreement is one of the most obvious and ambitious goals, an EU-China-India Agreement would cover a similar percentage of the global GHG emissions. Likewise, while a US-EU-India Agreement would be more geographically spread out, a China-India Agreement would cover slightly more GHG emissions than the former.

Why a regional trade agreement would foster future universal/multilateral discussions

The goal of the WTO and regional trade agreements is the same, namely the liberalization of trade. Their only difference is the level at which the removal of barriers to trade is imposed; while the WTO aims to remove trade barriers at a universal level, RTAs do so at the regional level, for which specific requirements must be met, in accordance with GATT Article XXIV.

Multilateralism and regionalism seem to be competing. In theory, regionalism should complement, not supplant, multilateralism.[10] It can be argued that RTAs play a role in promoting multilateral free trade if they manage to multilateralize regionalism. This argument is even recognized formally in the first sentence of Article XXIV(4) of GATT 1994. It is certainly easier and more manageable to negotiate among a small number of large players than it is among many small players. Therefore, the creation of clubs could be utilized as a first step toward universalism.

Technological cooperation

Much more global cooperation is necessary. Around 80% of all patent applications in clean energy technology are owned by Japan, the US, the EU (Germany and France), South Korea, and the UK.[11] However, many of the green technology patents have not been used due to either incapability of commercialization or patent suppression behavior. The latter phenomenon is particularly concerning even if it has been declared as legal by the US courts (*Continental Paper Bag Co. v Eastern Paper Bag Co.* 210 US 405 (1908)). Nonetheless, countries such as Brazil and South Africa have made their progression under the nationally determined contributions conditional on technological transfer and support.

Therefore, an important part of a climate club may be to foster this technological cooperation through a "limited" compulsory licensing among large GHG emitters. This idea may be more compelling for patent holders than making patents internationally available. In other words, green patents could become a climate club good to incentivize participation. This idea may be controversial as green patents should be ideally shared with least developed nations.

approaches. Equally, the level of economic development of the members of both mega-RTAs (RCEP and CPTPP) is vastly different.

10. *See* R. Leal-Arcas, "Proliferation of Regional Trade Agreements: Complementing or Supplanting Multilateralism?," 11:3 *Chicago Journal of International Law* 597-629 (2011).
11. https://www.irena.org/inspire/Intellectual-Property-Rights/Global-Patenting-Trends-in-RET.

However, if climate clubs' cooperation allows large emitters to reduce their emissions, least developed nations could still benefit, as opposed to maintaining the current deadlock on green patents. Moreover, there should be some limitations to countries leaving the club once they obtain a patent. This is where a corporate climate club may be preferable. On technological cooperation, such a club may establish a mega-joint venture scheme that will bind companies under contractual obligations, which do not apply to states.

A climate club of companies

Companies' involvement in climate change mitigation is fundamental, especially following important rising trends regarding corporation social responsibility and investor awareness/responsiveness toward climate change. Countries around the world are making laws that will make corporations accountable for environmental damage. Due to the increasing power of (major) companies, it would be difficult to have effective global emissions reduction without companies' support.

Moreover, there seems to be an apparent incompatibility between strict competition law and climate change law. It seems that competition law hinders effective cooperation between companies seeking to produce sustainable goods. A climate club agreement involving companies may seek to establish a clear and legitimate framework for state aid as well as vertical/horizontal cooperation among companies to promote emissions reduction. This has been, to some extent, already initiated under the European Green Deal.

Domestic policy preferences

Part of the justification/incentive to join a climate club constitutes a need of structural economic reform in face of crises (including COVID-19, but also financial and international crises). One of the reasons why the parties to the Trans-Pacific Partnership proceeded toward the conclusion of the CPTPP even after the US's withdrawal was their internal motivations beyond the benefits derived from access to the US market. An example is Japan's economic policy. Robert Falkner even notes that "getting a deal on internationally-agreed mitigation efforts is less a question of reducing the number of players than of the convergence of *domestic policy preferences* toward strong international action."[12] This change in domestic policy preferences toward reducing emissions is similar to actions taken in the face of a crisis, e.g.:

- Geostrategic crises relating to energy: Where a state decides to proceed with energy transition/sustainable energy development to reduce energy dependence, it is likely to also want to participate in a climate club since it will already be decarbonizing and able to sell emission units.
- Financial crises relating to economic concentration: There are states, such as the Gulf Cooperation Council (GCC) countries, that are in the process of

12. R. Falkner, "A Minilateral Solution for Global Climate Change? On Bargaining Efficiency, Club Benefits, and International Legitimacy," 14:1 *Perspectives on Politics* 87-101 (2016) [emphasis added].

economic diversification and structural economic reform with a view to becoming less dependent on fossil fuel exportation. These can also be incentives to link their efforts to a climate club.
- Green recovery and COVID-19: If anything, COVID-19 demonstrated the gaps in the high-emission global supply networks. It is an opportunity to increase regional ties and even support countries in localizing the essential food-water-energy[13] nexus. A climate club can be a platform setting the basis for doing so in a sustainable way.

Environmental justice

In the context of climate clubs, one obvious highlight is the concept of the free-rider theory as one of the main disincentives of climate action and of participation in a climate club. It could be argued that free-riding as a concept in itself is not a problem, but a prerequisite in the broader context of achieving environmental justice. In other words, there are states (e.g., least developed nations and small island developing states (SIDS)) that have not contributed in the past nor contribute very much in the present to climate change. Yet, they are the main victims of climate change created by emissions from the major emitters. Would it not be in line with the principles of environmental justice to accept that fairness mandates these states to be "free-riders" to the benefits of emissions reduction by major emitters simply because they are "captive riders" to negative climate change effects created by large emitters?

If yes, the main issue of a climate club is to distinguish the states that should legitimately be allowed to be free-riders and those that should not. Legitimate free-riders seem to be the states that are adversely affected by climate change, but do not significantly benefit from activities contributing to it. Evidently China, India, or the GCC countries, for example, do not meet the criterion, despite being developing nations. Hence, this perspective shifts the focus from optimal exclusion to optimal inclusion.

While it is accepted that not all the countries that should ideally take part in the climate club will actually do so, the above should at least provide a justification for why not all countries must take part.

II ANALYTICAL FRAMEWORK

We know the science of climate change; we know the economics of climate change; we also know the law of climate change. However, we do not know how countries may come together to cooperate on climate change mitigation. One way of doing so successfully is by putting together the climate regime with the international trading system via the creation of climate clubs, namely the coalition of the willing. These clubs can be parallel to, within, or outside the UNFCCC. The result may be a club of sorts: of national carbon taxes or carbon-market coalitions.

13. This third element links back to geostrategic crises and energy dependence.

While President Trump opted to ditch the Paris Climate Agreement, the European Union (EU) thinks there is now scope for a plethora of joint initiatives as President Biden reversed course by returning the US to the Paris Agreement on Climate Change.[14] These initiatives include cooperating on regulation for sustainable finance, building on the EU's so-called green taxonomy, forming a "green tech alliance" on clean technologies, collaborating on carbon pricing, and forming a Trade and Climate Initiative within the World Trade Organization.

The use of trade as a tool for climate action can go a long way. It is not just punitive. It can also be cooperative mechanisms where the environmental exceptions within Article XX of the GATT can be broadened to allow people to pursue more ambitious climate and environmental domestic policies, where one can work with countries and bilateral trade agreements to expand clean energy trade and investment as well as cooperation on technology.

If we are serious about deep decarbonization, we are going to need large amounts of clean energy trade around the world. We should think about countries that are very good at producing cheap batteries, solar panels, and other emerging technologies. We are going to need a lot more trade in those technologies around the world.

Instead of the Paris Agreement on Climate Change, some economists such as Scott Barret or William Nordhaus prefer a climate-club approach to climate-change mitigation, a climate club-type approach for more effective climate-change mitigation. This idea of a climate club is, among other things, about border adjustments to level the playing field. Beyond levelling the playing field, a climate club is intentionally punitive to incentivize countries to do things that their domestic political system will not support.

The climate-club concept is a very interesting idea for how one can come together with a set of countries and then encourage and force other countries to escalate their ambition. Three indicators seem relevant to give legitimacy to a climate club: (1) how much of the world's population it represents; (2) coverage of the world's GDP; and (3) level of GHG emissions' coverage. On this point of indicators, around 50% of humanity lives within the world's ten largest economies. Bringing together these economies would make an effective climate club. However, one needs to think carefully about what happens after that, how countries respond to it, and whether that may cause a breakdown of much of the international trading system that is so important and necessary to advance trade in clean technologies.

Three characteristics appear evident for the creation of a climate club:

(1) Most big GHG emitters need to be members of the club.
(2) Membership benefits are a must, and they should outweigh obligations.
(3) The club would need to be related to sanctions for noncompliance.

14. In fact, in April 2021, President Biden hosted the Leaders' Summit on Climate, where he invited 40 world leaders. *See* https://www.whitehouse.gov/briefing-room/statements-releases/2021/03/26/president-biden-invites-40-world-leaders-to-leaders-summit-on-climate/.

One would need to make sure that such sanctions would not violate international law and/or WTO legal rules. However, one should be concerned about the overuse of sanctions as a tool of diplomacy and economic statecraft. One should also take into account the response to retaliation and escalation resulting from sanctions in climate clubs. In William Nordhaus's analysis, the level of sanction of border adjustment required to do the job is not just on the carbon intensity of carbon-intensive goods and services; rather, it is across the entire economy.[15]

More than 100 countries/trading blocs, many of the biggest companies,[16] and 400 cities have promised to reach net-zero carbon emissions by 2050. The EU, the UK, South Korea, and Japan are among them. China said it will get there by 2060. This means that their economies will not put more carbon dioxide into the atmosphere than they take out, which will come at an enormous economic price. It remains to be seen whether voters in liberal democracies will accept higher energy prices to fix climate change. President Biden of the US is also keen to reach that target by 2050. Doing so puts all these economies in line with the Paris Agreement on Climate Change of limiting global warming to 2°C above preindustrial levels. Therefore, those six players, among others, are ideal candidates for a climate club.

Along the same lines, I would argue that (major emitting) companies should also create their climate club. Why? Because companies are on the front line on issues such as climate change and some have promised to be carbon-neutral by 2050. In addition, we could have a green recovery, turning the COVID-19 crisis into a climate opportunity,[17] whether in COP26 or China stating that it will reduce the carbon intensity of GDP by 18% between 2021 and 2025[18] and by 65% by 2030 (based on the 2005 levels), and go net-zero by 2060.[19]

The world has moved from hyperglobalization (in the words of Dani Rodrik)[20] to localization, characterized by national protectionism. Regionalism can be explained by the fact that globalization peaked in 2007-2008 and, since then, we have seen the rise of regional blocs, slowbalization, and localization. In the context of climate change mitigation, this reality explains why countries may be thinking about the creation of climate clubs (as examples of regional/plurilateral arrangements, as opposed to a global architecture).

15. William Nordhaus, "A New Solution: The Climate Club," *The New York Review of Books*, 2015, available at http://www.nybooks.com/articles/2015/06/04/new-solution-climate-club/.
16. https://carbon.ci/insights/companies-with-net-zero-targets/.
17. Restrictions related to COVID-19 caused a 6% reduction in energy-related global CO_2 emissions in 2020. The decline in road activity accounted for 50% of the fall in demand for oil and 35% came from aviation. The Economist, March 6, 2021, p. 7.
18. The Economist, March 13, 2021, p. 5.
19. The Economist, "Cleaning Up: Can China's Carbon Market Take Off?," February 27, 2021, p. 57.
20. D. Rodrik, *The Globalization Paradox: Democracy and the Future of the World Economy*. W. W. Norton & Co., 2011.

III STRUCTURE OF THIS BOOK

This book is divided into an introduction, six chapters, and a conclusion. After this introduction, Chapter 1 explains the main concepts and typological considerations regarding climate clubs. Chapter 2 provides a comprehensive examination of the greatest threat humanity faces today: climate change. It does so by looking at the legal, political, institutional, regulatory, and economic angles of climate change from the perspective of major emitters and the most relevant actors in the mitigation of climate change. This chapter offers a fresh understanding of the international trading system as a way to reach a prosperous, modern, and sustainable society that will help decarbonize the economy effectively. This chapter offers new ideas on the creation of a climate club linked to the international trade regime as an effective way to mitigate climate change. How should policies, regulations, laws, and agreements change to get us there? What incentives will be necessary to get there effectively? What has failed and why? We conclude that innovation and clean technology will be among the key elements of this equation, without which we will miss the goal, given that innovation is a major part of dealing with climate change. This chapter also investigates the potential role of international trade in reducing fossil-fuel consumption. The use of fossil fuels contributes to climate change, so there is no excuse for continued inaction. We argue that, in light of the lack of international consensus on how to effectively tackle climate change, an upstream carbon tax with reinvestment supported by border tax adjustments should be enacted unilaterally by the biggest emitters of greenhouse gases to reduce fossil-fuel consumption and incentivize global carbon abatement.

Chapter 3 explores emission units as a linkage in climate clubs in the hope of making a remarkable difference in climate change mitigation. It analyzes emission units trading in the context of regional trade agreements as a novel, promising, and effective way to mitigate climate change. It then examines Article XX of the General Agreement on Tariffs and Trade (GATT) as a potential and encouraging remedy for the development of climate clubs. Specifically, it explores the scope of application of Article XX and investigates to what extent it can be applied to climate concerns. It then provides an analysis of climate-club measures according to the jurisprudence of the WTO on GATT Article XX.

In Chapter 4, we argue that international trade law may be the solution to decarbonize the economy and invest in renewable energy, but there must be effective measures agreed by the international community. That would require agreeing on preferential trade agreements (PTAs) with environmental and sustainable development chapters on a unilateral, bilateral, or plurilateral basis, as multilateral agreements have proven to be increasingly difficult to negotiate. This chapter will assess the feasibility of including such chapters through an analysis of existing environmental and climate change-content in recent PTAs. Equally, recent developments in newer bilateral investment treaties (BITs) show that there is a positive shift toward explicit inclusion of sustainable development issues into their wording (e.g. states' right to adopt sustainable development-oriented regulation). Reference to sustainable development issues may appear in different parts of a BIT, e.g., as a general objective of the treaty in a preamble, as a separate obligation of the state parties or investors, or as an exception to the substantive protection standards, such as prohibition of unlawful indirect expropriation. Specific position and wording of sustainable development provisions determine to what extent such provisions will be binding and enforceable against state parties. Certain BITs contain only a broad reference to sustainable development without clear practical consequences, while others list sustainable development obligations that may be enforced in practice, e.g., by means of state-to-state dispute settlement mechanism or, if such obligations are imposed on foreign investors, as a counterclaim brought by a host state against an investor. Finally, we analyze how efforts to modernize the Energy Charter Treaty (ECT) may contribute to sustainable development.

Chapter 5 explores three case studies of "clubs" related to climate change, energy, and environmental protection to reach sustainability more effectively: the Generalized Systems of Preferences (GSP), trading energy in the Arab world, and EU-Russia energy trade relations:

a) Generalized System of Preferences

This case study explains two of the main GSP schemes—the US program and the EU's GSP. It indicates the reasons behind conditionality and possible future impacts. It discusses the EU's GSP, the objectives, the types of GSP offered and the various conditions to benefit from the scheme, focusing on environmental provisions. It then examines the US's GSP and the reasons for the lack of environmental conditionality. It then offers a discussion about the similarities and differences between the two systems' environmental conditions and an examination of what could be done better concerning the environmental conditionality in the GSP. The case study then provides some insights on how the COVID-19 pandemic may affect the GSP and globalization as we know it.

b) Trading energy in the Arab World

This case study contributes to the debate on energy security in the Arab world within the Middle East and North Africa (MENA) regions by highlighting various aspects of the energy sector, specifically electricity and gas, which are relevant to the discussion on regional integration and their convergence with sustainable development. Development is not possible without energy and sustainable development is not possible without sustainable energy. The case study discusses energy sustainability through the trading system and proposes regional trade agreements as means to further enhance peace and the energy security agenda of the region. The case study makes the case that, to achieve security of energy supply, two main factors are important: diversification to minimize risk and regional cooperation. It is suggested that the MENA region should reduce or eliminate the technical barriers to energy trade. The case study concludes with some recommendations on how to move forward.

c) EU—Russia energy trade

This case study aims to provide an analysis of the EU's external relations in the energy sector, focusing on EU-Russia energy trade relations and the Energy Community. Our hypothesis is that securing reliable flows of diverse sources of renewable energy would foster the EU's energy security and decrease its vulnerability toward major fossil fuel exporters. Section II provides an overview of pertinent data related to the EU's energy production and consumption, as well as current trends and challenges the EU faces in the energy realm. Section III covers related competence issues with regard to the EU treaties. Section IV examines the energy-related treaty regimes that the EU is a part of, whereas section V analyzes the EU's participation in the Energy Community and related issues. EU-Russia energy relations concerning trade in oil and gas are analyzed in section VI, which is followed by the conclusion with recommendations. The case study concludes that the EU energy structure is highly dependent on external supplies. This fact explains why external energy relations play a crucial role in foreign-policy development. The ability to accommodate an effective EU energy acquis among its neighboring states and the ability to ensure a stable relation with the biggest energy supplier are two of the major external goals of the Union. However, to do so efficiently while ensuring energy security requires certain policy solutions, which the EU has been implementing for some time: the diversifying energy supplies, e.g., by means of trade liberalization in green goods[21] and services, the promotion of renewables, as well as the lobbying of the ECT modernization.

Chapter 6 links international trade, climate change, investment and a prosperous future. We argue that a transformative, integrated, and holistic approach to sustainability is necessary to reach a prosperous future for all. With growing inequality in the world, demographic change, rapid technological development, and 40% of the world's population with no access to digital technology, access to digital technology for all and, ultimately, having a prosperous future are a must today. Doing so will give a voice to the voiceless. For that, governments, companies, and transnational institutions should invest as much as possible to make it happen and provide solutions that are rules-based, inclusive, innovative, disruptive, and simple. We the citizens should push for this human right to materialize. That way, everyone will benefit and no one will fall behind. Access to energy, mitigating climate change, and benefiting from international trade and investment all can be achieved thanks to technological advancement. These global challenges are complex, interconnected, multidimensional, interdependent, and multicausal. This chapter also explores the role citizens can play in shaping sustainable trade. Sustainability constitutes a principal concern for various regimes that, because of their nature, are in constant interaction with climate change.

21. The sales of electric cars are increasing. Liberalizing trade in such a green good would be a way forward in the fight against climate change. General Motors announced that it will phase out production of combustion-based vehicles by 2035 and has promised that 40% of its vehicles in the US will be electric by 2025. *See* The Economist, February 6, 2021, p. 6. Volvo also said that it would phase out petrol-powered cars by 2030 and produce only electric cars. *See* The Economist, March 6, 2021, p. 7.

CHAPTER 1

Climate Clubs: Key Conceptual and Typological Considerations

We know the science of climate change; we know the economics of climate change; we also know the law of climate change. However, we do not know how countries may come together to cooperate on climate change mitigation. One way of doing so successfully is by putting together the climate regime with the international trading system via the creation of climate clubs, namely the coalition of the willing. This chapter aims to explain that, by building climate clubs and making use of the international trading system, we can reach a better future for all.

I INTRODUCTION[1]

The shortcomings of multilateralism are well understood. Often treaty regimes, albeit by design, contain such limitations that consensus is necessary for their further development, including the subsequent imposition of binding obligations upon their parties. Consensus, on the other hand, is not easy to come by, given the complexities of negotiations among a plurality of states with extremely diverse interests and capabilities. This is even more so when it comes to climate mitigation, given the structural and other challenges of taking meaningful action. Case in point being the universal, multilateral climate governance regime *par excellence*, namely the United Nations Framework Convention on Climate Change (UNFCCC),[2] the overall governance of which is provided by its Conference of the Parties (COP) through its annual meetings ever since its advent. The UNFCCC is predicated on consensus-based approaches to rulemaking, resulting in standoffs between parties, given the disparity

1. The research input of Andrew Filis is acknowledged when writing this chapter.
2. United Nations Framework Convention on Climate Change, Durban, S. Afr., November 28-December 11, 2011, *Rep. of the Conference of the Parties*, U.N. Doc. FCCC/CP/2011/9/Add.1

of, among other things, interests, capabilities, abatement capacity, and development level.[3]

Moreover, the rise of greater protectionist reflexes during times of economic crisis (cf. the ongoing global recession following the 2008 financial crisis that is likely to be further compounded by the ongoing COVID-19 pandemic) have a corroding effect on multilateralism in various fields of international cooperation (cf., among other things, the European Union (EU) and Brexit,[4] the resetting of North American economic cooperation by the Trump administration with the replacement of the North American Free Trade Agreement (NAFTA) by the United States-Mexico-Canada Agreement (USMCA),[5] and currency and trade antagonisms between the United States (US) and China). Arguably, a further casualty of such global dynamics is the willingness of the international community to commit to *binding* and *enforceable* ambitious targets within the context of the UNFCCC and its successive agreements. To begin with, the agreements reached within the context of the UNFCCC have been relatively anemic given their modest goals and/or lack of enforcement and, thus, their limited capacity to meaningfully forestall potentially catastrophic climatic impacts caused by human activity. Such half-hearted and, in the event, fateful, multilateralism seems ripe for reconsideration. In the absence, therefore, of some effective nigh-universal binding *top-down* regime to meaningfully regulate emissions, it is vital to look at other means of achieving or, at the very least, approximating the necessary abatement targets.

What is more, a number of commentators, including Professor and Nobel Laureate William Nordhaus, consider a pernicious aspect of the existing multilateral order to be its scope for *free-riding*[6] and, therefore, that solutions must be designed to

3. J. Vogel, The Problem with Consensus in the U.N. Framework Convention on Climate Change. 32:2 *Philosophy & Public Policy Quarterly* 14-21 (2014).
4. In relation to its Brexit and collective climate abatement, following its departure from the EU on January 31, 2020, the United Kingdom (UK) committed to participate in the European Union (EU) Emissions Trading System (ETS) during the transition period until the end of 2020. UK installations therefore continued to face compliance obligations for their 2019 and 2020 emissions. In February 2020, the UK published its approach to negotiating the future relationship with the EU, in which the UK stated that its future carbon pricing initiative will support its net zero by 2050 target. It is also considering a link between any future UK ETS and the EU ETS, similarly to the Swiss-EU ETS linking model. This would allow allowances to be exchanged between the two systems. However, fallback options, such as a stand-alone UK ETS and a long-term carbon tax, are also being considered by the UK government has therefore legislated provisions for both a UK ETS and carbon tax as part of its 2020 legislative proposals. For more background information *see* World Bank. State and Trends of Carbon Pricing 2020 (May), World Bank, Washington, DC (39).
5. The USCMA entered into force on July 1, 2020. *See* the Office of the United States Trade Representative website for the US view on the USCMA's objectives: https://ustr.gov/trade-agreements/free-trade-agreements/united-states-mexico-canada-agreement. Also, for the text of USCMA, *see* https://ustr.gov/trade-agreements/free-trade-agreements/united-states-mexico-canada-agreement/agreement-between.
6. W. Nordhaus, "The Climate Club: How to Fix a Failing Global Effort." *Foreign Affairs* (May/June 2020). Nordhaus discusses such dynamic thus, "Free-riding is a major hurdle to addressing global externalities, and it lies at the heart of the failure to deal with climate change. Consider a voluntary agreement, such as the Kyoto Protocol or the Paris accord. No single country has an incentive to cut its emissions sharply. Suppose that when Country A spends USD 100 on abatement, global damages decline by USD 200 but Country A might get only USD 20 worth of the benefits: its national cost-benefit analysis would lead it not to undertake the abatement. Hence, nations have a strong incentive not to participate in such agreements. If they do participate, there

reduce or altogether eliminate such scope. However, as the notion of free-riding generally concerns the enjoyment of a benefit without contributing to its cost, it does not entirely tally with the realities of climate change, given that such free-riding has less to do with states *enjoying a benefit* cost-free, and more to do with states actually *not suffering potential anticompetitive effects* to their economies by not implementing meaningful carbon abatement measures. What is more, the global commons/public good to be defended within the context of multilateral climate-mitigation efforts remains somewhat elusive.[7] In that sense, it is not clear how free-riding arguments may be framed within the context of the enjoyment on the part of a state of some benefit cost-free when that *benefit* pertains to averting environmental degradation, which, after all, may impact states differently. For instance, the current state of play—i.e., the inadequacy of existing global abatement efforts for containing average global temperature increases by 2100 to no more than 2°C over preindustrial levels—[8] would have

is a further incentive to understate their emissions or to miss ambitious objectives. The outcome is a noncooperative free-riding equilibrium, in which few countries undertake strong climate change policies—a situation that closely resembles the current international policy environment." However, it is not clear what are the "global damages" that decline by USD 200, or how Country A only benefits by USD 20 vis-à-vis its USD 100 outlay while a free-riding Country B also benefits presumably by USD 20 while not spending a dime. Also *see* W. Nordhaus. "Climate Clubs: Overcoming Free-riding in International Climate Policy." 105:4 *American Economic Review* 1339-1370 (2015), where the issue of free-riding is analyzed in greater detail. In sum, the issue pertains to how abatement costs (often prohibitive and damaging for competitiveness) accrue nationally while the benefits of abatement accrue globally and are independent of where emissions take place. Moreover, given that cost of abatement, many states choose not to participate as the penalties are lower than such costs. This results in low participation rates but also low penalties given that so few states are in the climate club in question. Therefore, trade sanctions on nonparticipants seem to be indispensable, albeit highly controversial given their implications for global trade.

7. Things such as clean air, public health, water quality, the integrity of ecosystems, and so on, are public goods. Environmental degradation does not limit itself to national borders; that much is clear. Yet the notion of global commons/public goods remains elusive although that need not inhabit in the slightest the need for the international community to take meaningful action to address climate change, particularly when the current overall objective of ensuring that average global temperatures do not rise above 2°C by 2100 is sufficiently specific for states to consider how to go about collectively to achieve this common goal. In that sense, the *global public good* is to preserve the integrity of the global ecosystem to its projected state following an increase of no more than 2°C over preindustrial levels. The global public good in that sense accrues intergenerationally. Others have posited that "the stability of the Earth's climate is a global public good." *See* L. Weischer, J. Morgan, and M. Patel, *Climate Clubs: Can Small Groups of Countries make a Big Difference in Addressing Climate Change?*. 21:3 *RECIEL* 177 (2012).
8. It is worth noting that the end of 2020 marked the moment under the Paris Agreement "ratchet mechanism" when parties were expected to formally submit more ambitious emissions reduction commitments in relation to their 2030 targets. However, just 45 parties (44 countries, plus the EU *sui juris*) met this deadline. After a year disrupted by the global pandemic states responsible for only c. 28% of global emissions registered new or updated "nationally determined contributions" (NDCs) on the UN's official registry by the end of 2020 (cf. https://www4.unfccc.int/sites/NDCS taging/Pages/LatestSubmissions.aspx). Some emitters did register their NDCs in time, including the UK and the EU, but there were also major absences including China, India, and the US. Even among the new submissions many lacked increases in ambition since their first pledges made previously (2015), or even backtracked with scaled-back proposals. However, one expert, Professor Niklas Höhne, as reported by Josh Gabbariss for Carbon Brief, assesses collective plans to be still "totally off" what is required to achieve Paris Agreement targets. *See* J. Gabbariss, "Analysis: Which Countries Met the UN's 2020 Deadline to Raise 'Climate Ambition'?."

very different implications for Mauritius than for Russia, given that such failure would pose an existential threat to the former, while potentially advantageous to the latter.[9]

Disparities between parties to the UNFCCC, including their historical cumulative emissions, abatement capacity, and level of development, provide useful context as to why *differentiations in multilateral efforts* exist and may be entirely valid.[10] In fact, the notion of common but differentiated responsibility (CBDR) is a principle that is deployed across many multilateral environmental agreements, including the UNFCCC. Article 3 of the UNFCCC states that:

> parties should protect the climate system for the benefit of future and present generations of human kind on the basis of equity and in accordance with their common but differentiated responsibility and respective capabilities.

In that sense, CBDR affirms that, while it is a common responsibility of all states to protect the integrity of the environment, some should take stronger action than others. The justification for differentiated responsibilities is twofold: first, those that have most heavily contributed to (and benefited from) the accumulation of atmospheric greenhouse gas (GHG) emissions have a greater burden of responsibility and, second, those that have greater abatement capability, on account of access to resources, including wealth and technology, should contribute more. Against the above context, free-riding arguments may be falling short of properly reflecting the complexities of global climate policy.

That being said, the literature is rife with, among other things, not only game theoretical conceptualizations, but also scrutiny of trends in state practice, suggesting perhaps too great a role of the scope for free-riding in multilateralist failure concerning global abatement efforts. For instance, when one considers international cooperation

---

CarbonBrief (January 8, 2021). https://www.carbonbrief.org/analysis-which-countries-met-the-uns-2020-deadline-to-raise-climate-ambition. According to the analysis presented by Climate Action Tracker, just 8 parties (7 states plus the EU *sui juris*) have submitted stronger NDC targets than their previous commitments and 118 parties have not updated targets altogether. As of February 3, 2021, new NDC submissions covered c. 31.6% of global emissions and 25.2% of the global population. *See* https://climateactiontracker.org/climate-target-update-tracker/.

9. *See* M.F. Dellinger, "Narrowed Constellations in a Supranational Climate Change Regime Complex: The 'Magic Number' Is Three." 37 *Fordham International Law Journal* 373 (2014), 383, where it is stated that, "Russia's official attitude towards climate change is hesitant because of the perceived benefits to it of climate change and its self-proclaimed 'better' ability to deal with climate change than other nations."

10. Among other things, such differentiations in the distribution of obligations, albeit temporary in some cases, are witnessed not only in the UNFCCC's principle of common but differentiated responsibilities (CBDR) but also the notion of special and differential treatment (SDT) as reflected across various provisions within the context of the World Trade Organization (WTO), including the Decision on Differential and More Favorable Treatment, Reciprocity and Fuller Participation of Developing Countries (commonly referred to as the "Enabling Clause") that serves as the basis for the WTO's Generalized System of Preferences (GSP). For a fuller account of SDT provisions within the WTO order *see* Development: Trade and Development Committee (WTO) https://www.wto.org/english/tratop_e/devel_e/dev_special_differential_provisions_e.htm. *See also* the 27 Rio Principles found in the Rio Declaration on Environment and Development, signed at the 1992 Earth Summit in Rio de Janeiro. Principle 6 states: "The special situation and needs of developing countries, particularly the least developed and those most environmentally vulnerable, shall be given special priority."

in relation to environmental protection, it is apparent that the goal—namely the prevention of the sort of environmental degradation that would pose serious threats to human, animal, and plant life and health—actually pertains to a *transboundary, intergenerational common good.* Securing such an objective benefits the entire international community, including not only fully developed and post-industrial states—responsible for historical and ongoing emissions—but also developing states whose development is emissions-intensive (e.g. China and India, among others). It is not only those coastal including island states facing threats to their existence that are set to benefit from global temperature stabilization. However, those not facing existential threats are, therefore, tempted to shirk the likely economic burdens of abatement, particularly when the development of their economies is a higher priority.

Although, as discussed above, this is not *free-riding* in the sense of enjoying some existing benefit the cost of which burdens others, strictly speaking, such states are relieved of the current and ongoing cost of action necessary to safeguard an existing and future *benefit.* In future, should environmental collapse be averted, it is only then that such states would enjoy the benefits (or rather the felicitous preclusion of climatic collapse along with the rest of international community) of climatic integrity without having borne the economic, and potentially social and political, costs of the requisite abatement efforts. Notions from game theory offer perhaps cynical, albeit insightful, glimpses into the ways of statecraft. What we may witness is reverse Beggar-Thy-Neighbor effects,[11] whereby, assuming burdens accrued domestically (that is to say, the cost of abatement) actually translates into benefits to one's neighbors (that is to say, to those who bore no such costs).

Undoubtedly, this is all moot should failure to take meaningful action result in the sort of climatic degradation about which scientists have long warned the international community.

II CLIMATE CLUBS: MINILATERALIST PATHS TO CLIMATE MITIGATION

Nordhaus so aptly highlights how climate-mitigation solutions must *also* be sought outside a multilateralism that, under its current form, has persistently failed to produce a legally binding international agreement on climate change when one considers the last twenty-five years or so of its life, namely since the 1990s.[12] Such solutions must be well attuned to the realities of failed multilateralism, including the lack of adequate and

11. The concept of "Beggar-Thy-Neighbor" pertains to the likely negative implications of a state's economic policies for the economic interests of other states. This 18th-century notion originates in the writings of Adam Smith (*see An Inquiry into the Nature and Causes of the Wealth of Nations*, Book IV, Chapter III, W. Strahan and T. Cadell, London, 1776). Reverse Beggar-Thy-Neighbor effects would involve just the opposite—namely costly domestic measures on the part of one state resulting in free benefits to other states.
12. W. Nordhaus, "The Climate Club: How to Fix a Failing Global Effort." *Foreign Affairs* (May/June 2020). Nordhaus states that, "The bottom line is that climate policy has not progressed over the last three decades. The dangers of global warming are much better understood, but nations have not adopted effective policies to slow the coming peril."

binding targets, incentives, and penalties to achieve critical mass in global mitigation efforts, hence why Nordhaus calls for some complementary solution to multilateralism—namely the *climate club* model—that takes stock of existing pitfalls.

There is extensive literature exploring or advocating solutions beyond multilateralism, namely minilateralist solutions,[13] that is to say, solutions that involve at least three parties (be they jurisdictions or states).[14] The idea that groups of states/jurisdictions can come together to develop and pursue specific common objectives concerning their shared interests is not new. In fact, it is at least as old as the rise of empires in the Bronze Age, where separate realms would oscillate between warfare and peace to, among other things, advance their trade interests through coalitions and arrangements (cf. the *Amarna letters* containing diplomatic correspondence in Akkadian, the diplomatic lingua franca, between the Great Powers of the Ancient East Mediterranean/Near East and their vassals during the late half of 2 millennium BCE[15]) to the exclusion of others. In that sense, such modalities of inter-state cooperation that rely on obligations, sticks, and carrots are far from novel.

Terms of convenience encountered in the literature that pertain to such minilateralist solutions include *climate clubs, carbon clubs, carbon market clubs, club-like arrangements*, and *climate mitigation clubs*. However, the nomenclature is incidental as what actually matters is the content of such arrangements,[16] as is discussed throughout the present publication. What is crucial is whether such minilateralist

13. Among others, W. Nordhaus, "The Climate Club: How to Fix a Failing Global Effort." *Foreign Affairs* (May/June 2020); M.F. Dellinger, "Narrowed Constellations in a Supranational Climate Change Regime Complex: The 'Magic Number' Is Three." 37 *Fordham International Law Journal* 373 (March 1, 2014); L. Weischer, J. Morgan, and M. Patel, "Climate Clubs: Can Small Groups of Countries Make a Big Difference in Addressing Climate Change?." 21:3 *RECIEL*; J. Hovi, D.F. Sprinz, H. Sælen, and A. Underdal. "Climate Change Mitigation: A Role for Climate Clubs?." 2:16020 *Palgrave Communications* (2016); R. Falkner. "A Minilateral Solution for Global Climate Change? On Bargaining Efficiency, Club Benefits and International Legitimacy," 14:1 *Perspectives on Politics* 87-101 (2016). Centre for Climate Change. Economics and Policy. Working Paper No. 222, London, UK; Victor D G (2015) The Case for Climate Clubs. International Centre for Trade and Sustainable Development (ICTSD) and World Economic Forum: Geneva, Switzerland; J. Morgan and L. Weischer. *Two Degrees Clubs: How Small Groups of Countries Can Make a Big Difference on Climate Change* (World Resources Institute, October 29, 2012); J. Hovi, D.F. Sprinz, H. Sælen & A. Underdal, "The Club Approach: A Gateway to Effective Climate Co-operation?" 49 *B. J. Pol. S* 1071-1096 (2017); D.G. Victor, *The Case for Climate Clubs*. E15Initiative. Geneva: International Centre for Trade and Sustainable Development (ICTSD) and World Economic Forum, 2015. www.e15initiative.org/; W. Nordhaus "Climate Clubs: Overcoming Free-riding in International Climate Policy." 105:4 *American Economic Review* 1339-1370 (2015); and N. Martin and J.C.J.M. van den Bergh "A Multi-level Climate Club with National and Sub-national Members: Theory and Application to US States." 14:124049 *Environ. Res. Lett.* (2019).
14. *See* J. Hovi, D.F. Sprinz, H. Sælen, and A. Underdal, "The Club Approach: A Gateway to Effective Climate Co-operation?." 49 *B. J. Pol. S* 1071-1096 (2017) where they define a climate club "as any international actor (country) group that (1) starts with fewer members than the UNFCCC has and (2) aims to co-operate on climate change mitigation" (1072).
15. *See* S. Izre'el, "The Amarna Tablets for Background Information and Access to Particular Portions of the Corpus," https://www.tau.ac.il/humanities/semitic/amarna.html
16. *See* L. Weischer, J. Morgan, and M. Patel, "Climate Clubs: Can Small Groups of Countries make a Big Difference in Addressing Climate Change?." 21:3 *RECIEL* (2012), where the authors argue that the term "clubs" includes "any grouping that comprises *more than two and less than the full multilateral set of countries party to the UNFCCC* and that has not reached the degree of

formations pertain exclusively or partially to climate mitigation, are *impactful*, and are capable of bringing collective efforts on track to meeting global abatement objectives.

While minilateralist approaches are predicated on *coalitions of the willing*,[17] there is limited effectiveness in merely *clubbing* together the most ardent of environmentalist states if they do not include the most emitting or, at the very least, if they do not include such states with the necessary geopolitical and geoeconomic clout to draw in the most emitting players. For instance, the top four (by global share) emitters, namely China, the US, the EU27 plus the UK, and India, have collectively contributed to 55 per cent of total global GHG emissions over the last decade, while, collectively, G20 members are responsible for 78 per cent of total GHG emissions,[18] and, therefore, what these states, either separately or jointly, do to mitigate or compound climate change rightly attracts intense scrutiny. For instance, according to the latest United Nations Environmental Programme (UNEP) Emissions Gap report, collectively, G20 members are currently not on track to achieving their unconditional nationally determined contributions (NDCs), as some are on track, others are falling short, while for others, it is unclear.[19] What is clear, however, is that the *emissions gap*—i.e., the discrepancy between current commitments and those needed to limit average temperature rises by the end of the century to a maximum of 2°C—remains gaping.[20]

---

institutionalization of an international organization. While clubs may include other stakeholders, they are predominantly governed and funded by national governments. This broad definition can include very different degrees of formality and organization. Regular conferences and meetings, as well as groupings frequently referred to as "initiatives", "forums" or "partnerships", are, for the purposes of this chapter, all included in the "clubs" term. We exclusively consider clubs that discuss and promote greenhouse gas emissions reductions or a sub-issue directly relevant to climate change mitigation, such as the promotion of renewable energies." (emphasis added) (177).

17. *See* J. Bacchus, *The Willing World: Shaping and Sharing a Sustainable Global Prosperity*. Cambridge University Press, 2018, in which Bacchus makes the case for global action and multilateralism, based on the rule of law, on the part of willing actors in a manner that optimally promotes both environmental and economic objectives.
18. United Nations Environment Programme, *Emissions Gap Report 2020—Executive Summary*. 2020, v-vi. https://www.unep.org/emissions-gap-report-2020.
19. *Ibid.*, viii.
20. To track progress of global decarbonization efforts, UNEP collects data and compiles periodic reports to assess discrepancies between the purported goals and the situation on the ground. This discrepancy is known as the "emissions gap." Successive reports highlight the inadequacy of global efforts. To mark the ten-year anniversary since the first assessment, the UNEP published an assessment report taking stock of action over the ensuing decade, which makes for sober reading in that it finds that despite a decade of "increasing political and social focus on climate change and the milestone Paris Agreement, global greenhouse...emissions have not been curbed, and the emissions gap is larger than ever." What is more, it states that "... essentially, there has been no real change in the global emissions pathway in the last decade. The *effects* of climate policies have been *too small* to offset the impact of key drivers of emissions such as economic growth and population growth.... [T]he fact remains that at the global scale *we have failed to bridge or even narrow* the 2020 emissions gap ... *if the Paris goals are to be kept viable, the world cannot afford to lose another decade.* Altogether, ambition and action over the past decade have simply been inadequate, so *action now needs to be faster and more transformational.*" (emphasis added). *See* J. Christensen and A. Olhoff (2019). "Lessons from a decade of emissions gap assessments. United Nations Environment Programme." Nairobi (1, 3, and 4) https://www.unenvironment.org/resources/emissions-gap-report-10-year-summary.

Moreover, such minilateralism need not be antagonistic toward multilateralism,[21] and often minilateralist initiatives purport to be complimentary toward the latter, although, as some commentators have analyzed, minilateralism engenders fragmentation in global climate governance, with varied implications for multilateralism.[22] What is more, the involvement of particular states in such minilateralist arrangements may have considerable implications on their effectiveness, given the importance of such states when it comes to their GHG emissions and/or geoeconomic and geopolitical clout. For instance, the impact of the US withdrawal from the Paris Agreement under the Trump administration has been assessed as negative for multilateralism, given that the US had not merely gone from being leader to follower, but to actually becoming an outsider, thus undermining the extent of Paris Agreement abatement efforts.[23] It is worth noting that the US is the second largest emitter in absolute terms and was the twelfth in terms of emissions *per capita* in 2017,[24] and, therefore, its absence has implications globally. The Biden administration, however, by signaling on January 21, 2021 its acceptance of the Paris Agreement, has sought to reverse this situation, thus bringing the US back into the fold.[25]

21. *See* Victor (2015) *The Case for Climate Clubs.* E15Initiative. Geneva: International Centre for Trade and Sustainable Development (ICTSD) and World Economic Forum, 2015 (8) who states that minilateralist arrangements need not be taken as "replacements to the UN approach but as friendly competitors and complements."
22. F. Biermann, P. Pattberg, H. van Asselt, and F. Zelli "The Fragmentation of Global Governance Architectures: A Framework for Analysis," 9:4 *Global Environmental Politics* (2009) and Victor (2015) *The Case for Climate Clubs.* E15Initiative. Geneva: International Centre for Trade and Sustainable Development (ICTSD) and World Economic Forum, 2015.
23. D.F. Sprinz, H. Sælen, A. Underdal, and J. Hovi. "The Effectiveness of Climate Clubs under Donald Trump," 18:7 *Climate Policy* 828 (2018).
24. US total annual CO_2 emissions stand at c. 5.28 billion $MtCO_2$ (2019), annual emissions per capita stand at c. 16.16 $MtCO_2$ (2017). US total annual emissions represented c. 14.72% of global CO_2 emissions (2017), while US historical, cumulative CO_2 emissions stand at 410.24 billion $MtCO_2$ (up to 2019). For comparison, China's total annual CO_2 emissions stand at c. 10.17 billion $MtCO_2$ (2019); annual emissions per capita at c. 6.76 $MtCO_2$ (2017). China's total CO_2 emissions for 2017 represented 27.32% of global CO_2 emissions. China's historical, cumulative CO_2 emissions stand at c. 219.99 billion $MtCO_2$ (up to 2019). In terms of their respective share of global cumulative CO_2 emissions, the US share stands at 24.82% while China's at 13.31%. Based on 2017 figures, the US was the twelfth in terms of per capita emissions. Qatar, Kuwait, Bahrain, UAE, Saudi Arabia, Australia, Kazakhstan, Brunei, and Trinidad & Tobago, among others, actually preceded it. The above figures have been drawn from charts developed by Our World in Data (https://ourworldindata.org/co2/country/united-states?country=USA~CHN and https://ourworldindata.org/co2-emissions) and the World Resources Institute (https://www.wri.org/blog/2020/02/greenhouse-gas-emissions-by-country-sector), which, in turn, rely on a variety of datasets cited in their respective sources.
25. White House: Briefing Room (January 20, 2021) *Paris Climate Agreement* [Statement] https://www.whitehouse.gov/briefing-room/statements-releases/2021/01/20/paris-climate-agreement/. What is more, President Biden has alluded to his vision for the US to play a more leading role in climate diplomacy by stating that, "On day one, I signed the paperwork to rejoin the Paris Climate Agreement. We're taking steps led by the example of integrating climate objectives across all of our diplomacy and raise the ambition of our climate targets. That way, we can challenge other nations, other major emitters ... to up the ante on their own commitments. I'll be hosting climate leaders—a climate leaders' summit to address the climate crisis on Earth Day of this year. America must lead in the face of this existential threat. And just as with the pandemic, it requires global cooperation." *See* White House: Briefing Room (February 4, 2021) *Remarks by President Biden on America's Place in the World* [Speeches and Remarks] https://

It is also worth noting that the US has had an ambivalent approach toward multilateralist climate mitigation as this remains a controversial policy field at the domestic level. Historically, the US has consistently been reluctant to assume binding obligations where no such/analogous obligations are also assumed by parties with less developed economies, as had been the case under the Kyoto Protocol to the UNFCCC, which categorized parties into different groupings burdened by different obligations. Against that backdrop, US reticence and vacillation may well reappear in future, unless some monumental cultural shift takes place in the US to definitively catalyze public opinion in favor of multilateralism. That being said, the US has been involved in successful minilateralism when it comes to its geopolitical rival, China.

For instance, in 2013 China and the US agreed to bilaterally phase down a group of the most harmful pollutants, namely hydrofluorocarbons (HFCs) and, furthermore, to provide global leadership to that end;[26] an event that has been heralded as indicative of their "potential willingness to accede to an internationally binding climate change agreement which in turn may cause a watershed in negotiations and spur further action."[27] Interestingly, such bilateral cooperation took place in the midst of antagonisms on various fronts, including in connection to currency and trade disputes. This fact contradicts the notion that successful inter-state collaboration is predicated on cordial relations.[28] That being said, while the participation of the US is not crucial for the formation or eventual success of minilateralist solutions, such clubs are less likely to extend over more than 50% of global emissions and to thus secure substantial cuts.[29]

As mentioned earlier, the implications of minilateralist solutions for multilateralism should not be ignored. Such implications are neither always apparent nor uniform, given the variety of minilateralist initiatives at play. Sprinz et al. conclude that

---

www.whitehouse.gov/briefing-room/speeches-remarks/2021/02/04/remarks-by-president-biden-on-americas-place-in-the-world/.

26. White House: Office of the Press Secretary (June 8, 2013) *United States and China Agree to Work Together on Phasing Down of HFCs* [Press Release] https://obamawhitehouse.archives.gov/the-press-office/2013/06/08/united-states-and-china-agree-work-together-phase-down-hfcs, where it is stated, among other things, that, "President Obama and President Xi agreed on an important new step to confront global climate change. For the first time, the United States and China will work together and with other countries to use the expertise and institutions of the Montreal Protocol to phase down the consumption and production of hydrofluorocarbons (HFCs), among other forms of multilateral cooperation. A global phase down of HFCs could potentially reduce some 90 gigatons of CO_2 equivalent by 2050, equal to roughly two years' worth of current global greenhouse gas emissions."
27. M.F. Dellinger, "Narrowed Constellations in a Supranational Climate Change Regime Complex: The 'Magic Number' Is Three." 37 *Fordham International Law Journal* 373, 375 (2014).
28. In fact, Dellinger (*ibid.*, 391) offers examples of how treaty creation and collaboration may take place even at the height of antagonisms. For instance, the height of the Cold War did not prevent the erstwhile Soviet Union to negotiate and conclude an agreement on the governance of Antarctica with, among other states, its archenemy at the time, the US.
29. Sprinz, Sælen, Underdal, and Hovi have assessed the implications of US participation and absence from climate club initiatives, and their analysis indicates that the likely impact would depend on the US's initial role in such initiatives. For example, whether it had initiated/led, had been a follower, or had never participated. For the purposes of the present chapter, it would suffice to state that, predictably, climate club initiatives involving the most significant states in terms of emissions and/or geopolitical clout are likely to be the most impactful when it comes to climate mitigation. *See* D.F. Sprinz, H. Sælen, A. Underdal, and J. Hovi, "The Effectiveness of Climate Clubs under Donald Trump." 18:7 *Climate Policy* 835-836 (2018).

climate clubs, however, need not necessarily be antagonistic to multilateralism. While such minilateralism can provide alternatives to unsuccessful multilateralism—e.g. the Paris Agreement regime, should it fail to ensure that global efforts are on track to meet global objectives—they may actually be complimentary.[30]

1 Climate Club Minilateralism

The relevant literature contains much support for minilateralist approaches to climate governance, including for the purposes of surmounting existing consensus-based hurdles at the multilateral level. However, minilateralist solutions are not always successful and may actually undermine multilateralist efforts. For instance, a club of sorts, the erstwhile Asia-Pacific Partnership on Clean Development and Climate Change[31] (APP), to name but one example, sought to pursue voluntary climate-mitigation objectives outside the UNFCCC, despite the express claim that it was to complement the Kyoto Protocol regime,[32] but was unable to secure emissions reductions beyond what could probably have otherwise been achieved collectively under binding targets within the multilateralist system.[33] The 2009 Major Economies Forum on Energy and Climate[34] (MEF) and the 2012 Climate and Clean Air Coalition[35] are other examples of minilateralist arrangements in that regard.[36]

30. D.F. Sprinz, H. Sælen, A. Underdal, an J. Hovi, "The Effectiveness of Climate Clubs under Donald Trump." 18:7 *Climate Policy* 836 (2018).
31. The APP was established in July 2005 between Australia, China, India, Japan, South Korea, and the US (with New Zealand and Canada joining later) to pursue voluntary climate-mitigation objectives. The APP sought, among other things, to promote the take-up and use of energy efficient and cleaner technologies particularly on the part of less developed countries (India and China) on their path to industrialization and development. The APP adopted its Charter in Sydney in January 2006 and lasted until 2011, although various APP-related projects remain in operation. https://web.archive.org/web/20090920021825/http://www.asiapacificpartnership.org/pdf/resources/charter.pdf.
32. *See* APP Charter (*ibid.*) recitals where it is stated that, "Bearing in mind that the purposes of the Partnership are consistent with the principles of the United Nations Framework Convention on Climate Change and other relevant international instruments, and are intended to complement but not replace the Kyoto Protocol … ." *See also* APP Charter Annex I *Vision Statement of Australia, China, India, Japan, the Republic of Korea, and the United States of America for a new Asia-Pacific Partnership on Clean Development and Climate* of July 28, 2005, where it is stated that the APP "will be consistent with and contribute to our efforts under the UNFCCC and will complement, but not replace, the Kyoto Protocol."
33. *See* J. Hovi, D.F. Sprinz, H. Sælen and A. Underdal, "The Club Approach: A Gateway to Effective Climate Co-operation?" 49 *B. J. Pol. S* 1071-1096 (1091) (2017) who also suggest that club-like arrangements have been "no more effective" than the UNFCCC in mitigating climate change given that they neither provide exclusive member benefits nor make conditional emissions reduction commitments (1091).
34. Current MEF members: Australia, Brazil, Canada, China, the European Union (EU) *sui juris*, France, Germany, India, Indonesia, Italy, Japan, South Korea, Mexico, the Russian Federation (Russia), South Africa, the United Kingdom (the UK), and the US. President Obama announced the launch of the MEF on March 28, 2009 to facilitate 'candid' dialogue between the major developed and developing economies over clean energy and emission reduction/decarbonization objectives in light of the December 2009 UNFCCC Conference of Parties in Copenhagen. Denmark, in its capacity as President of the December 2009 Conference of the Parties, and the United Nations were also invited to participate. U.S. Department of State (2009)

Furthermore, it is crucial to analyze each minilateralist solution separately in assessing their particular content—particularly the nature, level, and bindingness of abatement ambition/targets—to assess whether they are in line with global climate-mitigation objectives. As Shuting Pomerleau highlights, minilateralist solutions are not a "silver bullet" for a variety of reasons, including the potential implications of measures for consumers in climate club member states.[37]

One such assessment is provided by Weischer, Morgan, and Patel (2012), who examine 17 clubs founded between 1974 and 2012 that, either exclusively or partially, concern climate mitigation and found that such arrangements generally fall within two categories: *dialogue forums* or *implementation groups*.[38] The former concern the type of club, where parties exchange information and gain a better understanding of their respective positions, whereas the latter collectively implement agreed policies. Examples of the former include the G8, G20, MEF, the Renewable Energy Policy Network for the 21st Century (REN21), and the Global Bioenergy Partnership, while examples of the latter include the Renewable Energy and Energy Efficiency Partnership (REEEP) and the Global Green Growth Institute. However, their analysis concludes that, while the clubs examined make "important contributions"—e.g., they facilitate a better

Major Economies Forum on Energy and Climate Change [Press Release] https://2009-2017.state.gov/e/oes/climate/mem/index.htm.

35. The Climate and Clean Air Coalition to Reduce Short-Lived Climate Pollutants, launched in 2012 by the UNEP, is a voluntary partnership of governments, intergovernmental organizations, businesses, scientific institutions, and civil society organizations aimed at improving air quality and environmental protection through actions to reduce short-lived pollutants. https://ccacoalition.org/en/content/about-0.
36. D.F. Sprinz, H. Sælen, A. Underdal, and J. Hovi, The Effectiveness of Climate Clubs under Donald Trump, 18:7 *Climate Policy* 830 (2018).
37. Shuting Pomerleau (Niskanen Center) argues that Nobelist William Nordhaus's alternative proposals hinge on the imposition of global carbon price and uniform import tariffs on non-members as a means to address free-riding which may actually hurt consumers in implementing members than those outside the bloc. Other difficulties include reaching consensus on carbon price and social costs and so on which is hard enough domestically. Nordhaus's proposal is not for a border carbon adjustment (BCA) contingent on emission content of goods but a uniform tariff on all imports from non-member states. Something along the lines of USD 25 per $MtCO_2$ and a 3% *ad valorem* tariff would provide strong incentive for non-members. Also, a *uniform* tariff imposed on non-members does not discriminate in favor of low-carbon imports, which is absurd and potentially counterproductive. This could also be said about the imposition of retaliatory measures by non-members. Solutions may exist in leveraging or amending current World Trade Organization (WTO) rules but this may be unfeasible given the complexities involved (e.g. unanimity of members to amend key norms). What is more, commenting on Nordhaus's Climate Club proposals, Pomerleau states that "... a climate club might not be a silver bullet to global climate-mitigation efforts. While the proposal is ambitious and backed by detailed modeling, it's not without its own challenges. Nordhaus admitted that "the international community is a long way from adopting a Climate Club or a similar arrangement to slow the ominous march of climate change." No country can solve climate change on its own. An effective and accountable international collaboration mechanism is overdue." *See* S. Pomerleau (2020, September 21), *Potential Challenges to a Climate Club*. Niskanen Center. https://www.niskanencentre.org/potential-challenges-to-a-climate-club/.
38. L. Weischer, J. Morgan, and M. Patel, "Climate Clubs: Can Small Groups of Countries make a Big Difference in Addressing Climate Change?." 21:3 *RECIEL* (2012). The membership of the clubs examined ranged from 7 (Asia-Pacific Partnership on Clean Development and Climate, which came to a close in 2011) to 73 (the REDD +). Most related to energy issues and were independent of the UNFCCC but a minority were connected to the implementation of UNFCCC COP decisions.

mutual understanding of the respective positions and interests of parties, facilitate the sharing of best practice, and help coordinate mitigation action—they are neither focused on significantly increased ambition[39] nor dramatically reduce GHG emissions at the scale and speed required to meet global mitigation targets (i.e. to ensure that average global temperatures do not exceed 2 °C by the end of the century).[40]

In response to what would make a climate club "transformational" and consistent with the 2°C target, the World Resources Institute (WRI) has arrived at the following four criteria:

1. ambitious vision;
2. clear membership conditions;
3. considerable member benefits; and
4. a clear pathway from immediate action to expansion (i.e., increase in ambition and potentially membership) over time.

More specifically, vision must be in line with climate science, and may relate to targets concerning emissions reductions, energy efficiency, renewable energy (RE) deployment, or price parity for RE technology. According to the WRI, ambition must reflect what is necessary to solve the problem.

Membership should be exclusive and only granted to parties that meet clear criteria that are consistent with the vision. Such conditions may be predicated on the track records, current or future targets, or other indicators of current and future commitment/action. Such criteria must be specific and measurable, and also reflect the willingness of parties to move forward but also raise ambition in future.

In terms of benefits, there must be strong incentives for joining so that prospective members will accept the ambitious conditions for membership. Therefore, the club has to create palpable member benefits in areas such as investment, technology sharing, and/or trade, which should not otherwise be available to those outside the club.

In terms of a clear pathway to current and future action, the club would need:

> to become operational quickly, so it is likely to start with relatively easy-to-implement activities, such as information sharing. However, a two degrees club should be set up in a way that allows it to address more difficult questions (e.g., around trade), grow in scope (e.g., expanding from renewable energy or efficiency to other areas), and increase the number of members over time. What starts as a small energy transformation club could turn into a "low-carbon union," providing

39. *Ibid.*, 192.
40. *Ibid.*, 187, and J. Morgan and L. Weischer, *Two Degrees Clubs: How Small Groups of Countries Can Make a Big Difference on Climate Change* (World Resources Institute (October 29, 2012). https://www.wri.org/blog/2012/10/two-degrees-clubs-how-small-groups-countries-can-make-big-difference-climate-change.

significant benefits across all sectors of the low-carbon economy to a growing number of member countries.[41]

A further review builds on existing analyses to find some merit in the pursuit of minilateralist solutions and to offer recommendations to that end. Hovi et al. suggest that starting small, namely with a small group of "enthusiastic" states to then seek to entice "reluctant" states may be "promising" for the purposes of climate mitigation. However, most importantly, the potential synergistic effects of minilateralist instances on multilateralism (namely the UNFCCC regime) may hold some positive implications for global abatement efforts, although the potential of climate clubs for being instrumental in global climate mitigation is not readily apparent in the scholarly literature.[42]

Falkner, in his analysis, concludes that minilateralism is unlikely to overcome structural barriers to a comprehensive and ambitious international climate agreement. That being said, minilateralist initiatives such as climate clubs may, among other things, enhance political dialogue within the context of existing multilateral negotiations and, thus, facilitate "great power bargaining,"[43] which, in itself, is a useful net gain, if no actual drawbacks to multilateralism are discernable.

Other studies seek to prescribe the *features* of effective minilateralism. For instance, Nordhaus provides extensive analysis on the pitfalls and likely facilitators of effective minilateralism by examining no less than 44 regimes across 15 regions.[44] Essentially, effective minilateralism seems to boil down to the presence of effective sticks and carrots to ensure the incremental increase in participation and ambition during the life of a club-like arrangement.

As discussed in greater detail in other parts of the present publication, examples would be regimes that contain trade sanctions or other trade barriers (e.g. taxes including carbon taxes, import tariffs, and/or carbon allowance requirements) for nonparticipants of a sufficient level that cannot be ignored.[45] For instance, when trade

---

41. J. Morgan and L. Weischer. *Two Degrees Clubs: How Small Groups of Countries Can Make A Big Difference on Climate Change* (World Resources Institute (October 29, 2012). https://www.wri.org/blog/2012/10/two-degrees-clubs-how-small-groups-countries-can-make-big-difference-climate-change.
42. J. Hovi, D.F. Sprinz, H. Sælen, and A. Underdal. "Climate Change Mitigation: A Role for Climate Clubs?." 2:16020 *Palgrave Communications* (2016).
43. R. Falkner, "A Minilateral Solution for Global Climate Change? On Bargaining Efficiency, Club Benefits and International LegitimacyLegitimacy," 14:1 *Perspectives on Politics* 87-101 (2016). Having analyzed the relevant literature and particular examples of clubs, Falkner concludes that minilateralism offers "no panacea for the ills of climate multilateralism. Most critically, climate clubs cannot pressurize or induce reluctant great powers to reduce their greenhouse gas emissions. However, a realistic approach to climate minilateralism, focused on coalitions of the willing, holds the promise of moving us beyond the current stalemate in international climate negotiations. The rise of minilateralism ... can be harnessed to strengthen an increasingly polycentric field of global governance. ... can inject political momentum into gridlocked international processes, provide new forms of collective leadership in a post-hegemonic world and reconcile existing multilateral regimes with shifts in the global power balance." (27).
44. W. Nordhaus, "Climate Clubs: Overcoming Free-Riding in International Climate Policy." 105:4 *American Economic Review* 1339-1370 (2015).
45. The appropriate level of a uniform *ad valorem* tariff on all imports from nonparticipants is not straightforward and would have to be conditional on other aspects of the climate club including its carbon pricing and particular membership. Nordhaus considers penalties for nonparticipants

sanctions are lower than the abatement costs necessary under a climate club, there is no tangible incentive (setting aside the need to meaningfully address climate change and avert further environmental degradation) for a nonparticipant state to join the club. In other words, if it costs less to abstain, states may choose to abstain. What is more, an effective climate club ought to contain consistent carbon pricing along with trade sanctions to induce substantial abatement. That said, Nordhaus suggests that the sustainability of such a regime is predicated on trade growth;[46] therefore, there seems to be insufficient evidence based on the modeling on whether effective carbon pricing can be sustained where there is no growth, given the implications of carbon pricing externalities for trade growth.

Following the analysis of minilateralist proposals and experiences, Weischer et al. also conclude on the sort of features that seem indispensable to optimal minilateralism, including the need for a shared vision and the offer of strong exclusive (primarily economic) incentives for members that are also compatible with existing international rules.[47]

Hovi et al. also conclude that effective minilateralism relies on the existence of a variety of factors, including conditional commitments (that is to say, that membership is conditional), exclusive *club goods* (that is to say, membership benefits), support from at least one global power (e.g. the EU or the US, and less so from China, which is less likely to be as enthusiastic), and higher marginal returns from participation.[48]

---

in the form of such tariffs varying from 0% to 10%. It is suggested that a rate of at least 4% could induce states to join a club which involves high-income states (given the implications for their trade exports), while a higher rate of 5% or more could be conducive to a club coming closer to achieving its abatement gains. However, even a tariff of 2% could induce a coalition that comes close to optimal abatement levels so long as the target carbon price is c. USD 50 per $MtCO_2$. *See* W. Nordhaus, "Climate Clubs: Designing a Mechanism to Overcome Free-riding in International Climate Policy." Presidential Address to the American Economic Association (January 4, 2014), (2, 24 & 29). This is also reflected in Nordhaus's subsequent publication, *see* W. Nordhaus, Climate Clubs: Overcoming Free-riding in International Climate Policy. 105:4 *American Economic Review* 1339-1370 (1356) (2015).

46. *See* W. Nordhaus. "Climate Clubs: Designing a Mechanism to Overcome Free-riding in International Climate Policy." Presidential Address to the American Economic Association (January 4, 2014), where it is stated that, "Assuming that international trade rises at the same time as world output and its composition is unchanged ... the optimal carbon price would need to rise by one-fifth relative to output and trade over the period to 2100 .... If the policy were to keep within a 2°C upper temperature limit, the target would [be around] $50 per ton CO_2 ... but it would grow more slowly than world GDP.... The modelling results indicate that modest trade penalties on non-participants can induce a coalition that approaches the optimal level of abatement as long as the target carbon price is less than $50 per ton range. The regime is *sustainable as long as world trade grows as fast as the optimal carbon price.* Such a regime would have incentives favorable for attracting a large majority of countries." (emphasis added) (32 & 33).
47. L. Weischer, J. Morgan, and M. Patel, "Climate Clubs: Can Small Groups of Countries Make a Big Difference in Addressing Climate Change?." 21 *RECIEL* 191, 192 (2012).
48. *See* J. Hovi, D.F. Sprinz, H. Sælen, an A. Underdal, "The Club Approach: A Gateway to Effective Climate Co-operation?." 49 *B. J. Pol. S* 1071-1096 (2017), where it is stated that, "First, the 'right' actors must be *enthusiastic.* In particular, the actors initiating the club must control a sufficiently large share of global emissions and income. The United States or the European Union (EU) (*but not China*) can under some conditions single-handedly initiate a club that can persist and attract other members. If both were enthusiastic, the prospects for at least moderate club success would appear bright, because in our model *the United States and the EU can in many cases entice China to join* by appealing to its self-interest." (emphasis added) (1073).

David Victor,[49] overall, also supports minilateralism and argues that there ought to be greater openness on the part of skeptics within the UN climate diplomatic community toward such solutions, given the limitations of UNFCCC-type multilateralism. Victor also argues that club-like arrangements should make specific pledges to the UNFCCC process and allow external scrutiny, both for the purposes of allaying fears concerning inter-regime antagonisms and to also encourage the sharing of best practice.[50] Moreover, Victor identifies two types of club member that effective clubs ought to draw in, given that their cooperation is likely to yield the greatest gains. On the one hand, there are those that are net contributors to global public goods by means of their considerable abatement efforts or targets (e.g., the EU, Norway, Sweden, and even the US), while, on the other, there is the type of club members that are reluctant "laggards" when it comes to assuming specific binding targets comparable to those of other members, and to allowing outside scrutiny (e.g. China).[51]

Among others, Brewer et al. have carried out a review of existing club-like arrangements, albeit in connection to carbon markets, to examine whether membership size is a determinant of success and whether club-like benefits can incentivize participation and compliance.[52] Among their findings is the notion that, while some climate clubs, when compared to multilateral organizations, may facilitate agreement, conditions on participation and compliance can still prevent participation and consensus building. In that sense, the sort of hurdles inherent to multilateralism may also be present in minilateralist solutions. As a starting point for potential participants to consider in designing a club arrangement, Brewer et al. refer to a number of issues that club rules may need to establish, including commitments to comparable targets, consistency across the membership (e.g. common definition of units, allowances, offsets etc.), the agreed scope of coverage, converging emissions-verification checks, compliance assurance, and a shared registry (or network of registries).[53]

Martin et al. list factors in assessing whether the top 15 global emitters are likely to participate in minilateralist arrangements, including their respective carbon dependence, domestic public opinion on climate change, official policy position, and involvement in climate coalitions.[54] They devise a "net likelihood" taxonomy that places each of the emitters under review in either the "likely" grouping or the "not likely." The former includes the EU as a bloc, three EU member states (France, Germany, and Italy), the UK, Canada, and Japan—which so far, represent six of the G7 members—along with Australia, Brazil, Mexico, and South Korea, while the latter

49. D. Victor. *The Case for Climate Clubs*. E15Initiative. Geneva: International Centre for Trade and Sustainable Development (ICTSD) and World Economic Forum, 2015 (8).
50. *Ibid.*, 5 & 6.
51. *Ibid.*, 9.
52. T.L. Brewer, H. Derwent, A. Błachowicz, and M. Grubb *Carbon Market Clubs and the New Paris Regime*. Networked Carbon Markets. World Bank, Washington, DC, 2016. World Bank. https://openknowledge.worldbank.org/handle/10986/25768.
53. *Ibid.*, 31.
54. N. Martin and J.C.J.M. van den Bergh, A multi-level climate club with national and sub-national members: theory and application to US states. 14:124049 *Environ. Res. Lett.* (2019).

grouping includes China and the US, along with India, Indonesia, Iran, Russia, Saudi Arabia, and South Africa.[55]

2 Climate Club Formation

While, undoubtedly, the formation of clubs—at the very least, coalitions of the *willing*—neither precludes nor, on the face of it, undermines concurrent or subsequent multilateralism, without the participation of the greatest culprits (i.e. highest GHG emitters), it simply cannot achieve the critical mass necessary to prevent temperature increases above the prescribed objectives (i.e. 1.5-2°C by 2100 relative to preindustrial temperature levels). The UNEP considers that the current emissions gap can be bridged, but that "unprecedented and immediate action" is necessary.[56] What this means is that the situation is so pressing that, at the very minimum, the top ten emitters must take serious abatement action in order to close the gap in time to avert potentially catastrophic knock-on effects to the climate. In that sense, there is no room for complacency and prevarication.

The science is settled. However, moral implorations and appeals to reason are clearly not enough as they have yet to yield significant abatement within the existing multilateralist setting. That being said, loci of minilateralism (even when they lack the highest emitters) could well serve as the nuclei upon which other states progressively latch, so long as effective incentives and penalties are present to incrementally lead to climate clubs of consequence when it comes to meeting global abatement goals. In fact, these are views supported by commentators who consider it useful to build clubs around enthusiastic and powerful states willing and able to contribute meaningfully.[57] Particularly fruitful would be concerted, meaningful action in relation to renewable energy storage, solar and wind energy, energy efficiency in appliances and passenger transport, afforestation, and halting deforestation.[58] However, the underlying assumption here is that:

> countries will act quickly and implement the most cost-effective measures in their national contexts [which] is evidently a very idealistic assumption, but it underlines the fact that the policies and technologies needed to bridge the gap are readily available and at limited costs.[59]

55. *Ibid.*, 4.
56. J. Christensen and A. Olhoff (2019). Lessons from a decade of emissions gap assessments. United Nations Environment Programme, Nairobi (6).
57. J. Hovi, D. F. Sprinz, H. Sælen, and A. Underdal, "The Club Approach: A Gateway to Effective Climate Co-operation?" 49 *B. J. Pol. S* 1091 (2017) and D. Victor *The Case for Climate Clubs*. E15 Initiative. Geneva: International Centre for Trade and Sustainable Development (ICTSD) and World Economic Forum, 2015 (3, 4, & 12).
58. *See* J. Christensen and A. Olhoff, Lessons from a Decade of Emissions Gap Assessments. United Nations Environment Programme, 2019, Nairobi. According to that report, these areas "present a combined potential of up to 21 $GtCO_2e$ per year by 2030, which is *more than sufficient* to get on a pathway to well below 2°C" (emphasis added). In fact, total emissions reduction could be as high as 38 $GtCO_2e$ per year (within an uncertainty range of 35-41 $GtCO_2e$) per year (6).
59. *Ibid.*, 6.

Furthermore, the viewpoint that *all* major emitters are needed for a climate club has been challenged by various commentators. Dillinger does not consider this essential and proposes a range or narrower constellations even arguing that *as few as two* parties could kickstart a new climate club: the obvious ideal being that of a China-US climate club. A bilateral precedent pertains to how initially intractable Japan-US negotiations on telecommunications actually resulted in the multilateralization of this policy area and the advent of the International Telecommunication Union. Another example of starting small is the current EU that had initially involved two (namely France and Germany), then six states with agreements on steel and coal, to then transform into the current global juggernaut presently comprising 27 member states. Other examples involve the 1947 General Agreement on Tariffs and Trade signed by 23 parties and how it resulted in the current World Trade Organization (WTO) multilateral trade order currently extending over 164 parties.[60]

Unsurprisingly, club efficiency is inescapably predicated on the *quality* and *quantity* of prospective members and their agreed abatement targets. While the most immediate thought may be on how to coopt many, if not all, of the largest emitters, it may be more instructive to focus on coopting a mixture of states that, among themselves, have much else to offer, including abatement capability, financing, geoeconomic clout, and leadership. Some have argued that it may be fruitful to start with a small group of leadership states. While this alone is unlikely to reduce the current global emissions gap, what it may achieve, among other things, is to facilitate technological development and transfer, "provide momentum," and become "transformational" by providing "proof-of-concept" for nonparticipants to emulate. Evidently, this would be contingent on the necessary political will on the part of climate leaders being forthcoming.[61]

For instance, a trilateral arrangement between China, the EU, and the US would obviously be ideal, given the fact that it involves the top three GHG emitters in absolute terms,[62] but also involves states representing the most responsible, the most capable, and the lesser developed. Dellinger further concludes that the "magic number" could be as little as three:

> The "magic number"—to recall, "the smallest possible number of countries needed to have the largest possible impact on solving a particular problem" or at least instigating crucial action towards a resolution—is, in the climate change context, three: the United States, the European Union, and China. This constellation would account for 49% of the world's total GHG emissions. And with Brazil as an increasingly likely fourth player, 52% of global GHGs would be accounted for. Scientifically, this is arguably not a high enough percentage to solve the ultimate problem, but ... broader multilateral action can be derived progressively from narrower beginnings, so even a coalition of "just" three or four initial parties

60. M.F. Dellinger, "Narrowed Constellations in a Supranational Climate Change Regime Complex: The 'Magic Number' Is Three." 37 *Fordham International Law Journal* 373 (401) (2014).
61. J. Morgan and L. Weischer. *Two Degrees Clubs: How Small Groups of Countries Can Make A Big Difference on Climate Change* (World Resources Institute, October 29, 2012).
62. *See* interactive chart at the World Resources Institute website using 2018 figures https://www.wri.org/blog/2020/12/interactive-chart-top-emitters.

> may be enough. So far, the international community has gained literally nothing from attempting the broadest possible solution; near-global treaty participation.[63]

As briefly mentioned earlier, given the Biden administration's re-accession to the Paris Agreement and US presidential statements expressing intention to provide climate leadership, coupled with EU enthusiasm to provide leadership and to further explore climate cooperation with China, such a prospect seems increasingly less chimeric. As the third highest emitter (and second highest GDP), the EU is evidently among the most appealing contenders for optimal climate club formation. What is more, it falls within the "likely" grouping under the Martin et al. taxonomy of prospective participants, mentioned earlier. Moreover, the EU currently pursues trade diplomacy with three other states that fall within the "likely" grouping, namely Japan, Mexico, and South Korea, and with one state from the "not likely" grouping, namely South Africa.[64] Additionally, the EU has a long history of being a climate leader with its ambitious (yet perhaps unrealistic, given the political and institutional realities)[65] targets. In that sense, it has the capacity to provide strong leadership globally, but also within an effective minilateralist climate club arrangement.

Other club-like arrangements could involve states such as India, Russia, Japan, South Korea, and Brazil, which could provide the right mix between capability, responsibility, and need/vulnerability when it comes to the configuration of members.[66]

Furthermore, discussion around climate clubs is not limited to national actors, and does extend to subnational, but also sectoral, with the obvious examples being the energy-intensive trade-exposed (EITE) sectors (including steel, aluminum, and concrete)[67] and other sectors (fashion, international maritime, and aviation).

In conclusion, arriving at the "right" configuration of participants to club-like climate arrangements is far from straightforward. Enticing the most polluting has not

63. *See* M.F. Dellinger, "Narrowed Constellations in a Supranational Climate Change Regime Complex: The 'Magic Number' Is Three." 37 *Fordham International Law Journal* 373 (438) (2014).
64. N. Martin and J.C.J.M. van den Bergh, "A Multi-level Climate Club with National and Sub-national Members: Theory and Application to US States." 14:124049 *Environ. Res. Lett.* (4) (2019).
65. J. Pisani-Ferry, a senior fellow at the Peterson Institute for International Economics and a senior fellow at the Brussels-based Bruegel think tank, in his Memorandum to the President of the European Commission on Concrete Initiatives for a More Outward-Looking, Geopolitical Europe, lauding aspects of President von der Leyen's State of the Union 2020 Address, highlights how the EU, in its present form, is "not wired" for the EU's climate agenda given its institutional limitations—after all, it remains a rules-based institution chiefly inward-looking with regard to its politics and its need to reach and sustain agreements between a disparate set of member states. *See* J. Pisani-Ferry, *Memo to the European Commission on Concrete Initiatives for a More Outward-Looking, Geopolitical Europe* (PIIE) (December 2, 2020). https://www.piie.com/blogs/realtime-economic-issues-watch/memo-european-commission-concrete-initiatives-more-outward.
66. Such configurations are explored at length by Dellinger. *See* M.F. Dellinger, "Narrowed Constellations in a Supranational Climate Change Regime Complex: The 'Magic Number' is Three." 37 *Fordham International Law Journal* 373 (412-420, 435-438) (2014).
67. Cement accounts for around 8% of global GHG emissions and steel for 7%-9%. *See* The Economist, "Climate change and innovation: Greenbacks for greenery," October 31, 2021, pp. 55-57, at 57. The fashion industry is responsible for around 15% of global GHG emissions, whereas aviation around 5% and shipping 3%-5%.

worked within multilateralist settings due to political and other pitfalls. Those are not necessarily absent in smaller settings. Nor is there a certain path to success. As Nordhaus puts it:

> An important question is, how would ... a Climate Club get started? Who would define the regime? Would it begin with a grand Bretton-Woods-type conference? Or would it evolve from a small number of countries who see the logic, define a regime, and then invite other countries to join?
>
> There are no clear answers to these questions. International organizations evolve in unpredictable ways. Sometimes, it takes repeated failures before a successful model is developed. The histories of the gold and dollar standards, cholera conventions, the WTO, the European Union, and the Internet all emphasize the unpredictability in the development of international regimes.... *The destination of a Climate Club is clear, but there are many roads that will get there* (emphasis added).[68]

3 Climate Clubs and the UNFCCC (Minilateralism Vis-à-Vis Multilateralism)

It should be noted from the outset that the UNFCCC contains provisions (cf. Articles 4(2) and 12) that expressly recognize the right of parties to come together to pursue particular policies and measures that contribute to the objectives of the UNFCCC. Moreover, provisions within the Paris Agreement reflect this openness toward minilateralist cooperation in recognizing that parties may choose to pursue voluntary cooperation in pursuit of their intended nationally determined contributions (INDCs) to allow for higher ambition in their mitigation efforts, but to also allow for reaching economies of effort (cf. Article 6).[69] In that sense, minilateralist arrangements are, on the face of it, compatible with the UNFCCC. However, this is not to say that such arrangements necessarily have positive implications for UNFCCC multilateralism.

It is crucial, therefore, to analyze each minilateralist initiative on its own merits to discern its likely impact on multilateralism. To that end, Biermann et al. have systematically not only analyzed examples of minilateralist climate-mitigation initiatives, but also conducted a systematic review of the relevant literature, to assess implications for global climate governance (and its cohesion) and how this issue is

68. W. Nordhaus, "Climate Clubs: Overcoming Free-riding in International Climate Policy." 105(4) *American Economic Review* 1339-1370, (1351-1352) (2015).
69. *See* The Economic Potential of Article 6 of the Paris Agreement and Implementation Challenges (2019) IETA, University of Maryland and CPLC. Washington, DC (15) which extensively analyzes the scope of Article 6 of the Paris Agreement for parties to work together to either lower the costs of achieving their pledges or to increase their ambition. Since parties have different abatement capabilities and global emissions shares, emissions trading under Article 6 may be necessary to incentivize parties with excess abatement capacity to become net-negative emitters to "compensate" for other parties unable currently to reach carbon neutrality. Also *see* World Bank. State and Trends of Carbon Pricing 2020 (May), World Bank, Washington, DC (49) and UNFCCC, Draft CMA Decision on the Rules, Modalities and Procedures for the Mechanism Established by Article 6, para. 4, of the Paris Agreement, December 15, 2019, for further detail.

examined in the literature.[70] They conclude that minilateralist initiatives may be initially assessed to establish whether they contribute to fragmentation in global climate governance.[71] Subsequently, those found to contribute to fragmentation may then be further categorized according to their "degree of fragmentation," i.e., on whether, overall, they have *synergistic*, *cooperative*, or *conflictive* fragmentative effects on global climate governance,[72] with the first two evidently being the least harmful.

Biermann et al. conclude that, for the most part, global climate governance appears to exhibit cooperative fragmentation, with its institutional hub/core being the UNFCCC, which lays down the foundations for multilateralist governance, while also recognizing that state parties with disparate realities (e.g. emissions history, development level, means, capabilities, and so on) may have common but differentiated responsibilities and respective capabilities (CDR-RC), thus acknowledging that global objectives may have to also rely on minilateralist solutions.

For instance, the Kyoto Protocol within the context of the UNFCCC provided for differentiation of obligations depending on the level of development of states, but also contained features (e.g. mechanisms such as emissions trading, funding arrangements, and the Clean Development Mechanism (CDM)) that go beyond the UNFCCC and do not apply to all its parties (e.g. the US is party to the UNFCCC, but not the Protocol, and this has fragmentative implications). The EU's Emissions Trading System (ETS), while in line with the UNFCCC and Kyoto Protocol, is systemically independent of those regimes and has clear collaborative effects vis-à-vis UNFCCC multilateralism. In that sense, fragmentation in global climate governance may provide abatement solutions not currently attainable within the existing multilateralist setting by facilitating the inclusion of more relevant actors and sectors than may be possible through a more

---

70. F. Biermann, P. Pattberg, H. van Asselt, and F. Zelli. "The Fragmentation of Global Governance Architectures: A Framework for Analysis." 9:4 *Global Environmental Politics* (14 & 15) (2009).
71. *Ibid.*, 19, where it is stated that "[I]n global environmental politics more than 700 multilateral agreements are in force. Most have evolved independently, cover different geographic and substantial scopes, and are marked by different patterns of codification, institutionalization, and cohesion. A decade ago, in response to this fragmentation, a United Nations task force recommended stronger cooperation between multilateral environmental institutions to facilitate synergies and promote policy coherence. This fragmentation has also been at issue in numerous policy proposals that call for the clustering and integration of environmental institutions, notably through establishment of a world environment organization, to create a less fragmented architecture in this field. While some observers support a world environment organization to tackle fragmentation in global environmental governance, others oppose the idea of 'organizational tinkering' and emphasize the benefits of a more fragmented architecture." *Ibid.*, 19.
72. *Ibid.*, 20-22. According to Biermann et al. an example of *synergistic fragmentation* would be the Vienna Convention for the Protection of the Ozone Layer (Vienna, March 22, 1985) and its Montreal Protocol on Substances that Deplete the Ozone Layer (Montreal, September 16, 1987) plus their subsequent amendments. An example of *cooperative fragmentation* would be in relation to a common issue but were there are different institutions, e.g., the UNFCCC and its Kyoto Protocol. An example of *conflictive fragmentation* would be where regimes are hardly connected and have conflicting sets of underlying objectives and principles, e.g., the 1995 Agreement on Trade-Related Aspects of Intellectual Property (TRIPS) (within the context of the World Trade Organization—NB. TRIPS is a "covered" WTO agreement binding on all WTO parties) and how aspects of it may interact with the 1992 Convention on Biological Diversity (CBD) (broadly within the context of the United Nations (UN)) in relation to how, among other things, genetic resources are fairly and equitably exploited.

"integrated but static architecture." Case in point is, for instance, how the Kyoto Protocol (or the Paris Agreement on Climate Change, for that matter) had not required emissions reductions from, among others, international aviation, yet the EU bloc, through the EU ETS, sought to cover this sector.[73]

However, other instances may have had more harmful implications. For instance, Weischer et al. consider that minilateralist initiatives such as climate clubs or coalitions were originally formed to compete with the UNFCCC, but that nowadays, for their most part, they complement the latter.[74] Moreover, Biermann et al. consider the now-defunct APP to be a departure from key UNFCCC regime features, including the differentiation between industrialized and developing states. Even though not comparable to the UNFCCC regime, they consider that the APP:

> still provide[d] an alternative to international climate action that may reduce incentives for complying with, or signing up to, international legally binding commitments.... Importantly, these instances of fragmentation in climate governance [were] intentional." The Asia-Pacific Partnership and similar proposals—backed by the United States—were created not out of ignorance of the climate regime but *because* of it, at a time when the climate convention and the Kyoto Protocol were well established and in force.[75]

This also chimes with the views of other commentators. Bäckstrand et al. have also suggested that the APP and other minilateralist initiatives had actually been designed as "alternative models or rival forums" to the UNFCCC. That being said, they find that successive waves of minilateralism have led to most of this initial inter-club "rivalry to have vanished," though they state it is too early to assume that this trend is irreversible.[76] Such fragmentation need not be *conflictive*; however, were an important actor such as the US to seek to form some minilateralist, even bilateral, arrangement with, say, an equally important actor such as China, where the abatement, etc. ambitions are substantially above what may currently be achieved multilaterally within the UNFCCC context.[77]

As Morgan and Weischer argue, an effective "transformational" climate club along the lines they envisage need not aim to replace the UNFCCC regime, which, after all, remains the one platform where *global* ambition and other key considerations, including equity, can keep being discussed and negotiated. In that sense, climate clubs could complement the UNFCCC regime, as they would enable their respective

73. *Ibid.*, 28 & 29. Biermann et al. also state that a fragmented architecture "might offer solutions that are specifically tailored for specific regions and thus increase equity by better accounting for special circumstances ... . Yet again, serious problems may outweigh benefits. First, conflictive fragmentation, where different actors pull in different directions, may complicate linkages with other policy areas."
74. L. Weischer, J. Morgan, and M. Patel "Climate Clubs: Can Small Groups of Countries make a Big Difference in Addressing Climate Change?." 21 *RECIEL* (191) (2012).
75. F. Biermann, P. Pattberg, H. van Asselt, and F. Zelli, "The Fragmentation of Global Governance Architectures: A Framework for Analysis." 9:4 *Global Environmental Politics* (22-24) (2009).
76. K. Bäckstrand, F. Zelli, and P. Schleifer. "The Legitimacy and Accountability in Polycentric Climate Governance." In A. Jordan, D. Huitema, H. van Asselt, and J. Forster (eds), *Governing Climate Change: Policentricity in Action*. Cambridge University Press, 2018, 338-356 (350).
77. F. Biermann, P. Pattberg, H. van Asselt, and F. Zelli, "The Fragmentation of Global Governance Architectures: A Framework for Analysis." 9:4 *Global Environmental Politics* (26 & 27) (2009).

memberships to assume more ambitious commitments than they would alone, and pull other countries with them. Furthermore, how efforts are coordinated between the UNFCCC and minilateralist arrangements such as climate clubs should not be neglected, not least to ensure that information and best practice sharing remain effective.[78] Furthermore, Falkner lauds Eckersley's climate club proposed model as the "most fully developed" in relation to formalizing links to the UNFCCC regime.[79] Other examples include the Non-state Actor for Climate Action (NAZCA), which is a platform of several minilateral arrangements over which the UNFCCC Secretariat has the possibility to keep track.[80]

Other analyses of global climate governance conclude that it is characterized by polycentrism, where there is an identifiable epicenter, namely the UNFCCC that provides the overarching normative and governance context, but where there are other loci of climate governance including public and private actors in relation to, among other things, market involvement (e.g. emissions trading, finance, etc.) and the Kyoto Protocol's CDM.[81] In that sense, again, minilateralist solutions involving institutional fragmentation may also involve interdependence among actors and could be seen as actually advancing UNFCCC multilateralist goals. Bäckstrand et al. also view that the multilateralist UNFCCC, through the subsequent Paris Agreement, institutionalized "hybrid multilateralism," where one witnesses an intensified interplay between multilateral and minilateral/transnational climate cooperation, whereby the UNFCCC Secretariat plays a coordinating role.[82]

---

78. *See* J. Morgan and L. Weischer. *Two Degrees Clubs: How Small Groups of Countries Can Make a Big Difference on Climate Change* (World Resources Institute (October 29, 2012), where it is stated that "Defining the relationship of such clubs to the UNFCCC could become an important issue in the ongoing negotiations—the clubs could formally or informally report to the UNFCCC. Going further, the actions of the club members could also be negotiated, recognized, and monitored within the UNFCCC." Also *see* L. Weischer, J. Morgan, and M. Patel, "Climate Clubs: Can Small Groups of Countries Make a Big Difference in Addressing Climate Change?." 21(3) *RECIEL* (2012), where the authors state that UNFCCC multilateralism should not be replaced given that "climate change is caused by the cumulative effects of all global emissions, all major emitters need to be brought into an agreement to avoid free rider problems that would undermine the effectiveness of any solution. Furthermore, the impacts of climate change are affecting all countries, especially the poorest and most vulnerable, so they need to have a voice in the decision-making process to ensure an ambitious outcome that will be accepted as legitimate." (178).
79. *See* R. Falkner, "A Minilateral Solution for Global Climate Change? On Bargaining Efficiency, Club Benefits and International LegitimacyLegitimacy," 14:1 *Perspectives on Politics* 87-101 (2016) and R. Eckersley, "Moving Forward in the Climate Negotiations: Multilateralism or Minilateralism?." 12(2): *Global Environmental Politics* 24-42 (33) (2012).
80. K. Bäckstrand, F. Zelli, & P. Schleifer (2018). The Legitimacy and Accountability in Polycentric Climate Governance. In A. Jordan, D. Huitema, H. van Asselt, & J. Forster (Eds.), *Governing Climate Change: Policentricity in Action.* Cambridge: Cambridge University Press. (351).
81. K. Bäckstrand, F. Zelli, & P. Schleifer (2018). The Legitimacy and Accountability in Polycentric Climate Governance. In A. Jordan, D. Huitema, H. van Asselt, & J. Forster (Eds.), *Governing Climate Change: Policentricity in Action* (338-356). Cambridge: Cambridge University Press.
82. *Ibid.* (345), where the authors state that, "the UNFCCC Secretariat tak[es] a role as facilitator or orchestrator of transnational climate action.... The Lima-Paris Action Agenda (which later morphed into the Marrakech Partnership for Global Climate Action) and the Non-state Actor Zone for Climate Action (NAZCA) were launched to galvanize the groundswell of actions on climate change mitigation and adaptation from cities, regions, businesses and civil society organizations."

Lastly, empirical insights suggest that minilateral approaches attract less support among the general public and the climate diplomatic community when compared to UNFCCC multilateralism, particularly when climate clubs fail to cover large shares of global emissions. That being said, support increases when other club-design factors such as differentiation in commitment levels, membership benefits/club goods, and sanctions for non-members are present.[83]

4 Legitimacy and Accountability

Questions of *legitimacy* and *accountability* often arise in relation to global climate governance, including minilateralist initiatives and polycentrism.[84] For instance, the UNFCCC regime is predicated, for the most part, on the voluntarist nature of international law. Parties voluntarily accede to it and assume the (general) international obligations under that treaty. It is then open to them to consent to the imposition of subsequent international obligations. In that sense, the UNFCCC Secretariat enjoys legitimacy and is accountable to the parties of the UNFCCC. This is a view also reflected empirically.[85] This could also be said for minilateralist initiatives such as climate clubs comprising states as members, but this is less so with transnational climate clubs that, in addition, comprise non-state actors such as industry/sectoral representatives and so on. Therefore, the implications such initiatives may have for UNFCCC multilateralism should also be examined in relation to how they actually interplay with the latter and, more specifically, whether and to what extent they provide scope for UNFCCC scrutiny.[86]

What is more, Bäckstrand et al. highlight how legitimacy may be intricately connected to questions of representation, and that climate minilateralism has exhibited "considerable lack of inclusiveness" of poorer states, such as small island states and

83. R. Gampfer, UNFCCC v. Minilateralism: Effects on Agreement Design Features on Support for Global Climate Governance Architectures, ETH Zürich, Center for Comparative and International Studies, 2014. Gampfer surveyed two constituencies: climate policymakers—namely, UNFCCC delegates—and the US general public.
84. For a discussion on polycentrism and governance across complex human systems, *see* Nobel Laureate E. Ostrom (December 8, 2009) *Beyond Markets and States: Polycentric Governance of Complex Economic Systems* (Nobel Lecture), among other writings by Ostrom.
85. *See* R. Gampfer, UNFCCC v. Minilateralism: Effects on Agreement Design Features on Support for Global Climate Governance Architectures, ETH Zürich (2014), Center for Comparative and International Studies (3) where survey conclusions include the notion that climate clubs lack the sort of legal legitimacy that UNFCCC possesses given the role and symbolism of the UN when it comes to multilateral negotiations over global issues.
86. *See* K. Bäckstrand, F. Zelli, and P. Schleifer, "The Legitimacy and Accountability in Polycentric Climate Governance," in A. Jordan, D. Huitema, H. van Asselt, and J. Forster (eds.), *Governing Climate Change: Policentricity in Action*. Cambridge University Press, 2018, 338-356, where it is stated "The problem of accountability at the transnational level is amplified in polycentric climate governance, where actors are both regulators and regulated, be it cities, intergovernmental agencies, carbon market actors or standard-setting organisations … . The range of accountability and legitimacy challenges is so varied in polycentric climate governance that the key task becomes one of analysing the dynamics and logics of legitimacy and accountability in each. Polycentricity includes governance arrangements requiring top-down, hierarchical accountability as well as horizontal, non-hierarchical (market, peer and reputational accountability)." (344).

least-developed economies, not only in relation to clubs such as those within the "global" context (e.g., G7, G8, G20), but also from more technology-specific initiatives, thus failing to reflect a more balanced participation in relation to abatement capability and climate vulnerability.[87] Furthermore, empirical evidence would suggest that:

> polycentric governance can equally be non-transparent and exclusive in providing closed venues for coalition building, trust and bargaining between powerful elites from government, market and civil society. This image is far from the normative ideal in polycentric theory: of multiple platforms and domains actively facilitating dialogue and deliberation between political decision-makers and affected stakeholders.[88]

Moreover, for Eckersley, "to restrict the negotiations of any anti-pollution treaty to the biggest polluters and to exclude victims of pollution simply because their pollution contribution is negligible" seems elitist and would run against "the basic principles of communicative justice."[89] To address such concerns, Eckersley's proposed model of climate club is also aimed at widening participation by seeking to also establish a "Climate Council" comprising major emitters and other members drawn from among the "most capable," "most responsible" and the "most vulnerable" groupings of states.[90]

Weischer et al. also highlight that lack of representation in minilateralist initiatives could result in privileging "the voices of those within clubs at the expense of those outside, reproducing existing international hierarchies," which could also mean that the abatement goals and ambition remain too weak.[91]

According to Bäckstrand et al., the picture has improved somewhat with what they term "third wave minilateralism," which, they assess, has had positive implications for inclusiveness and accountability. They cite a 2017 systematic comparison of 38 clubs which indicates that 33 of them are more open and inclusive, having foregone the "by-invitation-only" approach in "first and/or second wave" minilateralist initiatives.[92]

---

87. *See* K. Bäckstrand, F. Zelli, and P. Schleifer, "The Legitimacy and Accountability in Polycentric Climate Governance," In A. Jordan, D. Huitema, H. van Asselt, and J. Forster (eds.), *Governing Climate Change: Policentricity in Action* (2018), where they state that, "Minilateral climate coalitions or clubs, that is initiatives predominantly governed by a limited number of governments, have multiplied since the mid- 2000s. Established by elected state governments on the one hand, but excluding a large group of countries on the other, they have distinctive legitimacy and accountability implications...." (348 & 349).
88. *Ibid.* 341.
89. R. Eckersley, Moving Forward in the Climate Negotiations: Multilateralism or Minilateralism?'" 12:2 *Global Environmental Politics* 24-42 (33) (2012).
90. *See* R. Falkner, "A Minilateral Solution for Global Climate Change? On Bargaining Efficiency, Club Benefits and International LegitimacyLegitimacy," 14:1 *Perspectives on Politics* 87-101 (2016) and for specific details, R. Eckersley, "Moving Forward in the Climate Negotiations: Multilateralism or Minilateralism?," 12:2 *Global Environmental Politics* 24-42 (33) (2012).
91. L. Weischer, J. Morgan, and M. Patel, "Climate Clubs: Can Small Groups of Countries Make a Big Difference in Addressing Climate Change?," *RECIEL* 21 (184) (2012).
92. K. Bäckstrand, F. Zelli, and P. Schleifer, "The Legitimacy and Accountability in Polycentric Climate Governance," in A. Jordan, D. Huitema, H. van Asselt, and J. Forster (eds.), *Governing Climate Change: Policentricity in Action* (2018), where they mention that as a consequence "more than 120 countries are members of climate clubs today."

Notwithstanding the foregoing, it should be noted that more accountable initiatives do not necessarily result in optimal abatement outcomes.[93] Less "democratic" or "inclusive" institutions could, conceivably, be more effective as ambition would be less vulnerable to the vagaries of politics inherent to multilateralism. Dillinger makes this point to illustrate that "exclusive" minilateralism—i.e., climate club arrangements which exclude non-governmental organizations (NGOs) and the 175 or so states with the least emissions—may actually lead to the remaining parties to be better able to establish small, more responsive institutions/climate clubs. Moreover, there is precedent concerning "clubs within clubs" in the form of concentric circles, including the EU and the Eurozone, as well as the United Nations (UN) and the UN Security Council. In that sense, climate clubs need not be inclusive per se to attain optimal outcomes. As Dillinger puts it:

> One of the perhaps most prevalent premises promoted in climate discourse is the notion that a new treaty must be highly inclusive for reasons of procedural fairness and general democracy. This premise is proving dated or even outright false. While many legal scholars still refer to "all" major emitters, this notion denotes the ideal of international inclusiveness that has proved unrealistic so far.[94]

III CONCLUDING REMARKS

This chapter has provided an overview of the key considerations in the relevant discourse exploring or advocating minilateralist approaches to global climate governance. Analysis of the literature in connection to the minilateralist landscape suggests that existing climate clubs function either as *dialogue* or *implementation* forums, and that their implications for multilateralism vary. Some club-like arrangements may have *synergistic* fragmentative effects, others *cooperative*, while yet others may have *conflictive* effects for global climate governance.

Furthermore, they vary in their content: some may be predicated on carbon market arrangements, while others may have more specific policies to encourage financing, increased energy efficiency, or other modalities explored throughout this publication.

Moreover, the multilateralist setting (UNFCCC and subsequent agreements) contains scope for minilateralism. This is apparent also in the bottom-up nature of the Paris Agreement[95] that encourages coalitions of the willing, including among states and

93. *Ibid.* 347, where it is stated that "A glance at the wider literature on accountability in global environmental governance reveals further complexities and contradictions…; the rapid proliferation of accountability mechanisms in this domain has done little to stop the environment from deteriorating." NB. K. Bäckstrand et al. are referring to the work of C. Brandi and S. Bauer, *Climate Clubs: Potentials and Pitfalls for Enhancing Legitimacy in Global Climate Governance*. German Development Institute, 2017.
94. M.F. Dellinger, "Narrowed Constellations in a Supranational Climate Change Regime Complex: The 'Magic Number' Is Three." 37 *Fordham International Law Journal* 373 (2014) (405-411).
95. *See* World Bank. State and Trends of Carbon Pricing 2020 (May), World Bank, Washington, DC. (87) for more detail on the bottom-up approach to addressing climate change within the context of Article 6 of the Paris Agreement.

private actors. That being said, the current minilateralist landscape does not appear to be achieving gains on a par or higher than what could have been achieved within the UNFCCC multilateralist regime. In that sense, climate club arrangements need to be impactful in order to secure buy-in and optimal abatement. However, that remains the greatest challenge: namely how to surmount the very dynamics that are present in multilateralism. How to coopt the most impactful actors has received considerable scholarly attention, as have questions of what may be the golden mean when it comes to the configuration of members. The findings are not straightforward, as they are contingent on a multiplicity of factors.

What is more, the literature, for the most part, contains prescriptive insights into the features and configuration of prospective members for optimal club-like arrangements. The obvious features would be substantial ambition over and above current commitment levels. However, incentivizing participation to ambitious abatement commitments potentially harboring costly and anticompetitive effects relies on striking a shrewd balance between, on the one hand, exclusive club-like benefits for members and, on the other, impactful penalties for non-members. Sufficient benefits draw in those reluctant states that face high abatement costs. Penalties also incentivize buy-in when they are sufficiently impactful.

However, as is to be explored in subsequent parts of the present publication, there are considerable administrative and legal intricacies that would need to be addressed when it comes to the imposition of conditions or penalties (e.g. through the requirement for emission allowances and/or the imposition of levies, including import tariffs or carbon taxation affecting the end price of the goods or services from non-members). Concerns include the erosion of the current multilateral trade order—itself a considerable multilateralist achievement of the post-WWII order—what with the inherent threat of countermeasures when it comes to the unilateral imposition of uniform tariffs or carbon taxes to non-member exports. In that sense, minilateralist solutions need to address and explore all known-knowns and unknown-knowns, respectively, when it comes to blowback to the multilateralism concerning global climate governance and global trade governance.

What is more, it is no less an intricate matter to arrive at the precise configuration of participants for optimal outcomes. That being said, the literature suggests that one need not obsess with including all or any of the highest emitters from the outset. While optimal emissions coverage is the ideal[96] —an ideal after all that multilateralism has historically failed to attain—it need not be the starting point for club formation. Focusing on how to design and promote an effective, incrementalist solution could yield results in terms of advancing climate diplomacy, including facilitating "great power" dialogue, research and development, investment, and informational and technological exchange.

96. *See* R. Gampfer, UNFCCC v. Minilateralism: Effects on Agreement Design Features on Support for Global Climate Governance Architectures, ETH Zürich, 2014, Center for Comparative and International Studies (4). Gampfer lists a series of hypotheses including that, support for minilateralism in prospective member as well as non-member countries is higher the larger the share of global emissions regulated under a climate club arrangement.

While the relevant literature, for the most part, expresses an optimistic albeit cautious view on climate clubs,[97] it is reasonable to argue that such minilateralist arrangements, when pertaining to more ambitious policies within a small yet impactful coalition of the mighty and the willing, could certainly catalyze the existing state of global climate diplomacy by providing not only economic pressure, but also moral leadership to coopt non-members and thus expand progressively toward multilateralism. While legitimacy, inclusiveness, and accountability are not essential for the success of such minilateralist initiatives, addressing those concerns by, for instance, sufficiently subjecting or otherwise linking to multilateralist (i.e. UNFCCC) scrutiny could go some way to allay related concerns and encourage public acceptance.

In sum, many misconceptions surround climate diplomacy discourse. States ought not to strive for ideals and perfection, but for realism and effect. Effective climate clubs may contain pathways to optimal abatement efforts at the inter-state and transnational levels.

Let us know analyze climate clubs in the context of international trade.

97. Aptly summed up by Falkner when stating that "minilateralism offers *no panacea for the ills of climate multilateralism*. Most critically, climate clubs cannot pressurize or induce reluctant great powers to reduce their greenhouse gas emissions. However, a *realistic approach to climate minilateralism*, focused on coalitions of the willing, *holds the promise* of moving us beyond the current stalemate in international climate negotiations. The rise of minilateralism, often decried as a sign of the disintegration of the postwar multilateral order, *can be harnessed* to strengthen an increasingly polycentric field of global governance." (emphasis added) R. Falkner, "A Minilateral Solution for Global Climate Change? On Bargaining Efficiency, Club Benefits and International LegitimacyLegitimacy," 14:1 *Perspectives on Politics* 87-101 (2016). Also *see* A. Hagen and K. Eisenack, *International Environmental Agreements with Asymmetric Countries: Climate Clubs v. Global Cooperation*, 20th Coalition Theory Network Workshop (March 19-20, 2015), *Fondazione Eni Enrico Mattei*, Venice, Italy, where it is stated that it can generally be shown that "climate clubs are at least not detrimental to global cooperation...we need to conclude that the idea that climate clubs do benefit global climate protection has to be taken with precaution." (18).

CHAPTER 2

Climate Clubs and International Trade

This chapter provides a comprehensive examination of the greatest threat humanity faces today: climate change. It does so by looking at the legal, political, institutional, regulatory, and economic angles of climate change from the perspective of major emitters and the most relevant actors in the mitigation of climate change. This chapter offers a fresh understanding of the international trading system as a way to reach a prosperous, modern, and sustainable society that will help decarbonize the economy effectively. This chapter offers new ideas on the creation of a climate club linked to the international trade regime as an effective way to mitigate climate change. How should policies, regulations, laws, and agreements change to get us there? What incentives will be necessary to get there effectively? What has failed and why? We conclude that innovation and technology will be among the key elements of this equation, without which we will miss the goal.

I INTRODUCTION[1]

Academic scholarship has debated at length whether the international community is on the right track to fight climate change.[2] Droughts resulting from climate change are making people migrate. There are projections that the sea level will continue to rise.

1. An earlier version of this chapter was published as R. Leal-Arcas, "Climate Clubs and International Trade across the European and International Landscape," 29:3 *European Energy and Environmental Law Review* 72-88 (2020).
2. D. Esty (ed.) *A Better Planet: Forty Big Ideas for a Sustainable Future*. Yale University Press, 2019; Johan Rockström et al., "A Safe Operating Space for Humanity," 461 *Nature* 472-475 (2009), available at https://www.nature.com/articles/461472a; Garrett Hardin, "The Tragedy of the Commons," 162 *Science* 1243-1248 (1968); David Wallace-Wells, "The Uninhabitable Earth, Annotated Edition," *New York Magazine*, July 14, 2017, available at http://nymag.com/intelligencer/2017/07/climate-change-earth-too-hot-for-humans-annotated.html; Elizabeth Kolbert, *Field Notes from a Catastrophe*, Bloomsbury (2006), Updated 2015, chapters 1, 2, and 5; Kerry Emanuel, *What We Know About Climate Change*. MIT Press, 2007, pp. 3-93; Markowitz and Shariff, "Climate Change and Moral Judgment," 2 *Nature Climate Change* (April 2012); Will Steffen et al., "Planetary Boundaries: Guiding Human Development on a Changing Planet," *Science* (2015).

We argue that the climate-change problem is bigger than we thought and it is a threat multiplier. It has the potential of putting off the progress of humanity. Therefore, effective collective action is imperative.

Following the theory of the second-best option, the answer to uncertainty in climate-change science[3] cannot be inaction, and, therefore, we argue that it is better to do something about it than nothing. We also argue that, to have the climate-change issue under control, we would need a sustainable climate policy, not one that changes with the color of the flag depending on what voters decide in democratic capitalist systems.[4] This is so because the timescale of the climate-change problem is very long and the cost-benefit ratio tends to be negative in that the costs tend to be very high and the immediate benefits often minimal.

We make the claim that renewable energy is the foundation for climate-change policy. The world is asking for a carbon-free electricity.[5] Renewable energy still has the limitation of storage and the fact that the sun does not always shine (which is an issue for solar energy) and the wind does not always blow (which is an issue for wind energy).

Economic growth remains priority number one in many countries, as is energy security and technology transfer. Low-cost energy is key to the success of any economy, which might be achieved thanks to technology and innovation. Technology and innovation are imperatives in the fight against climate change. However, technology is also part of the challenge because no one knows what that future technology may be. We claim that innovation is part of the solution to a decarbonized economy in every sector, especially transportation, via electrification. We need new sources of

3. For an overview of the science of climate change, *see* Intergovernmental Panel on Climate Change, Report on Global Warming of 1.5 degrees C: Summary for Policymakers (2018), https://www.ipcc.ch/sr15/chapter/summary-for-policy-makers/; U.S. Global Change Research Program, Executive Summary, Climate Science Special Report: 4th National Climate Assessment (NCA4), Vol. 1, https://science2017.globalchange.gov/downloads/CSSR_Executive_Summary.pdf; Scripps CO2 Program, "Keeling Curve Lessons," http://scrippsco2.ucsd.edu/history_legacy/keeling_curve_lessons; Chris Mooney, "The Hockey Stick: The Most Controversial Chart in Science, Explained." *The Atlantic* (May 10, 2013), https://www.theatlantic.com/technology/archive/2013/05/the-hockey-stick-the-most-controversial-chart-in-science-explained/275753/; Zachos et al., "Trends, Rhythms, and Aberrations in Global Climate 65 Ma to Present," 292 *Science* 686 (2001).
4. A conservative approach to democratic capitalism like that of Sir Keith Joseph in the UK would argue the Government is the problem and that we should let the market run. For analyses on this point, *see* James Patterson, *Restless Giant: The United States from Watergate to Bush v. Gore*. Oxford University Press, 2005, Chapters 4-5; Paul Pierson, *Dismantling the Welfare State? Reagan, Thatcher, and the Politics of Retrenchment*, Cambridge University Press, 1990, Chapters 1-3, 6; Kevin Phillips, *The Emerging Republican Majority*, Arlington House, 1969.
5. *See* generally R. Leal-Arcas, A. Filis, M. Peykova, and M. Greger, "Towards a Carbon-Free, Decentralized, and Democratized System of Energy Generation," 35:2 *Connecticut Journal of International Law* 127-193 (2020); R. Leal-Arcas et al., "Electrifying the Energy Sector: The Case of Slovakia and the Czech Republic," 12 *Kentucky Journal of Equine, Agriculture, & Natural Resources Law* (2020); R. Leal-Arcas, A. Filis, M. Peykova, and M. Greger, "Towards a Carbon-Free, Decentralized, and Democratized System of Energy Generation," 35:2 *Connecticut Journal of International Law* 127-193 (2020); R. Leal-Arcas, M. Greger, and H. Knigge "Towards Energy Democratization," 31:1 *Colorado Natural Resources, Energy, and Environmental Law Review* (2020); R. Leal-Arcas, M. Peykova, V. Nalule, and P. Kara, "Decarbonizing the Energy Sector," XV *Journal of Animal and Natural Resource Law* 173-271 (2019).

energy supply that are cleaner and cheaper. Such objective can be achieved, among other things, through subsidies, a traditional funding mechanism. In the past, however, governments have not always chosen the right companies to make this goal a reality.

To reach these goals, this chapter will examine whether the creation of a climate club linked to the international trading system could be an effective platform to deal with all the challenges mentioned above: climate action, renewable energy, storage, cheaper energy, technology as well as innovation. This question matters because, if it is possible to achieve what we propose in this chapter, we would solve the main deficiency of the current legal system on climate-change mitigation, namely not being enforceable or legally binding, which has consequences for the effective mitigation of climate change.

After this introduction, the chapter first offers possible solutions to mitigate climate change effectively based on different levels of ambition. After making the case that there are currently unsustainable levels of consumption and production that need to end and be replaced with clean economic growth, this chapter provides a fresh and transformative understanding of the international trading system for the promotion of a sustainable and prosperous future. Understanding the trading system as an enabler of social justice, this chapter then provides specific architectures for the creation of a climate club linked to the international trade regime as an effective way to mitigate climate change and have a prosperous future for all.

Finally, this chapter investigates the role of international trade in reducing fossil-fuel consumption as a means toward sustainability. The use of fossil fuels contributes to climate change. So there is no excuse for inaction. According to The Economist, "by 2100, if the Paris agreement's preferred target to keep warming below 1.5 [degrees] C relative to preindustrial levels were met, sea levels would rise by 50 cm from today, causing worldwide damage to property equivalent to 1.8 % of global GDP a year. Failure to enact meaningful emissions reductions would push the seas up by another 30-40 cm, and cause extra damage worth 2.5 % of GDP."[6] The right investment in climate resilience[7] today can save a lot of money in losses tomorrow.[8] In other words, the cost of inaction will be greater than that of action.

6. The Economist, "Rising Seas: A World Without Beaches," August 17, 2019, pp. 11-12, at 11.
7. On climate resilience, *see* William Nordhaus, *Climate Casino*. Yale University Press, 2013, Chapter 13; Gernot Wagner and Martin Weitzman, *Climate Shock: The Economic Consequences of A Hotter Planet*. Princeton University Press, 2015, Preface, Chapters 1-3; "What do adaptation to climate change and climate resilience mean?" *UNFCCC*, available at https://unfccc.int/topics/adaptation-and-resilience/the-big-picture/what-do-adaptation-to-climate-change-and-climate-resilience-mean#eq-1; IPCC, "Climate Change 2014: Impacts, Adaptation, and Vulnerability," 2014, available at https://www.ipcc.ch/pdf/assessment-report/ar5/wg2/ar5_wgII_spm_en.pdf; Brad Plumer, "Five Big Ways the United States Will Need to Adapt to Climate Change," *New York Times* (November 26, 2018), available at https://www.nytimes.com/2018/11/26/climate/adaptation-us-climate-change.html; "Local Climate Adaptation and Resilience Plans," *Institute for Local Government*, available at https://www.ca-ilg.org/post/local-climate-adaptation-resilience-plans; "World Bank Group Announces $50 Million over 5 Years for Climate Change Adaptation and Resilience," World Bank, (January 15, 2019), available at https://www.worldbank.org/en/news/press-release/2019/01/15/world-bank-group-announces-50-billion-over-five-years-for-climate-adaptation-and-resilience; Ed King, "What Does It Mean to Be Climate Resilient?," *Climate Home*, November 17, 2014; Emily Boyd, "Climate Adaptation: Holistic Thinking Beyond Technology," *Nature Climate Change* (2017), available at https://www.nature.com/articles/ncl

We argue that, in light of the lack of international consensus on how to effectively tackle climate change, an upstream carbon tax with reinvestment (CTR) supported by border tax adjustments (BTAs) should be enacted unilaterally by the biggest emitters of greenhouse gases (GHG) to reduce fossil-fuel consumption and incentivize global carbon abatement. This upstream carbon tax should include reinvestment into renewable energies[9] and sustainable development. Although one cannot control GHG emissions outside one's jurisdiction via national legislation, BTAs may be a useful policy instrument. While carbon pricing alone cannot solely tackle the negative impacts of climate change, it is a key part of the solution.[10] It is also a simpler first step toward mitigating climate change than doing research on more cost-effective technology regarding carbon capture and sequestration[11] or solar energy.[12]

---

imate3211; World Resources Institute, "Climate Watch," available at https://www.climatewatchdata.org/; Amdavad Municipal Corporation, "Ahmedabad Heat Action Plan," 2018, available at https://www.nrdc.org/sites/default/files/ahmedabad-heat-action-plan-2018.pdf.

8. According to The Economist, "governments investing $1 in climate resilience today will save $5 in losses tomorrow." The Economist, "Rising Seas: A World Without Beaches," August 17, 2019, pp. 11-12, at 12.
9. On renewable energy, *see* Michael B. Gerrard, "Legal Pathways for a Massive Increase in Utility-Scale Renewable Generating Capacity," July 2017, available at https://scholarship.law.columbia.edu/cgi/viewcontent.cgi?article=3046&context=faculty_scholarship; Richard Kauffman, "Obstacles to Renewable Energy and Energy Efficiency," *Silos to Systems: Issues in Clean Energy and Climate Change*, available at http://environment.research.yale.edu/publication-series/documents/downloads/0-9/03-Kauffman.pdf; Dave Roberts, "Is 100% Renewable Energy Realistic? Here's What We Know," *Vox* (2018), available at https://www.vox.com/energy-and-environment/2017/4/7/15159034/100-renewable-energy-studies; Varun Sivaram, *Taming the Sun*. MIT Press, 2018, pp. 1-54; Jeff Ball et al. "The New Solar System," Steyer-Taylor Center for Energy Policy and Finance, March 2017, available at https://www-cdn.law.stanford.edu/wp-content/uploads/2017/03/2017-03-20-Stanford-China-Report.pdf; Jeffrey Ball, "The New Age of Renewable Energy," *The Cairo Review of Global Affairs*, 2018, available at https://www.thecairoreview.com/essays/the-new-age-of-renewable-energy/; Emma Foehringer Merchant, "US Renewable Energy Generation Now 'Within Striking Distance' of Nuclear," *Greentech Media*, 2018, available at https://www.greentechmedia.com/articles/read/renewable-energy-generation-nuclear-bnef#gs.HIB=PaI; Callaway et al. "Location, Location, Location: The Variable Value of Renewable Energy and Demand-Side Efficiency Resources," *Journal of the Association of Environmental and Resource Economists* (2017); William Moomaw, "The EPA Says Burning Wood to Generate Power Is 'Carbon-Neutral.' Is That True?," *The Conversation*, 2018, available at https://theconversation.com/the-epa-says-burning-wood-to-generate-power-is-carbon-neutral-is-that-true-95727.
10. OECD *Effective Carbon Rates: Pricing CO_2 Through Taxes and Emissions Trading Systems*. OECD Publishing, 2016, 15.
11. On carbon sequestration and high-tech carbon removal, *see* Elizabeth Kolbert. "Can Carbon-Dioxide Removal Save the World?," *New Yorker*, 2017, available at https://www.newyorker.com/magazine/2017/11/20/can-carbon-dioxide-removal-save-the-world; Bronson Griscom et al. "Natural Climate Solutions," 114:44 *PNAS* (2017), available at https://www.pnas.org/content/114/44/11645.short; Oscar Venter and Lian Pin Koh, "Reducing Emissions from Deforestation and Forest Degradation (REDD+): Game Changer or Just Another Quick Fix?," 1249 *Annals of the New York Academy of Sciences* 137-150 (2012); Peter Irvine et al. "Halving Warming with Idealized Solar Geoengineering Moderates Key Climate Hazards," 9 *Nature Climate Change* 295-299 (2019); Joshua B. Horton et al. "Solar Geoengineering and Democracy," 18:3 *Global Environmental Politics*5-24 (2018); David Wallace-Wells, "Solar Geoengineering May Be Our Last Resort for Climate Change. What If It Doesn't Work?," *New York Magazine* (August 8, 2018), available at http://nymag.com/intelligencer/2018/08/solar-geoengineering-climate-change.html; "Deforestation and Greenhouse Gases." U.S. Congressional Budget Office, 2012, available at https://www.cbo.gov/sites/default/files/112th-congress-2011-2012/reports/

It is essential that the energy trading system be at the heart of the global effort to reduce fossil-fuel consumption and mitigate climate change because the purely environmental efforts are lacking in rigor. This is highlighted by the US's intention to withdraw from the Paris Agreement on Climate Change. As of 2018, there were 51 carbon pricing initiatives that were in implementation or scheduled; 25 of these were emissions trading systems operating cap-and-trade policies mostly at the subnational level, and 26 of these were carbon taxes, primarily at the national level.[13] While these efforts show recognition of the importance of carbon pricing, we argue that reducing fossil-fuel consumption can be better achieved through a carbon tax with reinvestment.

II AIMING HIGH: LOOKING FOR SOLUTIONS

As early as 1981, the UK showed a TV documentary that warned about global warming.[14] Since the early 1990s, we have been observing the worst climate-change situation. It has been reported that "temperatures in the Arctic are warming twice as fast as the global average"[15] and the Arctic is warming much faster than everywhere else. Water trapped on land in Greenland increases the sea level when it melts into the Atlantic Ocean. As a result, Greenland's ice could raise the sea level by seven meters.[16] This is one of many risks of inaction on climate change. Others are extreme heat, severe inundations, and droughts. The countries with the highest levels of risk are mostly in South Asia, equatorial Africa, and small island states. To avoid catastrophic consequences resulting from climate inaction, countries throughout the world are trying to have climate policies to tackle such crude reality.[17]

1-6-12-forest.pdf; Jon Gertner, "Is It O.K. to Tinker with the Environment to Fight Climate Change?," *The New York Times*, 2017, available at https://www.nytimes.com/2017/04/18/magazine/is-it-ok-to-engineer-the-environment-to-fight-climate-change.html; Ingrid Visseren-Hamakers et al. "Trade-offs, Co-benefits and Safeguards: Current Debates on the Breadth of REDD+," *Current Opinion in Environmental Sustainability* (2012); David Victor et al., "The Geoengineering Option: A Last Resort Against Global Warming?," *Foreign Affairs*, 2009, available at https://www.foreignaffairs.com/articles/arctic-antarctic/2009-03-01/geoengineering-option; Margaret Skutsch, "The Evolution of International Policy on REDD+," *Climate Science* (2017).

12. The Economist, "Green New Democrats," June 15, 2019, p. 37.
13. World Bank Group, 2018. *State and Trends in Carbon Pricing*, Washington DC, May 2018: 39. In Australia, there was a carbon tax introduced by the Labor Government, which former Prime Minister Tony Abbott later axed. The rationale for the carbon tax was that, relative to its population, Australia produces more GHG emissions than almost any other rich country.
14. https://www.carbonbrief.org/warming-warning-1981-tv-documentary-warned-climate-change.
15. The Economist, "Ice Would Suffice," September 21, 2019, p. 93.
16. *Ibid*.
17. For instance, the Global 2000 Report to the President, commissioned by US President Jimmy Carter and published in 1980, is arguably the beginning of US policy on climate change. *See* https://www.cartercenter.org/resources/pdfs/pdf-archive/global2000reporttothepresident--enteringthe21stcentury-01011991.pdf.

1 Net-Zero Carbon Emissions by 2050 as the Landing Target

According to The Economist, "some 57% of Americans, and 84% of self-declared Democrats, say climate change is a big threat."[18] As a result of this threat in the US[19] and throughout the world, the European Commission,[20] several countries (mainly in Europe—which tend to be more committed to nature than other parts of the world[21]—and small island states[22]) as well as rich cities and regions (such as California) are aiming high and therefore setting new targets to mitigate climate

18. The Economist, "Climate-Change Targets: Omissions," October 19, 2019, pp. 12-13, at 12.
19. For analyses of 21st-century US climate change policy, *see* David Roberts, "The Green New Deal: Explained," *Vox* (March 30, 2019), available at https://www.vox.com/energy-and-environment/2018/12/21/18144138/green-new-deal-alexandria-ocasio-cortez; Jason Bordoff, "Getting Real about the Green New Deal," *Democracy: A Journal of Ideas* (March 25, 2019), available at https://democracyjournal.org/arguments/getting-real-about-the-green-new-deal; James A. Baker III et. al., "The Conservative Case for Carbon Dividends," *Climate Leadership Council* (February 2017), available at https://www.clcouncil.org/media/2017/03/The-Conservative-Case-for-Carbon-Dividends.pdf; Ernest Moniz, "Innovating a Green Real Deal," *Science* (June 14, 2019); Julia Pyper, "The Business Case for Carbon Pricing," *Greentech Media* (October 3, 2018), available at https://www.greentechmedia.com/articles/read/the-business-case-for-carbon-pricing#gs.43v321.
20. European Commission, "Green Deal," COM(2019) 640 final, December 11, 2019, available at https://eur-lex.europa.eu/resource.html?uri=cellar:b828d165-1c22-11ea-8c1f-01aa75ed71a1.0002.02/DOC_1&format=PDF.
21. Germany, for instance, despite having a conservative government at the moment, has a new climate-change package that aims to reach carbon neutrality by 2050 by putting a carbon price for sectors not included in the European Union's (EU) current emissions trading system, such as buildings and transport. *See* The Economist, "Germany: Not Good Enough," September 28, 2019, p. 49. The UK, France, Denmark support the idea of net-zero emissions by 2050.
22. Although they count for less than 1% of the world's GDP, population, territory and greenhouse gas (GHG) emissions, small island states have become a main player in climate change negotiations for the past three decades. *See* The Economist, "Small Islands and Climate Diplomacy," September 21, 2019, pp. 66-67, at 66. For an analysis of climate change in small island states, *see* Brad Plumer and Lisa Friedman, "Island Nations, with No Time to Lose, Take Climate Response Into Their Own Hands," *The New York Times*, 2017, available at https://www.nytimes.com/2017/11/17/climate/islands-climate-change-un-bonn.html; Timothée Ourbak and Alexandre K. Magnan, "The Paris Agreement and Climate Change Negotiations: Small Islands, Big Players," Regional Climate Change (November 17, 2017), available at https://link.springer.com/article/10.1007/s10113-017-1247-9; Adaptation Fund in the Small Island Developing States (November 3, 2017), available at https://www.adaptation-fund.org/wp-content/uploads/2017/11/Adaptation-Fund-in-the-SIDS-1.pdf; Global Environment Facility, "LDCs," available at https://www.thegef.org/topics/least-developed-countries-fund-ldcf; "Inside Kiribati: The Island Being Erased by Climate Change," *Al Jazeera Documentary* (October 10, 2017), available at https://www.youtube.com/watch?v=plzBaNaCIv4; Justin Worland, "The Leaders of These Sinking Countries Are Fighting to Stop Climate Change. Here's What the Rest of the World Can Learn," *TIME* (June 13, 2019), available at https://time.com/longform/sinking-islands-climate-change; Peter Yeung, "Climate Change Is Already Battering this West African City," *CityLab* (April 7, 2019), available at https://www.citylab.com/environment/2019/04/climate-change-rising-sea-levels-west-africa-coastal-cities/586651/; Tara John, "While the Rich World Braces for Future Climate Change, the Poor World Is Already Being Devastated By It," *CNN*, (April 1, 2019), available at https://www.cnn.com/2019/03/31/africa/poorest-hit-the-hardest-climate-change-mozambique-intl/index.html.

change.[23] Shopping malls, major energy companies,[24] corporations,[25] airports, and offices are following suit.[26] Although much of recent climate action is happening at the local/state level,[27] rather than the national/international level, the international dimension is necessary because climate-change policy is an issue of global commons.[28] In fighting climate change, all these actors are aiming at net-zero carbon emissions by 2050 as the landing target.[29]

These new targets, however, seem to be at odds with subsidizing fossil fuels, which is expensive, regressive, and damages the environment.[30] One deficiency of these targets is that they tend to ignore the impact of international trade on climate

23. For instance, Slovakia, Germany and other countries joined an alliance in 2019 to stop the construction of coal plants. *See* The Economist, "Climate Action: The Day after Tomorrow," September 28, 2019, pp. 56-57, at p. 56. *See also* W. Leal and R. Leal-Arcas (eds.) *University Initiatives in Climate Change Mitigation and Adaptation*, Springer, 2018.
24. In February 2020, BP's CEO pledged net-zero emissions by 2050. *See* A. Raval, "New BP Boss Bernard Looney Pledges Net Zero Carbon Emissions by 2050," *Financial Times*, February 12, 2020, available at https://www.ft.com/content/e1ee8ab4-4d89-11ea-95a0-43d18ec715f5?emailId=5e43e413ded3600004d25293&segmentId=3d08be62-315f-7330-5bbd-af33dc531acb.
25. At the 2019 UN climate summit, companies such as Nestle' and Salestorce pledged to reach net-zero GHG emissions by 2050. Amazon promised to buy 100,000 electric vehicles to reach net-zero GHG emissions by 2040. *See* The Economist, "Climate action: The day after tomorrow," September 28, 2019, pp. 56-57, at 56. Even major energy companies pledged investing in carbon capture and sequestration. *Ibid.* at p. 57. Which raises the following question: how do you transport the carbon from where it has been produced to where it will be captured?
26. The Economist, "Climate-Change Targets: Omissions," October 19, 2019, pp. 12-13, at p. 13.
27. *See* the example of the United States, where, in addition to the federal level, much climate-change governance and action is taking place at the subnational level. Nathan Rott, "Trump Moves to Let States Regulate Coal Plant Emissions," *NPR*, 2018, available at https://www.npr.org/2018/08/21/639396683/trump-moves-to-let-states-regulate-coal-plant-emissions; The New York Times, "What Is the Clean Power Plan, and How Can Trump Repeal It?," *The New York Times*, 2017, available at https://www.nytimes.com/2017/10/10/climate/epa-clean-power-plan.html; Umair Irfan, "Trump's EPA Just Replaced Obama's Signature Climate Policy with a Much Weaker Rule," *Vox* (June 19, 2019), available at https://www.vox.com/2019/6/19/18684054/climate-change-clean-power-plan-repeal-affordable-emissions; *Massachusetts v. EPA*, 549 U.S. 497 (2007), available at https://supreme.justia.com/cases/federal/us/549/497/; "Climate Change: An Evangelical Call to Action," The Evangelical Climate Initiative, 2006, available at https://www.npr.org/documents/2006/feb/evangelical/calltoaction.pdf?t=1578129166653; Bill McKibben, *The Global Warming Reader*. Penguin Books, 2011; US EPA, "Regulatory Impact Analysis for the Review of the Clean Power Plan: Proposal," 2017, https://www.epa.gov/sites/production/files/2017-10/documents/ria_proposed-cpp-repeal_2017-10.pdf; US EPA, "FACT SHEET: Proposed Affordable Clean Energy Rule—Overview," 2018, https://www.epa.gov/sites/production/files/2018-08/documents/ace_overview_0.pdf; Emily Holden, "Carbon Emissions up as Trump Agenda Rolls Back Climate Change Work," *The Guardian*, January 2019, https://www.theguardian.com/environment/2019/jan/08/carbon-emissions-trump-agenda-climate-change; Coral Davenport and Lisa Friedman, "How Trump Is Ensuring That Greenhouse Gas Emissions Will Rise," *New York Times*, November 2018, https://www.nytimes.com/2018/11/26/climate/trump-greenhouse-gas-emissions.html.
28. R. Leal-Arcas, "Sustainability, Common Concern and Public Goods," 9:4 *The George Washington International Law Review* 801-877 (2017).
29. The idea of net-zero carbon emissions is to take "as much carbon dioxide out of the atmosphere as you put in." *Ibid.* at 12. That said, some major GHG emitters (such as China, the US, India, and Russia) have not made such commitment to reach net-zero emissions by 2050.
30. The Economist, October 19, 2019, p. 8.

change[31] and do not always reflect the total greenhouse gas (GHG) emissions created by products consumed in one country but produced elsewhere.[32] A case in point is the products produced in China,[33] but consumed in other countries.[34]

31. According to The Economist, around 25% of global greenhouse gas emissions are related to trade. *See* The Economist, "Climate-Change Targets: Omissions," October 19, 2019, pp. 12-13, at 13. What mode of transport is used for trading goods or services (whether by air, train, road or sea) makes a difference for climate change, in that transport by sea is 70 times more carbon-efficient than by air. *Ibid.* and p. 73. So electrifying transportation will certainly help, though it is still difficult when it comes to transport by air or sea. Beyond transportation, the fashion and food industries produce far more GHG emissions than aviation. For a comprehensive analysis on the links between climate change and international trade, *see* R. Leal-Arcas, *Climate Change And International Trade*, Edward Elgar Publishing Ltd, 2013.
32. According to The Economist, the measurement gap between consumption and production emissions is very large. The UK consumes around 40% more carbon emissions than it produces and, in the case of the European Union, it is 19%. *See* The Economist, "Trade and Emissions," October 19, 2019, pp. 72-73, at 72.
33. For an analysis of China's climate change policy, *see* Manish Bapna and Lailai Li, "China's Climate Leadership: 3 Signs to Watch in the Year Ahead," *World Resources Institute*, 2017, available at http://www.wri.org/blog/2017/11/chinas-climate-leadership-3-signs-watch-year-ahead; Zhang Chao, "The Climate Change Promise of China's Belt and Road Initiative," *The Diplomat*, 2018, available at https://thediplomat.com/2018/01/the-climate-change-promise-of-chinas-belt-and-road-initiative/; Isabel Hilton, "How China's Big Overseas Initiative Threatens Global Climate Progress," Yale E360, 2019, available at https://e360.yale.edu/features/how-chinas-big-overseas-initiative-threatens-climate-progress; Edward Wong, "Can China Take the Lead on Climate Change? That Could Be Difficult," *The New York Times*, 2017, available at https://www.nytimes.com/2017/06/02/climate/china-climate-change-trump-paris-accord.html; Joanna I. Lewis and Kelly Sims Gallagher, "How China's Domestic Energy and Environmental Challenges Shape Its Global Engagement," in Regina S. Axelrod and Stacy D. VanDeveer (eds.), *The Global Environment: Institutions, Law, and Policy*. Sage Publishing (August, 2019); Geoff Dernbicki, "The Timely Disappearance of Climate Change Denial in China," *Foreign Policy* (May/June, 2017), available at https://foreignpolicy.com/2017/05/31/the-convenient-disappearance-of-climate-change-denial-in-china/#; Jennifer Duggan, "China Makes Carbon Pledge Ahead of Paris Climate Change Summit," *The Guardian*, 2015, available at https://www.theguardian.com/environment/2015/jun/30/china-carbon-emissions-2030-premier-li-keqiang-un-paris-climate-change-summit; Elizabeth Losos, Alexander Pfaff, and Lydia Olander, "The Deforestation Risks of China's Belt and Road Initiative," *Brookings*, 2019, available at https://www.brookings.edu/blog/future-development/2019/01/28/the-deforestation-risks-of-chinas-belt-and-road-initiative/; Javier C. Hernández and Adam Nagourney, "As Trump Steps Back, Jerry Brown Talks Climate Change in China," *The New York Times*, 2017, available at https://www.nytimes.com/2017/06/06/world/asia/xi-jinping-china-jerry-brown-california-climate.html; Yuli Shan et al. *Science Advances*, "City-Level Climate Change Mitigation in China" (June 27, 2018), available at https://advances.sciencemag.org/content/4/6/eaaq0390.full; Ye Qi and Jiaqi Lu, "China's Coal Consumption Has Peaked," Brookings Institution, January 22, 2018, available at https://www.brookings.edu/2018/01/22/chinas-coal-consumption-has-peaked/.
34. The Economist states that two thirds of China's GHG emissions exports end up in the OECD countries and that six main products are responsible for around 30% of trade-related GHG emissions, namely cares, electronics, textiles, iron, steel, and chemicals. *See* The Economist, "Trade and Emissions," October 19, 2019, pp. 72-73.

2 How to Get There?

2.1 Transition to Cleaner Energy

A way forward to reduce GHG emissions is by making a transition to cleaner sources of energy such as wind, solar or hydropower. As in any transition, there will be geopolitical winners and losers of a full transition to clean energy.[35] Moreover, when trying to effectively mitigate climate change, some people will be negatively affected, such as fossil fuel producers and coal miners, to name a few. Natural gas could be the transition fuel to renewable energy, since it is the least dirty of all fossil fuels, and fracking has provided easy and cheap access to it.[36] In fact, GHG emissions are decreasing in the US due to switching to natural gas, which is slowly replacing coal. As a result of fracking, there may be cheaper electricity, the coal industry is being negatively affected, and it is harder for renewable energy to compete. In other words, the dirtiest source of energy (i.e., coal) and the cleanest type of energy (i.e., renewable energy) are being affected by fracking, which provides access to natural gas.

Modern bioenergy, which is the production of sustainable biofuels, such as ethanol, biogas or biodiesel, can help mitigate climate change. These sustainable biofuels can play a role in decarbonizing sectors of the economy.[37] Big oil and gas corporations can also play a major role by cutting production by 40% by 2040 compared to 2019 levels.[38]

Equally important is investing in offshore wind, as the UK is doing very successfully, given that wind energy has a very small carbon footprint. In fact, for the first time since the industrial revolution, the UK generated in 2019 more of its energy

35. *See* for instance Indra Overland et al., "The GeGaLo Index: Geopolitical Gains and Losses after Energy Transition," 26 *Energy Strategy Reviews* 100406 (2019).
36. For further details on natural gas, *see* Megan O'Sullivan, *Windfall: How the New Energy Abundance Upends Global Politics and Strengthens America's Power*. Simon & Schuster, 2017, Chapters 3 and 7; Daniel Yergin, *The Quest: Energy, Security, and the Remaking of the Modern World*. Penguin Books, 2012, Chapter 16; Barth-Naffihan, Sohng & Saiers, "Methane in Groundwater Before, During, and After Hydraulic Fracturing of the Marcellus Shale," PNAS (2018); Julio Friedmann, Congressional Testimony, Senate Committee on Energy and Natural Resources, Enhancing the Future of CCUS (2019), available at https://www.energy.senate.gov/public/index.cfm/2019/5/full-committee-hearing-to-examine-ccus-and-to-receive-testimony-on-legislation; Great Plains Institute, "Three Things to Know about Changes to the 45Q Tax Credit for Carbon Capture and Storage Projects," March 2018, available at https://www.betterenergy.org/blog/three-things-know-about-changes-45q/.
37. On decarbonization, *see* R. Leal-Arcas, M. Peykova, V. Nalule, and P. Kara, "Decarbonizing the Energy Sector," XV *Michigan State Journal of Animal and Natural Resource Law* 173-271 (2019); R. Leal-Arcas, A. Filis, and V. Nalule, "Energy Decentralization and Decarbonization: The Case of Romania and Malta," 60:1 *Natural Resources Journal* (2020); R. Leal-Arcas, "Electric Vehicles, Sustainability, and Decarbonization," in A. Fortes Martin (ed.), *Movilidad Urbana Sostenible y Accion administrativa: Perspectiva social, estrategias jurídicas y políticas públicas de movilidad en el medio urbano*. Aranzadi, 2019.
38. Carbon Tracker, "To Deflate the Carbon Bubble and Protect Investors, Oil & Gas Companies Must Shrink," November 1, 2019, available at https://www.carbontracker.org/reports/balancing-the-budget/?segmentId=1d846c0f-a9f0-a1db-a1f9-95194756b089.

from zero-carbon sources than from fossil fuels.[39] When it comes to GHG emission mitigation,[40] there is potential for offshore renewable-energy technologies such as tidal, wave, thermal, and wind, to name but a few.[41] Wind and solar power are among the most attractive and fastest growing candidates of clean energy.[42] Currently, they represent 7% of global electricity generation, double the amount in 2013.[43] But for clean energy to be effective, it would need to be kept in storage, especially because the sun does not always shine (think of night-time) and the wind may not always blow.[44] Storage via better batteries would allow to release clean energy based on needs.[45] India, with its great engineers, could be a world leader in clean energy production engineering.

39. N. Asgari, "UK Hit Historic Renewable Power Milestone in 2019," *Financial Times*, January 1, 2020, available at https://www.ft.com/content/943ebb3c-2be6-11ea-a126-99756bd8f45e?segmentId = 3f81fe28-ba5d-8a93-616e-4859191fabd8.
40. For an analysis of the interactions between the various sustainable development goals, *see* P. Pradhan, L. Costa, D. Rybski, W. Lucht, and J.P. Kropp, "A Systematic Study of Sustainable Development Goal (SDG) Interactions," 5:1 *Earth's Future* 1169-1179 (2017).
41. On renewable energy, *see* Michael B. Gerrard, "Legal Pathways for a Massive Increase in Utility-Scale Renewable Generating Capacity," July 2017, available at https://scholarship.law.columbia.edu/cgi/viewcontent.cgi?article = 3046&context = faculty_scholarship; Richard Kauffman, "Obstacles to Renewable Energy and Energy Efficiency," *Silos to Systems: Issues in Clean Energy and Climate Change*, available at http://environment.research.yale.edu/publication-series/documents/downloads/0-9/03-Kauffman.pdf; Dave Roberts, "Is 100% Renewable Energy Realistic? Here's What We Know," *Vox* (2018), available at https://www.vox.com/energy-and-environment/2017/4/7/15159034/100-renewable-energy-studies; Varun Sivaram, *Taming the Sun*. MIT Press, 2018, pp. 1-54; Jeff Ball et al. "The New Solar System," Steyer-Taylor Center for Energy Policy and Finance, March 2017, available at https://www-cdn.law.stanford.edu/wp-content/uploads/2017/03/2017-03-20-Stanford-China-Report.pdf; Jeffrey Ball, "The New Age of Renewable Energy," *The Cairo Review of Global Affairs*, 2018, available at https://www.thecairoreview.com/essays/the-new-age-of-renewable-energy/; Emma Foehringer Merchant, "US Renewable Energy Generation Now 'Within Striking Distance' of Nuclear," *Greentech Media*, 2018, available at https://www.greentechmedia.com/articles/read/renewable-energy-generation-nuclear-bnef#gs.HIB = Pal; Callaway et al. "Location, Location, Location: The Variable Value of Renewable Energy and Demand-side Efficiency Resources," *Journal of the Association of Environmental and Resource Economists* (2017); William Moomaw, "The EPA Says Burning Wood to Generate Power Is 'Carbon-Neutral.' Is That True?," *The Conversation*, 2018, available at https://theconversation.com/the-epa-says-burning-wood-to-generate-power-is-carbon-neutral-is-that-true-95727.
42. The EU has its own clean energy regulation. *See* "Clean energy for all Europeans" package, https://ec.europa.eu/energy/en/topics/energy-strategy-and-energy-union/clean-energy-all-europeans.
43. The Economist, "Climate Change and Energy: A Bid for Better Batteries," November 30, 2019, p. 16.
44. For further details, *see* the work of ClearPath at https://clearpath.org/.
45. On electric cars and electrification more generally, *see* Paige Jadun et al., "Electrification Futures Study," NREL, 2018, available at https://www.nrel.gov/docs/fy18osti/70485.pdf; Ravi Manghani and Rory McCarthy, "Global Energy Storage: 2017 Year-in-Review and 2018-2022 Outlook," GTM Research, 2018, available at https://www.greentechmedia.com/research/report/global-energy-storage-2017-year-in-review-and-2018-2022-outlook#gs.sgLYFwQ; Chris Nelder, James Newcomb, and Garrett Fitzgerald, "Electric Vehicles as Distributed Energy Resources," Rocky Mountain Institute, 2016, available at https://www.rmi.org/wp-content/uploads/2017/04/RMI_Electric_Vehicles_as_DERs_Final_V2.pdf; Robert Walton, "Short Term Pain for Long Term Gain: EPRI Assesses Expanding US Electrification," Utility Dive, 2018, available at https://www.utilitydive.com/news/short-term-pain-for-long-term-gain-epri-assesses-expanding-us-electrificat/521496/; Bill Chappell, "Electric Cars Hit Record In Norway,

In terms of reliability, since solar and wind power are intermittent, they are currently not able to replace coal and gas power plants. One solution is better batteries that will enable to store clean electricity in larger quantities. Currently, only around 1 % of renewable energy is complemented by storage.[46] As demand for better batteries goes up, there will be production on a larger scale and the price will drop. Legislation can also play a major role in helping to reduce GHG emissions by mandating long-term storage.[47]

The implication of this transition to cleaner energy may be a stagnating or decreasing demand of fossil fuels in future.[48] Although that seems to be a good intention, predictions are that world will continue to demand oil up to at least 2050.[49] Even if 188 countries of the world pledged to curb their GHG emissions to avoid more than 2 °C of warming above preindustrial levels, studies show that the expected CO_2 emissions from fossil fuels over the next 20 years will be higher than is required to keep warming below 2 °C.[50] Just oil today is responsible for around 33 % of global energy use and around 30 % of CO_2 emissions.[51] To that, one should add that oil is constantly found in different parts of the world, which makes oil discoverers want to expand their oil production capacity.[52]

Making up Nearly 60 Percent of Sales in March," NPR, April 2019, available at https://www.npr.org/2019/04/02/709131281/electric-cars-hit-record-in-norway-making-up-nearly-60-of-sales-in-march; Kenneth Gillingham and James Sweeney, "Barriers to Implementing Low Carbon Technologies," *Climate Change Economics* (2012); Julia Pyper, "Utilities See Distributed Generation as a Challenge—and Owning It as the Solution," *Greentech Media* (2016), available at https://www.greentechmedia.com/articles/read/utilities-see-distributed-generation-as-a-challenge-and-owning-it-as-the-so#gs.KlLKMj0; Eric Wanless and Kelly Carlin, "Growing the Minigrid Market in Sub-Saharan Africa," *RMI* (2017), available at https://rmi.org/news/growing-minigrid-market-sub-saharan-africa/.

46. *See* The Economist, "Energy Storage: To Have and to Hold," November 30, 2019, pp. 60-61, at 60, citing Morgan Stanley, a bank.
47. For a comprehensive analysis in various EU countries, *see* R. Leal-Arcas et al, "Smart Grids in the European Union: Assessing Energy Security, Regulation & Social and Ethical Considerations," 24:2 *Columbia Journal of European Law* 291-389 (2018); R. Leal-Arcas, M. Peykova, V. Nalule, and P. Kara, "Decarbonizing the Energy Sector," XV *Michigan State Journal of Animal and Natural Resource Law* 173-271 (2019); R. Leal-Arcas, M. Greger, and H. Knigge "Towards Energy Democratization," 31:1 *Colorado Natural Resources, Energy, and Environmental Law Review* 101-185 (2020); R. Leal-Arcas, A. Filis and V. Nalule, "Energy Decentralization and Decarbonization: The Case of Romania and Malta," 60:1 *Natural Resources Journal* (2020).
48. For various analyses on energy, *see* R. Leal-Arcas (ed.), *Commentary on the Energy Charter Treaty*. Edward Elgar, 2018; R. Leal-Arcas and J. Wouters (eds.) *Research Handbook on EU Energy Law and Policy*. Edward Elgar, 2017; R. Leal-Arcas et al., *Energy Security, Trade and the EU: Regional and International Perspectives*. Edward Elgar, 2016; R. Leal-Arcas, *The European Energy Union: The Quest for Secure, Affordable and Sustainable Energy*. Claeys & Casteels Publishing, 2016; R. Leal-Arcas et al., *International Energy Governance: Selected Legal Issues*. Edward Elgar, 2014.
49. The Economist, "The Goliath," November 2, 2019, pp. 61-63, at 62.
50. The Economist, "Climate Hypocrisy," November 23, 2019, p. 73.
51. The Economist, "To the Last Drop," November 2, 2019, p. 11.
52. For instance, the United Arab Emirates said in November 2020 that it had found 24 billion barrels-worth of oil underneath Abu Dhabi and that it hopes production capacity will increase by 25 % to 5m barrels per day by 2030. In the US, the Energy Information Administration forecast that the US would increase its oil production from 11.2m barrels per day in November 2020 to 11.4m barrels per day by the end of 2021. *See* The Economist, "Oil production: Opening the taps," December 12, 2020, p. 66.

Other ways forward to reduce GHG emissions are by drastically lowering the use of coal (something Germany is failing to do).[53] This approach raises the question of what to do with people hired in the high-carbon industry, taking into account the no one left behind principle.[54] Equally, one could have a properly integrated energy grid.[55] Doing so will offer immediate and meaningful benefits to society at large such as improved health and cleaner air. In fact, the price of fossil fuels does not reflect the actual cost of climate change: there are health issues directly linked to fossil-fuel consumption. The digital revolution can certainly help lower GHG emissions. If the benefits of any policy are tangible, then policies and legislation via activism are easier to get through national parliaments. That said, one cannot control GHG emissions outside one's country using one's national legislation other than, perhaps, a border tax adjustment or taxing the consumer, rather than the producer.

2.2 Communication

To get to meaningful GHG emissions reduction will require good communication[56] to citizens about these issues via ecumenical dialogue, technological improvements

53. However, in the US, the coal industry is been replaced by renewable energy and natural gas. *See* The Economist, November 2, 2019, p. 9.
54. *See* for instance the movement "Just transition," at https://www.e3g.org/showcase/just-transition/.
55. *See* for instance Rafael Leal-Arcas, *Solutions for Sustainability: How the International Trade, Energy, and Climate Change Regimes Can Help*, Springer International Publishing, 2019, Chapters 6-9; R. Leal-Arcas, M. Greger, and H. Knigge "Towards Energy Democratization," 31:1 *Colorado Natural Resources, Energy, and Environmental Law Review* 101-185 (2020); R. Leal-Arcas, M. Peykova, V. Nalule, and P. Kara, "Decarbonizing the Energy Sector," XV *Michigan State Journal of Animal and Natural Resource Law* 173-271 (2019); R. Leal-Arcas, F. Leniewska, and F. Proedrou, "Smart Grids in the European Union: Assessing Energy Security, Regulation & Social and Ethical Considerations," 24:2 *Columbia Journal of European Law* 291-389 (2018).
56. For analyses of climate change perceptions, *see* Yale Program on Climate Change Communication, "Politics and Global Warming." Executive Summary (December 2018), available at http://climatecommunication.yale.edu/publications/politics-global-warming-december-2018/2/; William Nordhaus, *The Climate Casino*. "Climate Science and its Critics," Yale University Press, 2013, Chapter 24, pp. 293-326; Renee Lertzman, "In Climate Change, Psychology Often Gets Lost in Translation," *Pacific Standard*, 2015, available at https://psmag.com/environment/in-climate-change-psychology-often-gets-lost-in-translation; Cameron Brick and Sander van der Linden, "Yawning at the Apocalypse," *The Psychologist: Climate Change* (September 2018), pp. 30-34, available at https://thepsychologist.bps.org.uk/volume-31/september-2018/yawning-apocalypse; Hiroko Tabuchi, "In America's Heartland, Discussing Climate Change Without Saying 'Climate Change,"" *The New York Times*, 2017, available at https://www.nytimes.com/2017/01/28/business/energy-environment/navigating-climate-change-in-americas-heartland.html; Shannon Hall, "Exxon Knew about Climate Change Almost 40 Years Ago," *Scientific American* (October 26, 2015); Campbell and Kay, "Solution Aversion: On the Relation Between Ideology and Motivated Disbelief," *Journal of Personality and Social Psychology* 107, 809 (2014); Marx et al. "Experiential and Analytic Processing of Uncertain Climate Information," *Global Environmental Change* 17, pp. 47-58 (2007); Suzanne Goldberg, "Work of Prominent Climate Change Denier Was Funded by Energy Industry," *The Guardian*, 2015, available at https://www.theguardian.com/environment/2015/feb/21/climate-change-denier-willie-soon-funded-energy-industry; Julia Corbett, "Media Power and Climate Change," *Nature Climate Change* (2015); Maxwell Boykoff and Gesa Luedecke, "Elite News Coverage of Climate Change," *Climate Change Communication* (2016); Dan Kahan, "Climate-Science Communication and the

(thankfully, the technology is getting cheaper), and major investments throughout the world.[57] Major GHG emitters such as China and India[58] are now investing very much in renewable energy,[59] making sure that they uphold their pledges of the Paris Agreement[60] on Climate Change.[61] In fact, one-fourth of the world's solar panels and

---

Measurement Problem," 36 *Advances in Political Psychology* 1-43 (2015), available at https://ssrn.com/abstract=2459057; Connie Roser-Renouf et al., "Message Strategies for Global Warming's Six Americas," *Handbook of Environment and Communication*, 2014, Yale Program on Climate Change Communications, available at http://climatecommunication.yale.edu/publications/global-warmings-six-americas-book-chapter-preview/.

57. *See* the results of a study by Tsinghua University cited by The Economist, which states that without decarbonization, the Belt and Road countries will account for 66% of global GHG emissions by 2050, compared to 28% today. *See* The Economist, "Banyan: Cold Command," September 21, 2019, p. 46.
58. According to The Economist, "twelve of the world's 15 most polluted cities are in India ..., and the country ranks 120th of 122 on the global index of water quality. *See* The Economist, "Special report: India," pp. 1-12, at 10, October 26, 2019. For an analysis of India's climate change policy, *see* Varun Sivaram, "Still Shining? Our Third Annual Review on Solar Scale-Up in India," *Council on Foreign Relations*, 2018, available at https://www.cfr.org/blog/still-shining-our-third-annual-review-solar-scale-india; Damien Carrington and Michael Safi, "How India's Battle with Climate Change Could Determine All Our Fates," *The Guardian*, (November 6, 2017), available at https://www.theguardian.com/environment/2017/nov/06/how-indias-battle-with-climate-change-could-determine-all-of-our-fates; Adil Najam, "The View from the South: Developing Countries in Global Environmental Politics," in Regina S. Axelrod and Stacy D. VanDeveer (eds.), *The Global Environment: Institutions, Law, and Policy*, Sage Publishing, August, 2019; Aaron Atteridge, Manish Kumar Shrivastava, Neha Pahuja, and Himani Upadhyay, "Climate Policy in India: What Shapes International, National and State Policy?," AMBIO, 2012, available at http://link.springer.com/article/10.1007%2Fs13280-011-0242-5; Tom Osborn, "Why Developing Countries are Disproportionately Affected by Climate Change—and What Can They Do About it," *Huffington Post*, 2015, available at https://www.huffingtonpost.com/tom-osborn/why-developing-countries-_b_6511346.html; Barbara Buchner, "Innovative Climate Finance: A Spotlight on India," *Climate Action*, 2016, available at http://www.climateactionprogramme.org/climate-leader-papers/innovative_climate_finance_a_spotlight_on_india; Mendelsohn et al. "The Distributional Impact of Climate Change on Rich and Poor Countries," 11 *Environment and Development Economics* 159-178 (2006).
59. That said, in the first half of 2019, China's investment in renewable energy fell by around 40%. *See* L. Hook, "Climate Change: How China Moved from Leader to Laggard," *Financial Times*, November 25, 2019, available at https://www.ft.com/content/be1250c6-0c4d-11ea-b2d6-9bf4d1957a67?segmentId=6132a895-e068-7ddc-4cec-a1abfa5c8378.
60. For an analysis of the Paris Agreement and climate change in general, *see* William Nordhaus, *Climate Casino*, Yale University Press, 2013; ES. "History of UN Climate Change Talks, available at https://www.c2es.org/content/history-of-un-climate-talks/; IPCC, *Fifth Climate Change Assessment Report: Synthesis Report*. 2014, available at http://www.ipcc.ch/pdf/assessment-report/ar5/syr/AR5_SYR_FINAL_SPM.pdf; Daniel Bodansky and Lavanya Rajamani, "Evolution and Governance Architecture of the Climate Change Regime," in Detlef Sprinz and Urs Luterbacher (eds.), *International Relations and Global Climate Change: New Perspectives*, 2nd edition. MIT Press, 2016; Daniel Esty and Peter Boyd, "To Move Paris Accord Forward, Bring Cities and Companies on Board." *Yale Environment 360*, 2018, available at https://e360.yale.edu/features/to-move-paris-accord-forward-bring-cities-and-companies-on-board; Daniel Esty and Dena Adler, "Changing International Law for a Changing Climate," *American Journal of International Law*, 2018, available at https://www.cambridge.org/core/services/aop-cambridge-core/content/view/DB50564B97DEE8620D818BFFCB9BBC94/S2398772318000764a.pdf/changing_international_law_for_a_changing_climate.pdf; Decision of the Parties to the 2015 Paris Conference of the Parties (COP21), available at https://unfccc.int/sites/default/files/english_paris_agreement.pdf; UN Environment, "Emissions Gap Report." 2018, available at https://www.unenvironment.org/resources/emissions-gap-report-2018; "Essential Elements of the Paris 'Rulebook'," Center for Climate and Energy Solutions, November 2018, available at https://

one-third of the world's electricity-generating capacity from wind are currently in China.[62] Norway, although not a major GHG emitter, is a good example of a country whose electricity is relatively cheap thanks to renewable energy, specifically hydro-power. Meaningful GHG emissions reduction will also require the political and social dimension of climate change to come together.

2.3 *Putting a Price on Carbon*

Further ways in which climate change can be mitigated is by having carbon taxes[63] or other forms of putting a price on carbon[64] that may result in consumers wanting less

---

www.c2es.org/site/assets/uploads/2018/11/essential-elements-paris-rulebook.pdf; Ted Nordhaus, "The Two-Degree Delusion: The Dangers of an Unrealistic Climate Change Target" *Foreign Affairs*, February 8, 2018, available at https://www.foreignaffairs.com/articles/world/2018-02-08/two-degree-delusion; Jules Kortenhorst, "The Paris Agreement: 10 Key Takeaways For the Global Energy Landscape," Rocky Mountain Institute, 2015, available at https://rmi.org/news/blog_2015_12_18_the_paris_agreement_10_key_takeaways/; "Adapt, Curb, Engage: 21 Solutions to Protect Our Shared Planet," Cities4Climate, 2015, available at https://envirocenter.yale.edu/sites/default/files/files/21%20Solutions%20to%20Protect%20Our%20Shared%20Planet.pdf.

61. In November 2019, the US formally notified the United Nations of its intention to withdraw from the Paris Agreement, in accordance with its Article 28. The US will therefore leave the Paris Agreement in November 2020, one year after formal notification. The Trump administration argues that the Agreement hurts American businesses.
62. The Economist, "Climate Change Targets: From Smog to Slog," September 21, 2019, pp. 47-48, at 47.
63. Ian Parry and William Pizer, "Emissions Trading Versus CO2 Taxes Versus Standards" Resources for the Future, 2007, available at https://www.rff.org/publications/issue-briefs/emissions-trading-versus-co2-taxes-versus-standards/; Kotchen et al., "Public Willingness to Pay for a US Carbon Tax and Preferences for Spending the Revenue," *Environmental Research Letters* (2017), DOI: https://doi.org/10.1088/1748-9326/aa822a.
64. For analyses on carbon pricing, *see* William Nordhaus, "Revisiting the Social Cost of Carbon." *PNAS* (2017), available at https://doi.org/10.1073/pnas.1609244114; "The true cost of carbon pollution." EDF, available at https://www.edf.org/true-cost-carbon-pollution; Gillingham, Carattini, and Esty, "Lessons from First Campus Carbon-Pricing Scheme," *Nature* (2017), available at https://www.nature.com/news/lessons-from-first-campus-carbon-pricing-scheme-1.22919; World Bank Group, "State and Trends of Carbon Pricing 2017," 2017, available at https://openknowledge.worldbank.org/bitstream/handle/10986/28510/wb_report_171027.pdf?sequence=7; "Economists' Statement on Carbon Dividends," *Wall Street Journal* (2019), available at https://www.clcouncil.org/media/EconomistsStatement.pdf; Karen Palmer, Anthony Paul, and Amelia Keyes, "Changing Baselines, Shifting Margins: How Predicted Impacts of Pricing Carbon in the Electricity Sector Have Evolved Over Time," Resources for the Future, 2018, available at http://www.rff.org/files/document/file/RFF%20WP%2018-15.pdf; Brad Plumer and Nadja Popovich, "These Countries Have Prices on Carbon. Are They Working?" *The New York Times*, April 2, 2019, available at https://www.nytimes.com/interactive/2019/04/02/climate/pricing-carbon-emissions.html?mtrref=www.google.com; Kotchen et al. "Public Willingness to Pay for a US Carbon Tax and Preferences for Spending the Revenue," Environmental Research Letters (2017), available at https://doi.org/10.1088/1748-9326/aa822a; C2ES, "The Business of Pricing Carbon: How Companies Are Pricing Carbon to Mitigate Risks and Prepare for a Low-Carbon Future," 2017, available at http://www.c2es.org/document/the-business-of-pricing-carbon-how-companies-are-pricing-carbon-to-mitigate-risks-and-prepare-for-a-low-carbon-future/; "How Can Carbon Prices and Policies Be Effectively Aligned?," Carbon Pricing Leadership Coalition, 2016, available at http://pubdocs.worldbank.org/en/221021478831141991/CPLC-Executive-Brief-Policy-Alignment-Nov2016-FINAL.pdf;

stuff.[65] The economic rationale for putting a price is that the price of almost everything is too low, which explains why there are large quantities of so many things. So how can we get a price signal strong enough to change behavior and to make sure we limit GHG emissions or pay for the harm caused? Using a carbon tax, which is a moving target, has the advantage that countries control the price of carbon, but it also implies that countries will not know by how much GHG emissions are being reduced, whereas a cap-and-trade system (such as the emissions trading system in different parts of the world)[66] enables countries to cap emissions up to a limit. If the cap is too loose, one can always tighten it to get more change-behavior. However, the disadvantage of a cap-and-trade system is that countries do not know how much revenue they will generate or what the price of carbon will be.

What would be the right price on carbon? A High-Level Commission on carbon prices produced a report to provide some clarity on the matter.[67] An example of carbon pricing is the Carbon Offsetting Scheme for International Aviation (CORSIA).[68] That said, there are several obstacles to carbon pricing. One is unfairness, in that the economic burden will fall mainly on the poorest people.[69] Another obstacle is the status

---

"Carbon Pricing: Gaining a Competitive Edge in a Climate-Constrained World," available at https://cbey.yale.edu/programs-research/carbon-pricing-gaining-competitive-edge-in-climate-constrained-world.

65. *See* the seminal work by Martin Weitzman on how regulators should deal with pollution, whether via a price on carbon or tradable pollution permits.
66. On January 1, 2020, Switzerland became the first country to link its GHG emissions trading system (ETS) with the EU ETS. *See* Council Decision (EU) 2017/2240 of 10 November 2017, https://eur-lex.europa.eu/legal-content/EN/TXT/HTML/?uri = OJ:L:2017:322:FULL&from = EN. China and other countries have their own ETS.
67. High-Level Commission on Carbon Prices (2017), Report of the High-Level Commission on Carbon Prices, Washington, DC: World Bank, available at https://static1.squarespace.com/static/54ff9c5ce4b0a53decccfb4c/t/59b7f2409f8dce5316811916/1505227332748/CarbonPricing_FullReport.pdf.
68. For further details on CORSIA, *see* https://www.iata.org/contentassets/fb745460050c48089597a3ef1b9fe7a8/corsia-fact-sheet.pdf.
69. For further analysis on climate-change justice, *see* Michael Gerrard and Edward McTiernan, "Patterns of Climate Change Litigation During Trump Era," *New York Law Journal*, 2018, available at http://columbiaclimatelaw.com/files/2018/03/070031820-Arnold.pdf; "Climate Change and Future Generations Lawsuit in Colombia: Key Excerpts from Supreme Court Decision," *Dejusticia*, (April 13, 2018), available at https://www.dejusticia.org/en/climate-change-and-future-generations-lawsuit-in-colombia-key-excerpts-from-the-supreme-courts-decision/; Angelique Chrisafis, "Who are the gilets jaunes and what do they want?" *The Guardian* (December 7, 2018), available at https://www.theguardian.com/world/2018/dec/03/who-are-the-gilets-jaunes-and-what-do-they-want; Anabella Rosemberg, "Embedding Just Transition in Long-term Decarbonization Strategies: What, Why, and How," *World Resources Institute*, available at https://www.wri.org/climate/expert-perspective/embedding-just-transition-long-term-decarbonization-strategies-why-what; Laura Parker, "143 Million People May Soon Become Climate Migrants," *National Geographic*, (March 19, 2018), available at https://news.nationalgeographic.com/2018/03/climate-migrants-report-world-bank-spd/; Umair Irfan, "Playing Hooky to Save the Climate: Why Students are Striking on March 15," *Vox* (March 14, 2019), available at https://www.vox.com/2019/2/21/18233206/march-15-climate-strike; Brooke Jarvis, "Climate Change Could Destroy His Home in Peru. So He Sued an Energy Company in Germany," *The New York Times* (April 9, 2019), available at https://www.nytimes.com/interactive/2019/04/09/magazine/climate-change-peru-law.html?smid = nytcore-ios-share; George David Banks, "Poorer and Richer States Should Be Held to the Same Greenhouse Standards" *The Hill* (July 22, 2018), available at http://thehill.com/opinion/energy-environment/398239-poor

quo, in that people tend to be uncomfortable with change. A third obstacle is competitive disadvantage, in that those affected will feel they are being burdened, whereas others are not. All these obstacles raise the following question: how can we build a political coalition to work together on carbon pricing?

2.4 Policy Choices and Eco-Labelling

Moreover, what sorts of incentives are necessary for consumers to reach zero GHG emissions? Simple policy choices such as charging for the use of plastic bags are making a difference to GHG emissions reduction in various parts of the world. Also, better labelling will help consumers know the carbon footprint of what they buy and, therefore, can help mitigate climate change.[70] Consumers increasingly seek information on the production and process methods of the products they buy and request eco-labelling. But from an international trade perspective, labelling raises the following question: does it create obstacles to international trade? One way around this question is by reassuring consumers that the energy they buy is green.

2.5 Blue Economy

The blue economy is another opportunity to tackle climate change. It offers shipping companies the chance to develop zero-carbon fuels and new technologies that will cut CO_2 emissions from the shipping industry by 50% by 2050.[71] A blue paper commissioned by the High Level Panel for a Sustainable Ocean Economy states that countries could do much more regarding the blue economy to comply with the Paris Agreement.[72]

---

er-and-richer-states-shouldnt-be-held-to-the-same-green-house; Amity A Doolittle, "The politics of indigeneity: Indigenous strategies for inclusion in climate change negotiations," *Conservation & Society* (2010); Josephine Balzac, "Climate Justice: Promoting Corporate Responsibility Through the Divestment of Fossil Fuels and Socially Conscious Investment," *The Environmental Forum* (July/August 2018); Chukwumerije Okereke and Philip Coventry, "Climate Justice and the International Regime: Before, During, and After Paris," 7 *WIREs Climate Change* 834-851 (2016); "Principles of Climate Justice," Mary Robinson Foundation, available at https://www.mrfcj.org/wp-content/uploads/2015/09/Principles-of-Climate-Justice.pdf; World Bank (2010), Equity in Climate Change, available at http://elibrary.worldbank.org/doi/pdf/10.1596/1813-9450-5383.

70. Council of the European Union, "Labelling of tyres: Council endorses agreement to make labels more visible," Press release, November 22, 2019, available at https://www.consilium.europa.eu/en/press/press-releases/2019/11/22/labelling-of-tyres-council-endorses-agreement-to-make-labels-more-visible/?utm_source=dsms-auto&utm_medium=email&utm_campaign=Labelling+of+tyres%3a+Council+endorses+agreement+to+make+labels+more+visible#.
71. The Economist, "Climate-Change Risks and Opportunities for the Blue Economy," December 21, 2019, p. 108.
72. S. Gaines et al., "The Expected Impacts of Climate Change on the Ocean Economy," Washington, DC: World Resources Institute, 2019, available at http://oceanpanel.org/sites/default/files/2019-12/expected-impacts-climate-change-on-the-ocean-economy.pdf.

2.6 *Fashion, Shipping, and Aviation*

Neither shipping nor aviation is mentioned in the Paris Agreement. This is a deficiency of the Paris Agreement in that it only deals with countries when it comes to reducing GHG emissions, but not industries. In fact, some industries (for instance, fashion, aviation or shipping) emit more GHGs than many countries. Which is why, in addition to making sure countries decarbonize their economies, it is crucial to have a sector-specific approach to GHG emission reduction for transnational industries. Predictions are that, if we continue business as usual, GHG emissions from aviation and shipping will increase due to population growth and economic growth.

Shipping has been rather slow at capping its GHG emissions. Only in 2018 did the delegates of the International Maritime Organization announce that 170 of its members had agreed to reduce their emissions from shipping to just about 50% of the 2008-levels by 2050.[73] The Economist believes that "even with the agreed cuts, shipping [...] will probably see its share of emissions rise in coming years."[74] Equally important is making use of sustainable fuel in transportation, especially commercial aviation, to reduce GHG emissions from that industry. Having global standards on the measurement of sustainability would allow the aviation industry to get fuel certified as sustainable in different airports throughout the world.[75]

2.7 *Forests*

Forests, a natural carbon sequestration, are a further way to remove and store carbon from the atmosphere.[76] They absorb most when trees are young. However, deforestation,[77] done to expand agriculture, is an issue in some countries (such as Brazil,[78] Ecuador, Bolivia, Indonesia, Nigeria, Philippines, Thailand and Malaysia), which can

73. Josh Gabbatiss, "Carbon Emissions from Global Shipping to Be Halved by 2050, Says IMO," *The Independent*, April 13, 2018, available at https://www.independent.co.uk/environment/ships-emissions-carbon-dioxide-pollution-shipping-imo-climate-change-a8303161.html.
74. T. Standage (ed.), *Uncommon Knowledge: Extraordinary Things That Few People Know*, The Economist, p. 206, 2019.
75. *Ibid.*
76. Elizabeth Kolbert, "Can Carbon-Dioxide Removal Save the World?" *New Yorker*, 2017, https://www.newyorker.com/magazine/2017/11/20/can-carbon-dioxide-removal-save-the-world; Bronson Griscom et al., "Natural Climate Solutions," *PNAS* 114(44), 2017, https://www.pnas.org/content/114/44/11645.short.
77. Oscar Venter and Lian Pin Koh, "Reducing Emissions from Deforestation and Forest Degradation (REDD+): Game Changer or Just Another Quick Fix?," *Annals of the New York Academy of Sciences* 1249 (2012), pp. 137-150; "Deforestation and Greenhouse Gases," U.S. Congressional Budget Office, 2012, available at https://www.cbo.gov/sites/default/files/112th-congress-2011-2012/reports/1-6-12-forest.pdf; Jon Gertner, "Is It O.K. to Tinker with the Environment to Fight Climate Change?" *The New York Times*, 2017; Ingrid Visseren-Hamakers et al., "Trade-offs, co-benefits and safeguards: current debates on the breadth of REDD+," *Current Opinion in Environmental Sustainability* (2012); Margaret Skutsch, "The Evolution of International Policy on REDD+," *Climate Science* (2017).
78. In the academic literature, Brazil has been considered a pivotal state. *See* R. Chase, E. Hill, and P. Kennedy (eds.) *The Pivotal States: A New Framework for U.S. Policy in the Developing World*, W.W. Norton & Co. Inc., 1998.

be mitigated by a natural climate solution, namely the promotion of reforestation. Solar geoengineering, a type of high-tech carbon removal that manipulates the environment, is another way forward.[79] But it comes with governance issues attached, namely who gets to decide?

The consequences of climate change[80] will be visible in long-term economic growth[81] and social welfare. Therefore, a cleaner and more liveable planet is in the interest of everyone.

III ENDING UNSUSTAINABLE LEVELS OF PRODUCTION AND CONSUMPTION

An argument has been made that uncontrolled levels of economic growth and consumption would eventually exhaust the world's resources.[82] If so, such a situation raises the question whether the rich world would need to renounce some natural resources to allow the poorer world to develop. It would appear, however, that more and more natural resources are found since that argument was made in the 1970s. As a result, the burning of fossil fuels in the past 50 years has accelerated global warming.[83]

79. R. Leal-Arcas and A. Filis-Yeloghotis, "Geoengineering a Future for Humankind: Some Technical and Ethical Considerations," 6:2 *Carbon & Climate Law Review* 128-148 (2012); Peter Irvine et al., "Halving Warming with Idealized Solar Geoengineering Moderates Key Climate Hazards," 9 *Nature Climate Change*295-299 (2019); Joshua B. Horton et al., "Solar Geoengineering and Democracy," 18:3 *Global Environmental Politics*5-24 (2018); David Wallace-Wells, "Solar Geoengineering May Be Our Last Resort for Climate Change. What If It Doesn't Work?," *New York Magazine* (August 8, 2018); David Victor et al., "The Geoengineering Option: A Last Resort Against Global Warming?," *Foreign Affairs*, 2009.
80. For a policy history on climate change, *see* Cass R. Sunstein, *Worst-Case Scenarios*, 71-117 (2007); Ian Parry and William Pizer, *Emissions Trading Versus CO2 Taxes Versus Standards*, Resources for the Future (2007), https://www.rff.org/publications/issue-briefs/emissions-trading-versus-co2-taxes-versus-standards/; Kelly Levin & Benjamin Cashore, et al., *Overcoming the Tragedy of Super Wicked Problems: Constraining Our Future Selves to Emeliorate Global Climate Change, Policy Sci* 45 (2012); Nathaniel Rich, "Losing Earth: The Decade We Almost Stopped Climate Change," *The New York Times Magazine* (August 1, 2018), https://www.nytimes.com/interactive/2018/08/01/magazine/climate-change-losing-earth.html; David Roberts, "The Once and Future Democratic Consensus on Climate Change," *Vox* (November 18, 2017), https://www.vox.com/energy-and-environment/2017/11/18/16669094/democratic-consensus-on-climate-change; Justin Gundlach, "How Much Does the Existing Regulatory Patchwork Reduce U.S. Greenhouse Gas Emissions?," *Sabin Center for Climate Change, Columbia Law School* (2015), https://doi.org/10.7916/D8959HDJ; "The Waxman-Markey Bill: A Good Start, or a Non-Starter?," *Yale Environment* 360 (June 18, 2019), https://e360.yale.edu/features/the_waxman-markey_bill_a_good_start_or_a_non-starter; Amanda Reilly and Kevin Bogardus, *7 Years Later, Failed Waxman-Markey Bill Still Makes Waves*, E&E News (June 27, 2016), https://www.eenews.net/stories/1060039422.
81. There are economists that argue that the expansion of property rights is essential for modern economic growth, although privatizing property rights does not necessarily lead to a higher productivity boom. *See* The Economist, "Free Exchange: Common Sense," September 14, 2019, p. 72.
82. D. Meadows et al., *The Limits to Growth*, Earth Island, 1972.
83. The Economist, "Schumpeter: The Stuff Paradox," October 19, 2019, p. 66.

1 Clean Economic Growth

Rational choice dictates that people would prefer a better dishwasher and a bit of climate change caused than a not so good dishwasher but help save the planet. This is so because people care more about clean dishes than energy efficiency.[84] Further, there is clear evidence that people consume more as they become richer and as goods become cheaper, which means that the market for these goods explodes.[85] That said, even if people consume more, companies use fewer resources to make stuff,[86] the economy is becoming more efficient, and the world is moving beyond the era of resource-heavy goods.[87] These facts demonstrate that it is possible to grow micro- and macroeconomically and, at the same time, have GHG emissions controlled or even stagnated, as depicted in Figure 2.1 exemplifying the case of Russia.[88]

---

84. Michael Gerrard, *The Law of Clean Energy: Efficiency and Renewables*, ABA Book Publishing, 2011, pp. 1-22; Sheryl Carter, "Ramping Up Energy Efficiency Key to Address Climate Change," *Natural Resources Defense Council*, September 22, 2016; IEA, "Energy Efficiency 2017," 2017; Jim Tankersley, "Can We Nudge People Into Conserving Energy?" *Washington Post*, 2014; Kenneth Gillingham, David Rapson, and Gernot Wagner, "The Rebound Effect and Energy Efficiency Policy," *Review of Environmental Economics and Policy* (2015); Amory Lovins, "Energy Strategy: The Road Not Taken?," *Foreign Affairs*, 1976; Suh et al., "Life Cycle Environmental and Natural Resource Implications of Energy Efficiency Technologies," *Journal of Industrial Ecology* (2016).
85. According to The Economist, the number of cell phones between 1990 and 2011 went up from 11m to 6bn. *See* The Economist, "Schumpeter: The Stuff Paradox," October 19, 2019, p. 66.
86. *See* A. McAfee, *More from Less: The Surprising Story of How We Learned to Prosper Using Fewer Resources—And What Happens Next*, Simon & Schuster UK, 2019.
87. E. Brynjolfsson and A. McAfee, *The Second Machine Age: Work, Progress, and Prosperity in a Time of Brilliant Technologies*, W. W. Norton & Co., 2014.
88. Incidentally, until October 2019, Russia was the only major GHG emitter that had not ratified the Paris Agreement on Climate Change. It accepted the Paris Agreement on October 7, 2019. For an update on ratification, acceptance, and approval of the Paris Agreement, *see* https://treaties.un.org/Pages/ViewDetails.aspx?src=TREATY&mtdsg_no=XXVII-7-d&chapter=27&clang=_en#4.

Figure 2.1 Russia's GDP Versus CO_2 Emissions, 1998-2018

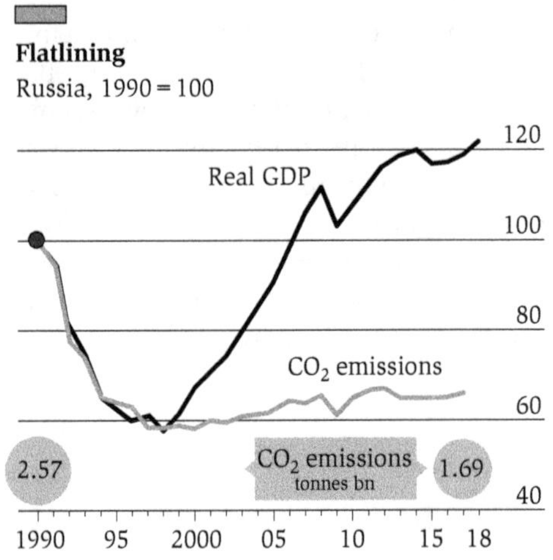

Source: The Economist, "Why Russia is ambivalent about global warming," September 19, 2019, available at https://www.economist.com/europe/2019/09/19/why-russia-is-ambivalent-about-global-warming.

2 Prosperity and Environmental Protection

Therefore, as they grow economically, businesses and countries can and are becoming greener/cleaner in their production of goods. In fact, the computer age promises to decouple environmental degradation from economic growth via dematerialization and competition.[89] This decoupling makes ever more sense as the threat of climate change increases.

That is what the European Commission is trying to demonstrate with its Green Deal:[90] that you can be prosperous and prioritize environmental protection. The EU Green Deal is not just about GHG emissions reduction; it is also about boosting innovation, green finance, modern mobility and much more. However, as is the case with environmental legislation, the issue is always enforceability. With its new Green Deal, by 2030 the European Union pretends to reduce its GHG emissions by 50-55% from 1990 levels,[91] which is a higher ambition than the existing target of 40% GHG emission reduction by 2030.[92] That target, however, does not seem enough for

89. The Economist, "Schumpeter: The Stuff Paradox," October 19, 2019, p. 66.
90. European Commission, "Green Deal," COM(2019) 640 final, December 11, 2019, available at https://eur-lex.europa.eu/resource.html?uri=cellar:b828d165-1c22-11ea-8c1f-01aa75ed71a1.0002.02/DOC_1&format=PDF.
91. *Ibid.* p. 4.
92. For an overview of the EU' climate change policy, *see* Dieter Helm, "The European Framework for Energy and Climate Policies," *Energy Policy* (2014); Lucas Merrill Brown, Alex Hanafi, and Annie Petsonk, "EU Emissions Trading Scheme Lessons Learned," *Environmental Defense Fund*,

environmental NGOs, which are demanding the EU cut its emissions by 65% by 2030 from 1990 levels because that is what environmental models suggest is required for the EU to keep warming below 1.5-2°C above preindustrial levels, as agreed by the Paris Agreement on Climate Change.[93]

One innovation of the EU's Green Deal is that it connects climate policies to the trading system, namely the EU will not conclude trade agreements with third parties that do not comply with the Paris Agreement's decarbonization requirement.[94] Equally, the EU's Green Deal also proposes a carbon border-adjustment mechanism,[95] which the EU views as a necessary scheme for EU companies to compete internationally on a level playing field due to the EU's climate-change mitigation ambitions.[96] Rather than embryonic, such initiatives should be dominant throughout the world.

Even if there is no institutionalized global environmental governance[97] per se because it poses a threat to national sovereignty, governments are now more active

---

2012, available at https://www.edf.org/sites/default/files/EU_ETS_Lessons_Learned_Report_EDF.pdf; Simon Evans, "The History of Energiewende," *Carbon Brief*, 2016, available at https://www.carbonbrief.org/timeline-past-present-future-germany-energiewende; "Spain Halts Feed In Tariffs," *Institute for Energy Research*, 2012, available at http://www.institutefor energyresearch.org/2012/04/09/spain-halts-feed-in-tariffs-for-renewable-energy; Nicola McEwen and Alexandra Raymond, "The Repatriation of Competences in Climate and Energy Policy after Brexit," *Brexit & Environment* (February 26, 2019), available at https://www.brex itenvironment.co.uk/2019/02/26/the-repatriation-of-competences-in-climate-and-energy-polic y-after-brexit/; Alissa J. Rubin and Somini Sengupta, "'Yellow vest' protests shake France. Here's the Lesson for Climate Change," *The New York Times* (December 6, 2018), available at https://www.nytimes.com/2018/12/06/world/europe/france-fuel-carbon-tax.html; Miguel Arias Cañete, "A new ambition for energy efficiency in Europe," *The European Files*, 2018, available at https://www.europeanfiles.eu/energy/new-ambition-energy-efficiency-europe-2; Alexandre Affre, Adrian van den Hoven, and Bruno Pedrotti, "A Competitive EU Energy and Climate Policy," *Business Europe*, 2013, available at https://bdi.eu/media/presse/publikatione n/klima-und-umwelt/20130618_FINAL_Brocure_2030_energy_and_climate_LOW_RESOLUTI ON.pdf.

93. The Economist, "The EU's Green Deal: The Way the Wind Is Blowing," December 14, 2019, pp. 23-24, at 23.
94. European Commission, "Green Deal," COM(2019) 640 final, p. 21, December 11, 2019, available at https://eur-lex.europa.eu/resource.html?uri=cellar:b828d165-1c22-11ea-8c1f-01aa75ed71a 1.0002.02/DOC_1&format=PDF.
95. *Ibid.* p. 5.
96. S. Fleming and C. Giles "EU Risks Trade Fight over Carbon Border Tax Plans," *Financial Times*, October 16, 2019.
97. All of this forces us to revisit the question of global environmental governance. For further discussion, *see* Katherine Michonski and Michael A. Levi, "Harnessing International Institutions to Address Climate Change," CFR, 2010, available at http://www.cfr.org/climate-change/harn essing-international-institutions-address-climate-change/p21609; Robert Keohane and David Victor, "Cooperation and discord in global climate policy," *Nature Climate Change* (2016); Michele M. Betsill and Desirée Fiske, "International Climate Change Policy: Complex Multilevel Governance," in Regina S. Axelrod and Stacy D. VanDeveer (eds.) *The Global Environment: Institutions, Law, and Policy*, Sage Publishing, August 2019; "Explainer: Dealing with the 'Loss and Damage' Caused by Climate Change," Carbon Brief, 2017, available at https://www.carbo nbrief.org/explainer-dealing-with-the-loss-and-damage-caused-by-climate-change; "Building Toward Breakthrough: Energizing the Paris 2015 Climate Negotiations and Post-Paris Action Agenda Through Broader Engagement," Yale Center for Environmental Law and Policy, (2015), available at https://envirocenter.yale.edu/sites/default/files/yale_climate_change_dialogue_w hite_paper.pdf; Daniel Esty and Maria Ivanova, "Revitalizing Global Environmental

than ever in loose global environmental governance,[98] which will help to expedite the dematerialization trend. In addition, people (especially young people) are more aware of the dangers of climate change. Martin Weitzman's theory that society should be prepared to pay unlimited amounts today for avoid the catastrophic consequences of climate change might be a stretch, but seems to be partly resonating with younger generations. So changing people's behavior by giving them incentives can be a game-changer in climate action.

Central bankers are also trying to contribute to climate change mitigation.[99] Christine Lagarde, President of the European Central Bank (ECB), wants to make sure that its monetary policy takes into account the impact of climate change.[100] There is a group of 51 central banks and supervisors (the Network for Greening the Financial System, NGFS) working on how central banks can contribute to green monetary policy.[101] Missing from the NGFS is the US Federal Reserve,[102] although it does

Governance: A Function-Driven Approach," in Daniel Esty and Maria Ivanova (eds.), *Global Environmental Governance: Options and Opportunities*, 2002.

98. In 1865, William Jevons published *The Coal Question: An Inquiry Concerning the Progress of the Nation, and the Probable Exhaustion of Our Coalmines*, Macmillan & Co. London, where he argued that more efficient use of fossil fuel leads to higher levels of consumption. Empirical evidence demonstrates that is not the case, as one can see with the decline in coal use in the United States even as coal became cheaper.
99. Mark Carney, Governor of the Bank of England, has argued that companies and investors need to plan for climate action. *See* V. Romei, "Carney Warns of Climate Change Hit to Corporate Assets," *Financial Times*, December 30, 2019, available at https://www.ft.com/content/f67833ba-2ad7-11ea-bc77-65e4aa615551?segmentId = 3f81fe28-ba5d-8a93-616e-4859191fabd8.
100. The Economist, "Green Central Banking: Carbon Capture," December 14, 2019, p. 64.
101. On financing climate action more generally, *see* "What Is the Financial Mechanism? What Are the Other Funds?," UNFCCC, available at http://bigpicture.unfccc.int/content/climate-finance/what-is-the-financial-mechanism-what-are-the-other-funds.html; Jacob Waslander and Niranjali Manel Amerasinghe. "Why Should Governments Refill the Green Climate Fund's Coffers?" WRI, 2019, available at https://www.wri.org/blog/2019/02/why-should-governments-refill-green-climate-funds-coffers; Daniel Esty and Todd Cort, "Corporate Sustainability Metrics: What Investors Need and Don't Get," *Journal of Environmental Investing*, Vol. 8, No. 1 (2017); Varun Sivaram, *Taming the Sun: Innovations to Harness Solar Energy and Power the Planet*, MIT Press, 2019, Chapter 4; Paul Griffin and Amy Myers Jaffe, "Are Fossil Fuel Companies Telling Investors Enough about the Risks of Climate Change?," *The Conversation*, 2017, available at http://theconversation.com/are-fossil-fuel-companies-telling-investors-enough-about-the-risks-of-climate-change-72562; "Financing Strategies: A Missing Link to Translate NDCs into Action," GIZ, 2017, available at https://www.giz.de/en/downloads/giz2017-en-climate-finance-strategies-ndcs.pdf; "Roadmap to US$100 Billion," White paper—Copenhagen Pledge, available at http://dfat.gov.au/international-relations/themes/climate-change/Documents/climate-finance-roadmap-to-us100-billion.pdf; NREL, Financing Renewable Energy Through Public Markets, available at http://www.nrel.gov/docs/fy13osti/58315.pdf; Whitney Leonard, "Clean Is the New Green: Clean Energy Finance and Deployment Through Green Banks," 33:1 *Yale Law & Policy Review* (2014); Coalition for Green Capital, "National Green Banks in Developing Countries," 2017, available at http://coalitionforgreencapital.com/wp-content/uploads/2017/07/Green-Banks-in-Emerging-Markets.pdf.
102. More broadly, for an analysis of US energy politics and policy history, *see* Michael Graetz, *The End of Energy*. MIT Press, 2011; David Anthoff and Robert Hahn, "Government Failure and Market Failure: On the Inefficiency of Environmental and Energy Policy," 26 *Oxford Review of Economic Policy* 197-224 (2010), available at http://oxrep.oxfordjournals.org/content/26/2/197.full.pdf+html; Trevor Houser, John Larsen, and Peter Marsters, "The Real Electricity Reliability Crisis," Rhodium Group, 2017, available at https://rhg.com/research/the-real-electricity-relia

recognize that climate-related disruption may affect the economy and productivity negatively.[103]

IV TRANSFORMING OUR UNDERSTANDING OF INTERNATIONAL TRADE TO PROMOTE SUSTAINABILITY

International trade has been responsible for much environmental harm. More trade implies more energy consumption and therefore more pollution. Now everyone assumes that trade damages the environment. But it does not have to be that way: international trade can be a very powerful tool for environmental protection.

1 International Trade as an Enabler of Social Justice

So why use international trade as a tool for change? Every country trades. Trade is a wealth generator. In 2015, trade made up 58% of global GDP.[104] Between 1950 and 2006, there was a 30-fold increase in international trade.[105] This trend is on the rise. Trade liberalization implies more trade. By the same token, trade leads to economic growth, and economic growth means that every country is better off.[106]

It is well known that international trade contributes to economic growth. A case in point is China's recent macroeconomic success since the late 1970s.[107] Thanks to the internet, international trade has been democratized, giving much potential for growth e-commerce in places like Africa, India, and China. Based on the rule of law,[108] the international community created an Appellate Body for the settlement of multilateral trade disputes. Yet, it ceased to exist in December 2019.

The main reason for the termination of the World Trade Organization's (WTO) Appellate Body is that the Trump administration blocked the nomination of new judges at the WTO Appellate Body because it does not agree with its interpretation of international law and claims the Appellate Body has gone, at times, beyond its remit.

---

bility-crisis-doe-nopr/; Stephen Lacey, "Energy Policy in Trump's Washington: Year One," *Greentech Media*, 2018, available at https://www.greentechmedia.com/articles/read/energy-policy-in-trumps-washington-year-one#gs.xpPu=1g; Daniel Yergin, *The Quest*, Penguin Books, 2011; Hannah Ritchie and Max Roser, "Energy Production & Changing Energy Sources," Published online at OurWorldInData.org (2018), https://ourworldindata.org/energy-production-and-changing-energy-sources.

103. The Economist, "Green Central Banking: Carbon Capture," December 14, 2019, p. 64.
104. https://www.ecb.europa.eu/pub/economic-bulletin/articles/2019/html/ecb.ebart201903_01~e589a502e5.en.html#toc1.
105. https://elibrary.worldbank.org/doi/abs/10.1596/27456.
106. For an analysis of international trade at various levels of governance, *see* R. Leal-Arcas, *International Trade and Investment Law: Multilateral, Regional and Bilateral Governance*, Edward Elgar, 2010; R. Leal-Arcas, *Theory and Practice of EC External Trade Law and Policy*, Cameron May, 2008.
107. R. Leal-Arcas, "The EU And China In A New World Order: Challenges Ahead," XIV:1 *The IUP Journal of International Relations* 1-20 (2020).
108. R. Leal-Arcas et al., "Aiming at Sustainable Trade in the Context of the Rule of Law: What Role for Citizens and How International Trade Can Help Reduce Fossil-Fuel Consumption," 11:2 *Trade, Law and Development* (2019); R. Leal-Arcas (ed.) *International Trade, Investment and the Rule of Law*. Eliva Press, 2020.

As a result, multilateral trade may become less predictable without the Appellate Body, and there will be a proliferation of bilateral and regional trade agreements.[109] Moreover, a rules-based system with an independent dispute settlement mechanism allows the poor and weak to be heard, whereas without that mechanism, the old might-is-right doctrine may prevail, allowing big country to potentially bully the rest.

It is also a well-known fact that international trade has lifted more people out of poverty in the past 30 years than in the entire history of humanity, benefiting many developing countries. Yet, the world is experiencing a trade war between the US and China that may cost the world economy USD 700 billion.[110] These two economies differ philosophically: one is based on exacerbated capitalism and the other combines a market with very strong state intervention. Since democratic systems (like that of the US) are incompatible with state-run economies (like that of China to a large extent), the trade tension between these two countries is not surprising. In the UN context, many countries support China since they see it as a great source of infrastructure, loans, and affordable technology. Moreover, many Chinese policymakers believe that the global liberal order established after 1945 and based on universal human rights, rules and regulations that bind the strong and weak alike goes against China's rise.[111] As one European diplomat put it, "China will not achieve its goals if it plays by the rules."[112]

Democratic systems have evolved over the years: one period of this new global order was from the late 18th century to the early 20th century, which was the case of the French Revolution and the US democracy; another period was the aftermath of World War II, which was the case of many African countries; and third period is post-1974, which was the case of Portugal, Spain, and many Latin American countries. This structural transformation happened thanks to intervention from outside, fighting from below against the elites (which was the case of the US, Romania and Portugal), or transformation by elites in power (which was the case of Spain, when the army gave up power to the people).

The US, like many other parts of the Western world, has a neoliberal approach to the economy, namely deregulation, privatization, and free trade, known as the Washington Consensus.[113] China, under its great leader Deng Xiaoping, opted for economic reforms,[114] not political reforms, which is why it clashes with the political structures of Western liberal democracies.[115] Nevertheless, China's economic success

109. R. Leal-Arcas, "Proliferation of Regional Trade Agreements: Complementing or Supplanting Multilateralism?," 11:2 *Chicago Journal of International Law* 597-629 (2011).
110. *See* The Economist, December 21, 2019, p. 8, citing the International Monetary Fund.
111. The Economist, "China Sees Its Moment," April 3, 2021, p. 48.
112. *Ibid.*
113. John Williamson, "A Short History of the Washington Consensus," in Narcis Serra and Joseph Stiglitz (eds.), *The Washington Consensus Reconsidered*: *Toward a New Global Governance*, Oxford University Press, 2008, Chapter 2. But *see also* Lucio Baccaro and Chris Howell, *Trajectories of Neoliberal Transformation: European Industrial Relations since the 1970s*, Cambridge University Press, 2017, Chapter 6.
114. Such reforms are known as the Beijing Consensus, which began to be instituted in the late 1970s.
115. *See* generally Edward Steinfeld, *Playing Our Game: Why China's Rise Doesn't Threaten the West*, Oxford University Press, 2010, Chapters 1 & 4; Yuhua Wang, *Tying the Autocrat's Hands: The Rise of the Rule of Law in China*, Cambridge University Press, 2015, Chapter 4.

has been largely due to better education, Confucian values of order, gradual reforms, a relatively good rule of law able to attract foreign direct investment, and an export-oriented model of economic growth. Lately, China has been able to seize its economic opportunity in Africa, realizing that African possesses a high number of natural resources, cheap labor and very cheap land.[116] China has also shown in recent times its assertiveness on the international stage by imposing sanctions on British, Canadian, and EU politicians, diplomats, lawyers, academics, and activists in retaliation for Western sanctions against Chinese officials accused of repressing the Chinese Muslim community.

The world is also experiencing high levels of protectionist and unilateral trade.[117] For instance, the Trump administration uses tariffs and other restrictive trade instruments to make sure other countries bring down their own trade barriers. If tariffs are conceived in a narrow manner, they invite consumers to purchase domestic products. But tariffs can backfire in that foreign countries can impose their own tariffs, thereby hurting exporters, or companies may also take their factories to other countries to avoid such tariffs.

Against the current trend of protectionist trade and unilateral actions, how can we have a trading system that is economically inclusive,[118] beneficial for all,[119] stays away from economic nationalism, avoids putting multilateral trade at risk, and is environmentally friendly? How can trade work for the environment? How can the trading system support the environment? Given that trade agreements affect our daily life, how democratic can trade policy become given how technocratic it is in nature? Given that local inequalities matter to people more than global inequalities, what can the trading system do about it? Surely trade policymakers need to listen to civil society

116. For a recount of China's economic position in Africa, *see* Deborah Brautigam, *The Dragon's Gift: The Real Story of China in Africa*, Oxford University Press, 2010, Chapters 1-5, 11; Irene Sun, *The Next factory of the World: How Chinese Investment is Reshaping Africa*. Harvard Business Review Press, 2017; *Dance of the Lions and Dragons*, McKinsey Report on China in Africa, 2017; Howard French, *China's Second Continent: How a Million Migrants Are Building a New Empire in Africa*. Vintage, 2015.

117. As the WTO panel report on *Section 301* highlighted, any unilateralist policy, such as a trade war like the one between the US and China, has "indirect effect (...) on individuals and the market-place, the protection of which is one of the principal objects and purposes of the WTO." *See* WT/DS152: United States—Sections 301-310 of the Trade Act 1974, para. 7.86.

118. European Commission, "Trade for all: Towards a more responsible trade and investment policy," 2014, available at https://trade.ec.europa.eu/doclib/docs/2015/october/tradoc_153846.pdf.

119. There is a vast body of literature that analyzes the reasons for economic inequality. *See* for instance some books published only in 2019: B. Sunkara, *The Socialist Manifesto: The Case for Radical Politics in an Era of Extreme Inequality*, Basic Books, 2019; P. Adler, *The 99 Percent Economy: How Democratic Socialism Can Overcome the Crises of Capitalism*. Oxford University Press, 2019; B. Milanovic, *Capitalism Alone: The Future of the System That Rules the World*, Belknap Press, 2019; C. Berry and J. Guinan, *People Get Ready! Preparing for a Corbyn Government*. OR Books, 2019. The argument on the conservative side of politics is that they do not promise to reduce inequality, but to increase salaries for everyone. *See* generally Paul Pierson, *Dismantling the Welfare State? Reagan, Thatcher, and the Politics of Retrenchment*. Cambridge University Press, 1990. On the anti-tax movement, *see* Michael Graetz and Ian Shapiro, *Death by a Thousand Cuts: The Fight over Taxing Inherited Wealth*. Princeton University Press, 2005.

about all the points raised above.[120] In turn, civil society's engagement in trade agreements enhances the legitimacy of any trade legislation, brings expertise to policy-making, and helps in the implementation process.

Given that environmental issues are present in our daily life and that trade has a role to play in sustainability,[121] the convergence of these two fields is what the Sustainable Development Goals are supposed to do, hoping to reach a better mutual understanding between trade and environment. The elimination or drastic reduction of technical barrier to trade in environmental goods and services is imperative, as is the avoidance of subsidies for fossil fuels. There are legal instruments that already bring trade and environment together: many free-trade agreements (FTAs) with sustainable-development chapters that aim to remove barriers to trade and investment in climate-change-friendly goods and services, unilateral actions such as the Generalized System of Preferences, as well as multilateral/plurilateral instruments such as the WTO's Committee on Trade and Environment, the OECD, or the Environmental Goods Agreement. A further instrument would be lower or no tariffs for environmental goods and certified products that comply with sustainability criteria.

In its recent free-trade agreements, the European Union (EU), like many other major actors in international trade, has imposed higher social and environmental standards on its trade partners.[122] That may come at the risk of less trade, but will certainly help protect the environment. China, a major country in international trade, is currently undertaking a major project called the Belt and Road Initiative (BRI), which aims to connect China to many parts of the world by investing in infrastructure and improving its trade relations. It would be very beneficial to the environment if the BRI were to be implemented in a manner that protects the environment through more renewable-energy investments and more energy-efficient industries. If China, the largest GHG emitter in the world largely due to coal consumption, gets on a sustainable trajectory, almost nothing else matters. However, if China does not, the world will most likely not be on the right track to mitigate climate change. This is so because China is fast growing, a very big population, and a very big country.

120. *See* the example of the European Commission (2018), "Feedback and way forward on improving the implementation and enforcement of Trade and Sustainable Development chapters in EU Free Trade Agreements," available at: http://trade.ec.europa.eu/doclib/docs/2018/february/tradoc_156618.pdf.

121. *See* for instance R. Leal-Arcas, *Solutions for Sustainability: How the International Trade, Energy and Climate Change Regimes Can Help*. Springer International Publishing, 2019; R. Leal-Arcas, "New Frontiers of International Economic Law: The Quest for Sustainable Development," 40:1 *University of Pennsylvania Journal of International Law* 83-132 (2018); R. Leal-Arcas, "Sustainability, Common Concern and Public Goods," 49:4 *The George Washington International Law Review* 801-877 (2017).

122. For a comprehensive analysis of EU trade law, *see* R. Leal-Arcas, *EU Trade Law*. Edward Elgar, 2019.

2 Enforceability of Environmental Clauses and the Creation of a Climate Club Linked to the International Trade Regime

Rules and regulations only become of any use if they can be enforced. Unlike trade laws and trade agreements—which are enforceable—the enforcement of international environmental laws and agreements has always been a deficiency of the legal system: if countries do not want to come along, there is nothing one can do. But when it comes to the enforcement of environmental provisions in FTAs, it is interesting to note that the Americans differ diametrically from the Europeans in that Americans use trade as a stick, whereas Europeans use it as a carrot.[123] The US approach is about legal sanctions if countries are not in compliance with the environmental clauses in FTAs between the US and those countries. The European approach is about dialogue, rather than punishment, if the Contracting Party is in breach of an environmental clause in an EU FTA.

The alternatives to the enforcement deficiency in environmental agreements are a name-and-shame approach, a carbon club to reduce GHG emissions, or regional groupings of countries that share an agenda. In all these options, trust and leadership are key elements.

Indeed, one way around this enforcement issue is the creation of a climate club of major GHG emitters linked to the international trade regime as an effective way to mitigate climate change, as two inter-related and inter-dependent regimes.[124] The rationale for a small club is that it is easier to negotiate among a small group of large countries (and major corporations) than it is among a large group of small countries (and companies). This bottom-up design of a club with incremental membership has been successful in other international regimes. For example, the GATT 1947 had only 23 parties, which eventually became the WTO and today has 164 members. The EU expanded from 6 to 28 members (but, since 2020, 27 members as a result of Brexit).[125] NATO enlarged from 12 to 29 members. The UN started with from 51 members and today has 193.

---

123. R. Leal-Arcas et al., "The Contribution of Free Trade Agreements and Bilateral Investment Treaties to a Sustainable Future," 1 *Zeitschrift für Europarechtliche Studien—ZEuS* (2020); R. Leal-Arcas et al., "Aiming at Sustainable Trade in the Context of the Rule of Law: What Role for Citizens and How International Trade Can Help Reduce Fossil-Fuel Consumption," 11:2 *Trade, Law and Development* (2019).
124. There is a growing body of academic literature on the links between international trade and climate change. *See* for instance, R. Leal-Arcas, *Climate Change and International Trade*. Edward Elgar, 2013; Daniel C. Esty. "Toward a Sustainable Global Economy: An Initiative for G20 Leadership," *Journal of Self-Governance and Management Economics* (2017); William Nordhaus. "A New Solution: The Climate Club," *The New York Review of Books*, 2015, available at http://www.nybooks.com/articles/2015/06/04/new-solution-climate-club/; Antoine Dechezleprêtre and Misato Sato, "The Impacts of Environmental Regulations on Competitiveness." *Review of Environmental Economics and Policy* (2017); Dani Rodrik, "Global Trade Needs Rules That Adapt to Economic Diversity," *Financial Times* (August 6, 2018) at p. 9; "Trade and Environment," World Trade Organization, available at https://www.wto.org/english/tratop_e/envir_e/envir_e.htm; Aaron Cosbey, "Climate Policies, Economic Diversification and Trade," 2017, available at http://dx.doi.org/10.2139/ssrn.3135359.
125. For an analysis of Brexit, *see* R. Leal-Arcas, "Three Thoughts on Brexit," *Queen Mary School of Law Legal Studies Research Paper 249/2016*, pp. 1-5.

In terms of members, this new climate regime would be of gradual and incremental membership. For this new architecture to work effectively, it would require only a few major GHG emitters and just three mega-regional trade agreements (RTAs) (i.e., the Transatlantic Trade and Investment Partnership,[126] the Comprehensive and Progressive Agreement for Trans-Pacific Partnership,[127] and the Regional Comprehensive Economic Partnership[128]) to make a great and meaningful contribution toward climate-change mitigation and sustainable energy.[129]

Let us suppose this club consists of the top 15 major GHG emitters. Out of the top 15 major GHG emitters—which, combined, are responsible for nearly 80% of global GHG emissions and most of the world's GDP—only Russia, Brazil, Iran, South Africa, and Ukraine are not parties to any of the three mega-RTAs mentioned above. Through a network of mega-RTAs/FTAs, these five major emitters, currently not parties to any of the three mega-RTAs mentioned above, could be linked to this new regime via FTAs they already have in place with the members of this club. For instance, the 2019 EU-Mercosur FTA would be a platform to link Brazil to this climate club. Equally, the EU has trade agreements with South Africa and Ukraine that would enable these two countries to be linked to this hypothetical club. One could be more ambitious and regionalize all these bilateral agreements; the next step would be to multilateralize all these regional agreements until we create a web of interlinked agreements that will eventually and gradually connect the world based on trade and climate obligations.

Alternative membership could be the 17 members of the Major Economies Forum on Energy and Climate (MEF)[130] or the G20.[131] But the political declarations that emanate from these two clubs are neither legally binding nor enforceable.

This new regime linking the governance of climate change to international trade would be legally binding and enforceable. The Paris Agreement on Climate Change is

126. The parties to this Partnership are the US and the EU.
127. The 11 Asia-Pacific nations that form this Partnership are Japan, Mexico, Canada, Australia, Malaysia, Chile, Singapore, Peru, Vietnam, New Zealand, and Brunei.
128. The 16 countries in this Partnership are in Asia and Oceania: the 10 members of the Association of Southeast Asian Nations (ASEAN) (Brunei, Cambodia, Indonesia, Laos, Malaysia, Myanmar, the Philippines, Singapore, Thailand, and Vietnam) and the six countries with which ASEAN has existing FTAs (Australia, China, India, Japan, South Korea, and New Zealand).
129. Eventually, a web of interconnected RTAs could be created to have a global reach on trade/climate action/sustainable energy. For the example of the Middle East and North Africa, *see* R. Leal-Arcas et al., "Energy Trade in the MENA Region: Looking Beyond the Pan-Arab Electricity Market," 10:6 *Journal of World Energy Law and Business* 520-549; Oxford University Press, 2017.
130. The MEF members include: Australia, Brazil, Canada, China, the EU, France, Germany, India, Indonesia, Italy, Japan, Korea, Mexico, Russia, South Africa, the UK, and the U.S. *See* Major Economies Forum on Energy and Climate, available at http://www.majoreconomiesforum.org/about/descriptionpurpose.html. For an analysis, *see* R. Leal-Arcas, "Alternative Architecture for Climate Change: Major Economies," 4:1 *European Journal of Legal Studies* 25-56 (2011); R. Leal-Arcas, "Small Is Beautiful: Why a Club Approach Is the Way to Go in Climate Change Mitigation," in W. Leal and D. Surroop (ed.), *The Nexus: Energy, Environment and Climate Change*, Springer, 2018, pp. 305-317.
131. The members of the G20 are Argentina, Australia, Brazil, Canada, China, the European Union, France, Germany, India, Indonesia, Italy, Japan, Mexico, Russia, Saudi Arabia, South Africa, South Korea, Turkey, the United Kingdom, and the United States. Incidentally, Saudi Arabia, Turkey and Argentina are the only G20 members that are not members of the MEF.

neither.[132] Absent a global, legally binding, and enforceable climate agreement, a climate club would make sense. It would provide economic incentives to the members of the club and be a more effective way to mitigate climate change. The incentives of membership of this club would need to be such that every country would want to join the club and no country would want to leave. In other words, becoming a member of a climate club would mean that you can check out any time you like, but you will never wish to leave.[133] Ultimately, this new architecture of governance of trade and climate change will also be beneficial toward geopolitics and national security in the context of climate change.[134]

132. R. Leal-Arcas and A. Morelli, "The Resilience of the Paris Agreement: Negotiating and Implementing the Climate Regime," 31:1 *Georgetown Environmental Law Review* 1-63 (2018); R. Leal-Arcas, "Re-thinking Global Climate Change: A Local, Bottom-Up Perspective," XX:1 *Seton Hall Journal of Diplomacy and International Relations* 4-20 (2018).
133. Here I am re-purposing the lyrics of the famous 1977 song "Hotel California" by the Eagles: "You can check out any time you like, but you can never leave!" My vision of a climate club is one where the benefits of this club should be so good that the members do not want to leave.
134. For further analysis on the links between climate change, national security, and geopolitics, *see* Jennifer Perron, "Fatih Birol and David Victor on the Geopolitics of Energy," *Brookings Institute* (July 2, 2018), available at https://www.brookings.edu/blog/planetpolicy/2018/07/02/fatih-birol-and-david-victor-on-the-geopolitics-of-energy/; Umair Irfan, "106 Lawmakers—Including 11 Republicans—Tell Trump Climate Change Is a National Security Threat," *Vox* (January 12, 2018), available at https://www.vox.com/energy-and-environment/2017/12/18/16791106/white-house-climate-change-national-security-strategy-threat-military; Tony Capaccio, Jennifer Dloughy, and Ari Natter, "Defense Department Warns about Climate Change Impacts to Armed Forces and Bases," *Time* (January 18, 2019), available at http://time.com/5507465/climate-change-impact-armed-forces-bases/; "Climate Change and Security: Strengthening Resilience to Climate-Fragility Risks," *UN Environment Programme*, available at https://postconflict.unep.ch/publications/ClimateChange_Security_twopager.pdf; Richard Morningstar, "Energy and Geopolitics: A Diplomat's View," *Atlantic Council (speech)* (October 12, 2018), available at https://www.atlanticcouncil.org/news/transcripts/energy-and-geopolitics-a-diplomat-s-view; Varun Sivaram, "Clean Energy Might Reduce Global Warming, But What Will It Do to Geopolitics?" *Council on Foreign Relations*, 2018, available at https://www.cfr.org/blog/clean-energy-might-reduce-global-warming-what-will-it-do-geopolitics; Ernest Moniz, "Shifting Dynamics Across the Energy Landscape" (August 6, 2018), available at http://energypolicy.columbia.edu/dr-ernest-moniz-shifting-dynamics-across-energy-landscape; Sarah Light, "Valuing National Security: Climate Change, the Military, and Society," *UCLA Law Review* (2014); Robinson Meyer, "Does Climate Change Cause More War?," *The Atlantic* (2018), available at https://www.theatlantic.com/science/archive/2018/02/does-climate-change-cause-more-war/553040/; Amy Larkin, "Energy Independence, National Security, and Another Reason for Sustainability," *The Guardian*, January 4, 2014, available at https://www.theguardian.com/sustainable-business/2014/jun/04/energy-independence-national-security-sustainability; Daniel Yergin, *The Quest*. Penguin Books, 2011, Chapters 13, 14, and 33; Henry Tricks, "The New Power Superpowers," in The Economist Special Report: The Geopolitics of Energy, *The Economist* (March 17, 2018), available at https://www.economist.com/special-report/2018/03/15/clean-power-is-shaking-up-the-global-geopolitics-of-energy.

V WHAT SHOULD BE THE ROLE OF THE TRADING SYSTEM IN REDUCING FOSSIL-FUEL CONSUMPTION?

1 Introduction[135]

Section V shall examine which carbon pricing method is better to achieve carbon abatement: cap-and-trade or a carbon tax.[136] The traditional carbon tax and cap-and-trade models shall be discussed, followed by an assessment of the benefits of a hybrid cap-and-trade with a price ceiling or price floor, and of CTR. Martin Weitzman of Harvard University made a great contribution to this debate by posing the following question: "how should regulators rein in pollution? Should they issue (tradable) pollution permits to firms, thereby picking a quantity? Or should they tax polluters, thereby picking a price?"[137] Which you should regulate depends on the relative costs of mistakes: if getting the quantity of pollution slightly wrong is costlier, then quantity should be pinned down; if, however, a slightly errant price does more damage, then a tax is a better option.[138] Based on the costs and benefits of acting to reduce GHG

135. A longer version of this section was published as R. Leal-Arcas et al., "Aiming at Sustainable Trade in the Context of the Rule of Law: What Role for Citizens and How International Trade Can Help Reduce Fossil-Fuel Consumption," 11:2 *Trade, Law and Development* 193-238 (2019).
136. For analyses on carbon pricing, *see* William Nordhaus, "Revisiting the Social Cost of Carbon," *PNAS* (2017), available at https://doi.org/10.1073/pnas.1609244114; "The True Cost of Carbon Pollution," EDF, available at https://www.edf.org/true-cost-carbon-pollution; Gillingham, Carattini, and Esty, "Lessons from First Campus Carbon-Pricing Scheme," *Nature* (2017), available at https://www.nature.com/news/lessons-from-first-campus-carbon-pricing-scheme-1.22919; World Bank Group, "State and Trends of Carbon Pricing 2017," 2017, available at https://openknowledge.worldbank.org/bitstream/handle/10986/28510/wb_report_171027.pdf?sequence=7; "Economists' Statement on Carbon Dividends," *Wall Street Journal* (2019), available at https://www.clcouncil.org/media/EconomistsStatement.pdf; Karen Palmer, Anthony Paul, and Amelia Keyes, "Changing Baselines, Shifting Margins: How Predicted Impacts of Pricing Carbon in the Electricity Sector Have Evolved Over Time," Resources for the Future, 2018, available at http://www.rff.org/files/document/file/RFF%20WP%2018-15.pdf; Brad Plumer and Nadja Popovich, "These countries have prices on carbon. Are they working?" *The New York Times*, April 2, 2019, available at https://www.nytimes.com/interactive/2019/04/02/climate/pricing-carbon-emissions.html?mtrref=www.google.com; Kotchen et al. "Public Willingness to Pay for a US Carbon Tax and Preferences for Spending the Revenue," Environmental Research Letters (2017), available at https://doi.org/10.1088/1748-9326/aa822a; C2ES, "The Business of Pricing Carbon: How Companies are Pricing Carbon to Mitigate Risks and Prepare for a Low-Carbon Future," 2017, available at http://www.c2es.org/document/the-business-of-pricing-carbon-how-companies-are-pricing-carbon-to-mitigate-risks-and-prepare-for-a-low-carbon-future/; "How Can Carbon Prices and Policies Be Effectively Aligned?" Carbon Pricing Leadership Coalition, 2016, available at http://pubdocs.worldbank.org/en/221021478831141991/CPLC-Executive-Brief-Policy-Alignment-Nov2016-FINAL.pdf; "Carbon Pricing: Gaining a Competitive Edge in a Climate-Constrained World," available at https://cbey.yale.edu/programs-research/carbon-pricing-gaining-competitive-edge-in-climate-constrained-world.
137. *See* The Economist, "Free Exchange: The Uncertainty of Genius," September 7, 2019, p. 68, referring to the work of Martin Weitzman.
138. *Ibid.*

emissions, Mr Weitzman argued in his dismal theorem that society should be willing to pay unlimited amounts of money today to avoid a catastrophic future.[139]

We argue that a carbon tax with reinvestment is the superior option because it is easier to implement and is aligned with the FASTER principles for successful carbon pricing.[140] We also argue that putting a price on carbon is better than climate change mitigation targets, but we acknowledge that getting people to pay for carbon in Western liberal democracies may imply losing the elections. This section will then turn to an analysis of why it is essential that carbon taxes be supported with BTAs to ensure competitiveness of national economies, prevent free-riding and induce a complete decarbonization of the international economy. We shall analyze how this can be achieved within the constraints of WTO law. We conclude that BTAs are fundamental to reduce carbon leakage and to incentivize less willing nations to turn toward carbon abatement.

2 Mitigating Climate Change at the International Level

As part of their climate-change mitigation strategy, quite a few countries have by now adopted some type of price on carbon—whether through a carbon tax or a cap-and-trade system.[141] In an interesting article in *The New York Times*, Brad Plumer and Nadja Popovich ask the question whether these schemes to tackle climate change in the various countries shown in Figure 2.2 are actually effective.[142]

139. *Ibid.*
140. OECD and World Bank Group, 2015. *The FASTER Principles of Successful Carbon Pricing: An Approach Based on Initial Experience,* September 2015: vii-viii.
141. The World Bank, "Carbon Pricing Dashboard," available at https://carbonpricingdashboard.worldbank.org/map_data.
142. B. Plumer and N. Popovich, "These Countries Have Prices on Carbon. Are They Working?" *The New York Times,* April 2, 2019.

Figure 2.2 Countries with a Price on Carbon

Source: B. Plumer and N. Popovich, "These Countries Have Prices on Carbon. Are They Working?" *The New York Times*, April 2, 2019.

Economists have long argued that increasing the cost of burning fossil fuels is a cost-effective way to mitigate climate change.[143] But even if all the major Western oil corporations decided to stop pumping oil and gas to reduce GHG emissions, global production would only shrink by 10%, since 90% of the fossil-fuel market is controlled by national oil companies.[144] So international efforts to tackle climate change began in 1992 with the UN Earth Summit, culminating in the production of the United Nations Framework Conventions on Climate Change. This legal text marked a first step in working toward solving climate change and provided a mitigation framework. Surprisingly quickly for international negotiations, this was extended in 1997 with the Kyoto Protocol, which set targets for industrialized nations to control their GHG emissions. It was perhaps the haste with which the Kyoto Protocol was drafted that led to its ill-thought-out provisions.[145] The Kyoto Protocol collapsed under the pressure of a lack of support from developed nations who objected to the absence of obligations for developing nations.[146] In fact, the US, Australia, Canada, Japan, and Russia all reneged on the obligations set out in the Protocol.[147]

A low point in the history of climate change mitigation came in 2008 at the 15th Conference of the Parties to the UNFCCC in Copenhagen, when it was impossible to

143. B. Plumer, "New U.N. Climate Report Says Put a High Price on Carbon," *The New York Times*, October 8, 2018, available at https://www.nytimes.com/2018/10/08/climate/carbon-tax-united-nations-report-nordhaus.html.
144. The Economist, "Ben van Beurden's Balancing Act," July 6, 2019, p. 62.
145. Victor. *The Collapse of the Kyoto Protocol and the Struggle to Slow Global Warming*, Princeton University Press, 2004, 26.
146. *Ibid*. 3-5.
147. M. Ivanova, "Politics, Economy and Society," in Klein et al. (eds.), *The Paris Agreement on Climate Change: Analysis and Commentary Analysis*. Oxford University Press, 2017, 19.

reach any sort of legally binding agreement.[148] In contrast, the Paris Agreement, adopted in December 2015, was seen as a great achievement in the international effort to mitigate climate change; in fact, Obama and Xi Jinping both demonstrated their unwavering commitment to climate change mitigating by vowing to promptly ratify the agreement, marking an important step for the two greatest emitters of GHG.[149] The substance of the Paris Agreement may be promising, but it is undermined by a lack of enforcement mechanisms; furthermore, the spirit of international climate change mitigation has been heavily dampened by the US's withdrawal from the agreement in 2017.

The current climate in international relations appears to be moving away from multilateral cooperation;[150] as a result, we argue that unilateral efforts to combat climate change should be made. The distinct lack of concrete progress in environmental agreements has led scholars to turn to trade law and policy as an effective means to combat climate change;[151] if the major GHG emitting nations were to unilaterally impose a carbon tax, then this could significantly contribute to global carbon abatement. One of the most important ways to mitigate climate change through trade policy is to effectively price carbon to induce producers and consumers to move toward more renewable and sustainable options.[152] We argue that if the largest emitters of GHG were to unilaterally price carbon effectively, and reinvest the revenues to support the development of renewable and sustainable energy, then this would reduce fossil-fuel consumption and lead to significant worldwide carbon abatement.

Carbon pricing is an effective mechanism through which fossil-fuel consumption can be reduced because it raises the price of carbon-based energy, which decreases the demand for it.[153] The higher the price of carbon, the more likely it is that sustainable alternatives will be sought; therefore, it is important that renewable energy policies are aligned with carbon pricing efforts.[154] The aim of a carbon pricing policy should be to incentivize emitters to invest in low-carbon technologies and to allow consumers to alter their choices toward more sustainable options in response to the price increases in GHG-intensive goods.[155]

The Paris Agreement, which entered into force in November 2016, may not be an international carbon pricing mechanism, but it does underline approaches toward emissions reductions through Article 6.[156] As of April 2019, there are 195 signatories to

148. *Ibid.* 17.
149. The New York Times, 2016. "Obama and President Xi of China to Sign Paris Climate Accord Promptly," March 31, 2016, https://www.nytimes.com/2016/04/01/world/asia/obama-and-president-xi-of-china-vow-to-sign-paris-climate-accord-promptly.html.
150. R. Leal- Arcas and A. Morelli, "The Resilience of the Paris Agreement: Negotiating and Implementing the Climate Regime," 31:1 *Georgetown Environmental Law Review* 62-63 (2018).
151. *Ibid.* 63.
152. N. Stern, *The Economics of Climate Change: The Stern Review*. Cambridge University Press, 2007, 349.
153. OECD. *Effective Carbon Rates: Pricing CO_2 Through Taxes and Emissions Trading Systems*. OECD Publishing Paris, 2016, 28.
154. *Ibid.* 29.
155. N. Stern, *The Economics of Climate Change: The Stern Review*, Cambridge University Press, 2007, 353.
156. World Bank Group. *State and Trends in Carbon Pricing*, Washington DC, May 2018, 33.

the Paris Agreement, and 185 ratifications;[157] yet, in the majority of these countries, there are no carbon pricing mechanisms in place. The Paris Agreement, while a great diplomatic achievement, contains legally binding as well as nonbinding instruments, and is devoid of enforcement mechanisms;[158] it is therefore necessary to turn to international trade in energy to seek to address climate change.[159]

There are various carbon pricing mechanisms in place across the globe, from cap-and-trade in the China[160] and New Zealand to carbon taxes in Argentina and South Africa; in the European Union,[161] the EU Emissions Trading System operates alongside

157. United Nations Treaty Collection, Chapter XXVII 7.d. Paris Agreement 4 November 2016 https://treaties.un.org/Pages/ViewDetails.aspx?src=TREATY&mtdsg_no=XXVII-7-d&chapter=27&clang=_en.
158. R. Leal-Arcas and A. Morelli, "The Resilience of the Paris Agreement: Negotiating and Implementing the Climate Regime," 31:1 *Georgetown Environmental Law Review* 3, 18, 61.
159. Leal-Arcas, "New Frontiers in International Economic Law: The Quest for Sustainable Development," 40:1 *University of Pennsylvania Journal of International Law* 85 (2018).
160. For an analysis of China's climate change policy, *see* Manish Bapna and Lailai Li, "China's Climate Leadership: 3 Signs to Watch in the Year Ahead," *World Resources Institute*, 2017, available at http://www.wri.org/blog/2017/11/chinas-climate-leadership-3-signs-watch-year-ahead; Zhang Chao, "The Climate Change Promise of China's Belt and Road Initiative," *The Diplomat*, 2018, available at https://thediplomat.com/2018/01/the-climate-change-promise-of-chinas-belt-and-road-initiative/; Isabel Hilton, "How China's Big Overseas Initiative Threatens Global Climate Progress," Yale E360, 2019, available at https://e360.yale.edu/features/how-chinas-big-overseas-initiative-threatens-climate-progress; Edward Wong, "Can China Take the Lead on Climate Change? That Could Be Difficult," *The New York Times*, 2017, available at https://www.nytimes.com/2017/06/02/climate/china-climate-change-trump-paris-accord.html; Joanna I. Lewis and Kelly Sims Gallagher, "How China's Domestic Energy and Environmental Challenges Shape Its Global Engagement," in Regina S. Axelrod and Stacy D. VanDeveer (eds.) *The Global Environment: Institutions, Law, and Policy*. Sage Publishing, August 2019; Geoff Dernbicki, "The Timely Disappearance of Climate Change Denial in China," *Foreign Policy* (May/June, 2017), available at https://foreignpolicy.com/2017/05/31/the-convenient-disappearance-of-climate-change-denial-in-china/#; Jennifer Duggan. "China Makes Carbon Pledge Ahead of Paris Climate Change Summit," *The Guardian*, 2015, available at https://www.theguardian.com/environment/2015/jun/30/china-carbon-emissions-2030-premier-li-keqiang-un-paris-climate-change-summit; Elizabeth Losos, Alexander Pfaff, and Lydia Olander, "The Deforestation Risks of China's Belt and Road Initiative." Brookings, 2019, available at https://www.brookings.edu/blog/future-development/2019/01/28/the-deforestation-risks-of-chinas-belt-and-road-initiative/; Javier C. Hernández and Adam Nagourney, "As Trump Steps Back, Jerry Brown Talks Climate Change in China," *The New York Times*, 2017, available at https://www.nytimes.com/2017/06/06/world/asia/xi-jinping-china-jerry-brown-california-climate.html; Yuli Shan et al. *Science Advances*, "City-Level Climate Change Mitigation in China" (June 27, 2018), available at https://advances.sciencemag.org/content/4/6/eaaq0390.full; Ye Qi and Jiaqi Lu, "China's Coal Consumption Has Peaked," Brookings Institution, January 22, 2018, available at https://www.brookings.edu/2018/01/22/chinas-coal-consumption-has-peaked/.
161. For an overview of the EU' climate change policy, *see* Dieter Helm, "The European framework for Energy and Climate Policies." *Energy Policy* (2014); Lucas Merrill Brown, Alex Hanafi, and Annie Petsonk, "EU Emissions Trading Scheme Lessons Learned," *Environmental Defense Fund*, 2012, available at https://www.edf.org/sites/default/files/EU_ETS_Lessons_Learned_Report_EDF.pdf; Simon Evans, "The History of Energiewende," *Carbon Brief*, 2016, available at https://www.carbonbrief.org/timeline-past-present-future-germany-energiewende; "Spain Halts Feed In Tariffs," *Institute for Energy Research*, 2012, available at http://www.instituteforenergyresearch.org/2012/04/09/spain-halts-feed-in-tariffs-for-renewable-energy; Nicola McEwen and Alexandra Raymond, "The Repatriation of Competences in Climate and Energy Policy after Brexit," *Brexit & Environment* (February 26, 2019), available at https://www.brexitenvironment.co.uk/2019/02/26/the-repatriation-of-competences-in-climate-and-energy-po

national carbon taxes in, for example, Norway and Sweden.[162] One of the key benefits of carbon pricing mechanisms is that they are dynamic; "all emitters have an ongoing incentive to reduce their carbon bill by finding new and cheaper ways to cut emissions."[163] Interestingly, while there is widespread agreement on the benefits of carbon pricing, there is even greater debate on which is the best mechanism.[164] We shall therefore now turn to analyze the arguments for and against the differing carbon pricing policies of cap-and-trade and carbon tax.

3 Cap-and-Trade System: A Viable Option for Reducing Fossil-Fuel Consumption?

One of the main attractions of a cap-and-trade system is its political feasibility. This is because it appeals to environmentalists, industries and politicians.[165] Crucially, for environmentalists, the cap-and-trade system is underpinned by the assumption that there should be a cap on emissions within a trading system because of their negative impact on climate change. This cap means that there is certainty about the environmental benefit of the measure.[166]

A potentially viable option to reduce fossil-fuel consumption would be an upstream cap-and-trade system, which would have an allowances system which can be traded within industries.[167] It, therefore, appeals to industry leaders because it is another source of potential revenue to firms which can successfully reduce their carbon emissions. If they do not need the allowances then they can trade them.[168] It is commonplace that such allowances are initially given out for free, and then are

licy-after-brexit/; Alissa J. Rubin and Somini Sengupta "'Yellow Vest' Protests Shake France. Here's the Lesson for Climate Change" (December 6, 2018) *The New York Times*, available at https://www.nytimes.com/2018/12/06/world/europe/france-fuel-carbon-tax.html; Miguel Arias Cañete, "A New Ambition for Energy Efficiency in Europe," *The European Files*, 2018, available at https://www.europeanfiles.eu/energy/new-ambition-energy-efficiency-europe-2; Alexandre Affre, Adrian van den Hoven, and Bruno Pedrotti, "A Competitive EU Energy and Climate Policy," *Business Europe*, 2013, available at https://bdi.eu/media/presse/publikation en/klima-und-umwelt/20130618_FINAL_Brocure_2030_energy_and_climate_LOW_RESOLU TION.pdf.

162. World Bank Group, *State and Trends in Carbon Pricing*, Washington DC, May 2018, 9.
163. OECD, 2016. *Effective Carbon Rates: Pricing CO_2 Through Taxes and Emissions Trading Systems*, OECD Publishing Paris: 29.
164. Goulder and Schein. "Carbon Taxes vs Cap and Trade: A Critical Review," Working Paper No. 19338, *NBER Working Paper Series*, National Bureau of Economic Research, 2013, 1-2.
165. Avi-Yonah and Uhlmann. "Combating Global Climate Change: Why a Carbon Tax is a Better Response to Global Warming than Cap and Trade," 28:3 *Stanford Environmental Law Journal* 5, 2009; OECD, *Effective Carbon Rates: Pricing CO_2 Through Taxes and Emissions Trading Systems*. OECD Publishing Paris, 2016, 24.
166. *Ibid.* 5, 36.
167. *Ibid.* 33.
168. *Ibid.* 5.

auctioned to create revenue, as in the EU Emissions Trading System (ETS). Cap-and-trade is therefore a significant policy tool, and makes the system a politically viable option.[169]

In fact, cap-and-trade systems are often the go-to policy for governments that are attempting to limit GHG emissions. The system allows governments to avoid using the word tax and shows them to be forward-thinking.[170] The administrative agency will initially allocate allowances for free in order to encourage participation in the scheme, but it is important that over time a monetary value is assigned to allowances in order to create revenue which can then be reinvested into renewable energy development. Furthermore, Carl and Fedor's analysis shows that an advantage of cap-and-trade is that the revenues are more likely to be spent on renewable energy initiatives.[171]

Overall, the disadvantages of cap-and-trade systems outweigh the advantages. In fact, Sewalk goes as far as to say that it is a failed policy.[172] While cap-and-trade ensures certainty about the environmental benefit of the policy, a key flaw of this method of carbon pricing is that there is much uncertainty about the price of emissions, and the price is likely to be incredibly volatile.[173] There are concerns that a cap-and-trade policy has poor economic efficiency and is therefore an ultimately untenable option to effectively reduce GHG emissions within an economic market.[174] Cap-and-trade policies may appeal to environmentalists because they produce emissions certainty; however, another underlying assumption of cap-and-trade is that carbon emissions below a certain level are acceptable.[175]

This is in stark contrast with the "polluter pays principle," which is a key element of the "Fairness" dimension to the "FASTER Principles for Successful Carbon Pricing" which were developed collaboratively by the OECD and World Bank.[176] It can be argued that cap-and-trade inherently suggests that there is a right to pollute; therefore, this mechanism of carbon pricing can create an incoherent message about what the goals of climate change policy.[177]

Cap-and-trade systems are difficult and costly to implement because it is complicated to set baselines for the emissions reduction targets, to allocate allowances, and

169. Carl and Fedor, "Tracking Global Carbon Revenues: A Survey of Carbon Taxes Versus Cap-and-Trade in the Real World," 96 *Energy Policy* 51 (2016).
170. Sewalk, "The EU-27, US, UK and China Should Dump Cap-and-Trade as a Policy Option and Adopt a Carbon Tax," XLVII *Suffolk University Law Review* 535-540 (2014).
171. Carl and Fedor, "Tracking Global Carbon Revenues: A Survey of Carbon Taxes Versus Cap-and-Trade in the Real World," 96 *Energy Policy* 51 (2016).
172. Sewalk, "The EU-27, US, UK and China Should Dump Cap-and-Trade as a Policy Option and Adopt a Carbon Tax," XLVII *Suffolk University Law Review* 525 (2014).
173. Goulder and Schein, "Carbon Taxes vs Cap and Trade: A Critical Review," Working Paper No. 19338, *NBER Working Paper Series*, National Bureau of Economic Research, 2013, 13.
174. Sewalk, "The EU-27, US, UK and China Should Dump Cap-and-Trade as a Policy Option and Adopt a Carbon Tax," XLVII *Suffolk University Law Review* 535-537 (2014).
175. *Ibid.* 540
176. OECD and World Bank Group, 2015. *The FASTER Principles of Successful Carbon Pricing: An Approach Based on Initial Experience,* September 2015: vii.
177. Sewalk, "The EU-27, US, UK and China Should Dump Cap-and-Trade as a Policy Option and Adopt a Carbon Tax," XLVII *Suffolk University Law Review* 542 (2014).

to decide on how to use offsets.[178] In fact, this is one of the reasons why Sewalk argues that the EU ETS cap-and-trade model is ineffective. The market is volatile which means that there have been high costs to consumers and it has failed to decrease GHG emissions.[179] Avi-Yonah and Uhlmann also argue that the EU system, the largest ETS in the world, has not been particularly successful.[180] This is because the cap-and-trade system has many complex features.

For example, the trading in allowances between firms needs to be established and then monitored by an administrative agency.[181] The failings of the EU ETS stem from problems related to the distribution of allowances. Too many were distributed which led to a collapse of the prices, and therefore cost uncertainty and volatility in the market.[182] Within cap-and-trade systems there are often provisions for the banking and borrowing of allowances, a mechanism which is designed to limit uncertainty within the system.[183] Banking allows firms within the market to "carry over unused quotas from one period to another," and borrowing allows actors within the market to "use or purchase quotas from a future period in the current period."[184] While this can improve flexibility, it means that cap-and-trade systems have lengthy and complex implementation periods.

One approach that has been proposed is a "hybrid cap-and-trade" system, which aims to solve the problems with price volatility by imposing either an allowance price floor or ceiling, or indeed both.[185] This policy prevents both emissions price volatility and carbon leakage, which is a key flaw of cap-and-trade policy, especially the EU ETS.[186] The hybrid policy allows the system to be more flexible and reactionary to changes: "enforcing the ceiling implies the introduction of extra allowances, while enforcing the floor can imply removal of some allowances that were in circulation."[187] Such caps and floors eliminate the risk of price spikes, and therefore increase the economic certainty of the cap-and-trade system.[188]

178. *Ibid.* at 541; Avi-Yonah and Uhlmann, "Combating Global Climate Change: Why a Carbon Tax Is a Better Response to Global Warming than Cap and Trade," 28:3 *Stanford Environmental Law Journal* 6 (2009).
179. Sewalk, "The EU-27, US, UK and China Should Dump Cap-and-Trade as a Policy Option and Adopt a Carbon Tax," XLVII *Suffolk University Law Review* 545 (2014).
180. Avi-Yonah and Uhlmann, "Combating Global Climate Change: Why a Carbon Tax Is a Better Response to Global Warming than Cap and Trade," 28:3 *Stanford Environmental Law Journal* 35 (2009).
181. *Ibid.* 38.
182. *Ibid.* 43.
183. *Ibid.* 39.
184. N. Stern, *The Economics of Climate Change: The Stern Review*. Cambridge University Press, 2007: 378; Sewalk, "The EU-27, US, UK and China Should Dump Cap-and-Trade as a Policy Option and Adopt a Carbon Tax," XLVII *Suffolk University Law Review* 539-541 (2014).
185. Goulder and Schein, "Carbon Taxes vs Cap and Trade: A Critical Review," Working Paper No. 19338, *NBER Working Paper Series*, National Bureau of Economic Research, 2013: 14.
186. *Ibid.* 29.
187. *Ibid.* 18.
188. N. Stern, *The Economics of Climate Change: The Stern Review*. Cambridge University Press, 2007, 376.

Despite this suggestion, the only concrete method of reducing the economic uncertainty associated with cap-and-trade is to incorporate a "safety valve" of additional allowances, but this would in turn sacrifice emissions reduction certainty.[189] The cap-and-trade system of carbon pricing is inherently complex and requires vast administrative support to ensure proper implementation. Even after implementation, the allocation and trading of allowances has to be monitored and regulated. Despite its political attractiveness as a policy, it is tainted by price uncertainty and does not satisfactorily fulfil the "polluter pays" principle.

4 Carbon Tax: The Superior Model for Carbon Pricing

In contrast to cap-and-trade, a carbon tax is a far simpler method of reducing GHG emissions. A carbon tax imposes a tax on each tonne of CO_2 emitted, therefore attributing a monetary value to the negative externality which GHG emissions have on the environment.[190] It is possible for a carbon tax to be established and become effective almost immediately, and it is therefore a far quicker method of inducing a reduction in fossil-fuel consumption.[191] An upstream carbon tax is the best option to induce carbon abatement because it prices carbon at the moment it enters the market; an upstream approach more efficiently distributes the costs across all sectors of the economy than a downstream approach.[192] A particular advantage of a carbon tax is that it provides cost certainty; the price is set to reflect the social cost of carbon, and it can be increased to produce the desired reduction in GHG emissions.[193] Furthermore, in contrast to the ambiguous environmental message which cap-and-trade suggests, the carbon tax is clear in its policy approach: "pollution imposes a negative externality on others and polluters should be forced to internalise that cost by paying the tax."[194]

Perhaps unsurprisingly, a carbon tax is less popular politically because of the opposition to policies which are simply labelled a "tax."[195] The significant advantage of the carbon tax however is that by definition it creates a revenue, and this can be invested into renewable and sustainable energy solutions. Furthermore, carbon pricing mechanisms, whether cap-and-trade or a carbon tax, are frequently faced with political opposition when they are to be enacted unilaterally by a state. Critics of unilateral carbon pricing argue that it will damage the competitiveness of domestic markets vis-à-vis their international competitors and will therefore drive industries abroad.

189. Avi-Yonah and Uhlmann, "Combating Global Climate Change: Why a Carbon Tax Is a Better Response to Global Warming than Cap and Trade," 28:3 *Stanford Environmental Law Journal* 43 (2009).
190. Sewalk, "The EU-27, US, UK and China Should Dump Cap-and-Trade as a Policy Option and Adopt a Carbon Tax," XLVII *Suffolk University Law Review* 547 (2014).
191. Avi-Yonah and Uhlmann, "Combating Global Climate Change: Why a Carbon Tax Is a Better Response to Global Warming than Cap and Trade," 28:3 *Stanford Environmental Law Journal* 7 (2009).
192. *Ibid.* 31.
193. *Ibid.* 32-38.
194. *Ibid.* 44.
195. Sewalk, "The EU-27, US, UK and China Should Dump Cap-and-Trade as a Policy Option and Adopt a Carbon Tax," (XLVII) *Suffolk University Law Review* 547 (2014).

In fact, the lack of border tax adjustment with the EU ETS meant that companies within the EU were at an economic disadvantage as a result of the cap-and-trade system.[196] It is challenging to adopt and incorporate a BTA into a cap-and-trade system, and Maruyama argues that "free emissions allowances rebates to trade-intensive industries represent a WTO-illegal export subsidy."[197] In contrast, border tax adjustments can be more easily incorporated into a carbon tax scheme.[198]

One weakness of a pure carbon tax system is that there is an inherent benefit uncertainty because there is no guarantee that the tax will lower emissions to the desired amount, and there may be intense political pressure against an increase in the tax to ensure less fossil-fuel consumption. Cap-and-trade might have the appeal of a greater environmental benefit certainty, but when coupled with price uncertainty and economic volatility, the system is inefficient, and ultimately does not induce significant carbon abatement. Instead, a carbon tax which is adjusted over time to fully decarbonize the economy is a far superior carbon pricing model. Furthermore, if reinvestment of the revenue is incorporated into this tax, (CTR), then the issues of environmental uncertainty surrounding a carbon tax are answered.[199]

Reinvestment of the revenue from a carbon tax is essential to ensuring that the carbon tax contributes effectively to the development of renewable energy sources. Taxes are administratively the simplest carbon pricing method to implement because they can be easily added on to existing national systems of taxation.[200] This means that policymakers can focus their efforts on how best to reinvest the tax revenues. Carl and Fedor's investigation into how carbon pricing revenues are spent by governments indicates that only a quarter is generally spent on developing green and renewable energy.[201]

If states are truly committed to completely decarbonizing the economy, then it is essential that climate change policy includes investment into the research and development of renewable and sustainable energy forms. As highlighted previously, carbon pricing alone cannot combat climate change. The FASTER principles of successful carbon pricing emphasize that effective carbon pricing is part of a "suite of measures"[202] to tackle climate change; it is paramount that the CTR model is used to ensure reinvestment into green energy.

An upstream carbon tax which is reinvested into developing renewable and sustainable energy sources is therefore the ideal carbon pricing model to decrease

196. *Ibid.* 546.
197. Maruyama, "Climate Change and the WTO: Cap and Trade Versus Carbon Tax?," 45:4 *Journal of World Trade* 681 (2011); Sewalk, "The EU-27, US, UK and China Should Dump Cap-and-Trade as a Policy Option and Adopt a Carbon Tax," XLVII *Suffolk University Law Review* 546 (2014).
198. *Ibid.* 681.
199. Sewalk, "The EU-27, US, UK and China Should Dump Cap-and-Trade as a Policy Option and Adopt a Carbon Tax," (XLVII) *Suffolk University Law Review* 553 (2014).
200. OECD, *Effective Carbon Rates: Pricing CO_2 Through Taxes and Emissions Trading Systems.* OECD Publishing Paris, 2016, 24.
201. Carl and Fedor, "Tracking Global Carbon Revenues: A Survey of Carbon Taxes Versus Cap-and-Trade in the Real World," 96 *Energy Policy* 51 (2016).
202. OECD and World Bank Group, *The FASTER Principles of Successful Carbon Pricing: An Approach Based on Initial Experience,* September 2015, vii.

fossil-fuel consumption and induce carbon abatement. The CTR model is flexible because it allows the country which is adopting it to choose which area of renewable energy to invest the majority of the tax revenue.[203] In this way, a carbon tax with reinvestment takes into consideration the different levels of development and industrialization in emerging and advanced economies.

Furthermore, border tax adjustments can be easily added on to a carbon tax with reinvestment, in order to create a level playing field between nations in the international trading regime.[204] From the perspective of the WTO's border tax adjustment rules, the simplest solution would be a carbon tax which is rebated on exports, and imposed in an equivalent amount on like imports, thus ensuring answering any international competitiveness concerns.[205] Sewalk argues that if a CTR system was implemented in the EU, US and China, this would incentivize other nations to move toward tougher GHG emissions policies.[206] Unilateral climate change policy is essential in an international community which is struggling to reach any concrete consensus on how to mitigate climate change. An ideal next step from unilateral implementation of a carbon tax with reinvestment could be a plurilateral coordination of such measures.[207]

We shall now turn to analyze how unilateral enactment of a carbon tax with reinvestment can be managed within the constraints of WTO law.

5 The Importance of Border Tax Adjustments for Effective Carbon Pricing Mechanisms

In the context of unilateral carbon taxation, the aim of a border tax adjustment on carbon is "to shift the economic burden of emission reduction to non-abating countries"[208] in order to tackle free-riding and induce international efforts to decrease fossil-fuel consumption. The Paris Agreement did not effectively tackle the issue of free-riding, but a WTO compliant border tax adjustment is a means of ensuring that unilateral climate change policy is not undermined by less climate-conscious states. Border tax adjustments are a fundamental part of the equation for effective carbon pricing mechanisms because they prevent carbon leakage through free-riding, they ensure the domestic competitiveness of industries in states who seek to unilaterally

203. Sewalk, "The EU-27, US, UK and China Should Dump Cap-and-Trade as a Policy Option and Adopt a Carbon Tax," XLVII *Suffolk University Law Review* 568-569 (2014).
204. *Ibid.* 569-570.
205. Maruyama, "Climate Change and the WTO: Cap and Trade Versus Carbon Tax?," 45:4 *Journal of World Trade* 682 (2011).
206. Sewalk, "The EU-27, US, UK and China Should Dump Cap-and-Trade as a Policy Option and Adopt a Carbon Tax," XLVII *Suffolk University Law Review* 571 (2014).
207. Rafael Leal-Arcas, "New Frontiers in International Economic Law: The Quest for Sustainable Development," 40:1 *University of Pennsylvania Journal of International Law* 116 (2018).
208. Böhringer et al., "The Role of Border Carbon Adjustment in Unilateral Climate Policy: Overview of an Energy Modelling Forum Study," 34 *Energy Economics* S97 (2012).

implement climate change policy, and they may give states leverage over non-carbon-abating states.[209]

The General Agreement on Trade and Tariffs (GATT) sets out the conditions in which WTO members can enact border tax adjustments in Article III.2. States may impose "an additional charge on an imported product on entry into their customs territory as long as it corresponds to an internal tax or charge that is being imposed on like domestic products."[210] GATT Article III, known as the "national treatment" article, requires that states do not discriminate between domestic and imported products. The destination-based charge of the border tax adjustment is therefore WTO compliant because the charge is levied once the product is within the customs territory, just like a domestic product. The BTA, therefore, levels the playing field in international trade and complements unilateral carbon pricing.[211]

WTO member states have the sovereign right to define their own climate change-motivated trade policies, but they must adhere to WTO balancing requirements.[212] The difficulty is that the issue of GHG emissions is transboundary, and when intending to induce international change through unilateral efforts, the state opens itself up to risks of a challenge at the WTO. Trachtman argues that it is possible to design an import border tax adjustment that has a low risk of a successful WTO legal challenge. Trachtman opines that "if the import tax adjustment is equal to the carbon tax of a domestic producer then it is compliant" with Article II(2) because this allows a charge equivalent to an internal tax.[213]

This is supported by the idea in *China—Auto Parts 2009*,[214] where it was held that a charge is lawful under Article II(2) if the charge is triggered by an internal factor, within the customs territory, and not because of the importation.[215] One important factor to consider is that if the border tax adjustment on carbon was to treat WTO member states differently, perhaps reflecting the differing needs of developing countries, then this could violate GATT Article I, i.e., the most-favored-nation principle, and bring into question the Rules of Origin principle. This is an important consideration because, at the Paris Conference of the Parties, it was agreed that developing countries should receive financial and technical assistance to facilitate the decarbonization of their economies.[216]

---

209. Condon and Ignaciuk, "Border Carbon Adjustment and International Trade: A Literature Review," *OECD Trade and Environment Working Papers* (2013/06): 5.
210. Maruyama, "Climate Change and the WTO: Cap and Trade Versus Carbon Tax?," 45:4 *Journal of World Trade* 690 (2011).
211. Böhringer et al., "The Role of Border Carbon Adjustment in Unilateral Climate Policy: Overview of an Energy Modelling Forum Study," 34 *Energy Economics* S98-9 (2012).
212. Will, "The Extra-Jurisdictional Effects of Environmental Measures in the WTO Law Balancing Process," 50:4 *Journal of World Trade* 616 (2016).
213. J. Trachtman, "WTO Law Constraints on Border Tax Adjustment and Tax Credit Mechanisms to Reduce the Competitive Effects of Carbon Taxes," 70:2 *National Tax Journal* 469-471 (2017).
214. *China—Auto Parts (2009)*.
215. J. Trachtman, "WTO Law Constraints on Border Tax Adjustment and Tax Credit Mechanisms to Reduce the Competitive Effects of Carbon Taxes," 70:2 *National Tax Journal* 471 (2017).
216. Odell, 2018. "Our Alarming Climate Crisis Demands Border Adjustments Now," *Climate and Energy*, International Centre for Trade and Development: 11.

Since *US—Shrimp*,[217] it is arguable that environmental policy is a more widely accepted reason to limit the WTO free trade rules.[218] While the WTO has not ruled on border tax adjustments for carbon, it has approved environmental measure which restrict trade, citing GATT Article XX(g), which provides an exception for the conservation of exhaustible natural resources.[219] The environmental measure either has to fulfil GATT XX(g) or the necessity test set out in subparagraphs (a) and (b) which analyzes whether the environmental measure is necessary, how it restricts trade and whether there are any less restrictive alternatives. Because of the regulatory autonomy which is left to states it is arguable that the environmental benefit which comes from pricing carbon can be argued to be legitimate.[220] While the domestic competitiveness of industries vis-à-vis their international competitors in states which are not enacting carbon-abatement policies is important, Odell argues that border tax adjustments on carbon should not be promoted as a protectionist strategy,[221] and instead should be seen primarily as an attempt to decarbonize the international economy.[222]

One key issue to consider for unilateral attempts to induce carbon abatement by national energy trade policy supported by border tax adjustments on carbon is to ensure such unilateral efforts do not lead to retaliation and trade wars.[223] Böhringer et al. suggest that "back-door" attempts to induce other nations to decarbonize their economy could backfire instead of strategically convincing other nations to comply.[224] Nevertheless, it is hoped that such measures would "incentivise conformity"[225] and lead governments to create their own carbon taxation policies in line with the nations who initiated the unilateral climate change effort.[226] The threat of trade wars and retaliation could be mitigated by a plurilateral effort to successfully price carbon among the key GHG emitting nations, but whether this is plausible considering the current negative climate between the two greatest emitters, i.e., the US and China, remains to

217. *US—Shrimp 1 (1998).*
218. Will, "The Extra-Jurisdictional Effects of Environmental Measures in the WTO Law Balancing Process," 50:4 *Journal of World Trade* 618 (2016).
219. Odell, "Our Alarming Climate Crisis Demands Border Adjustments Now," *Climate and Energy*, International Centre for Trade and Development, 2018, 12.
220. Will, "The Extra-Jurisdictional Effects of Environmental Measures in the WTO Law Balancing Process," 50:4 *Journal of World Trade* 619-620 (2016).
221. A note deserves to be made on global trade protectionism. It has been on the rise in recent years. Per data from the Global Trade Alert, over 10,000 protectionist measures were implemented by 19 members of the G20 between 2009 and 2018. The US ranked first in the world, with 1,693 of them. Germany is responsible for 1,225 measures in the same period. As a result of these measures, protectionism has become a major impediment to world trade. The G20 came up in 2019 with a Strategy for Global Trade Growth, available at http://trade.ec.europa.eu/doclib/docs/2016/july/tradoc_154789.pdf.
222. Odell, "Our Alarming Climate Crisis Demands Border Adjustments Now," *Climate and Energy*, International Centre for Trade and Development, 2018, 13.
223. T. Cottier et al., "The Potential of Tariff Policy for Climate Change Mitigation: Legal and Economic Analysis," 48:5 *Journal of World Trade* 1034 (2014).
224. Böhringer et al., "The Role of Border Carbon Adjustment in Unilateral Climate Policy: Overview of an Energy Modelling Forum Study," 34 *Energy Economics* S98 (2012).
225. Will, "The Extra-Jurisdictional Effects of Environmental Measures in the WTO Law Balancing Process," 50:4 *Journal of World Trade* 617 (2016).
226. Odell, "Our Alarming Climate Crisis Demands Border Adjustments Now," *Climate and Energy*, International Centre for Trade and Development, 2018: 15.

be seen. The most important benefit, and the one which should be promoted when states attempt to unilaterally enact carbon taxes supported by border tax adjustments, is the mitigation of carbon leakage, and the possibility of inducing other nations to enact similar policies.[227]

VI CONCLUSION

Many of the solutions to these global challenges are knowable, but not yet known. To paraphrase Immanuel Kant, policy without politics is empty, but politics without policy is blind.[228] Policy without politics is something that happens all the time (such as initiatives outside the political arena). Politics is about interests, institutions, and ideals. There have been recently political movements without policy (such as "Occupy Wall Street"). The two concepts (policy and politics) would need to converge to be meaningful. As Montesquieu put it over two hundred and fifty years ago, the trading system can be a great instrument for change and peace.[229] We have shown that trade can also help toward climate-change mitigation with the right policies in place (*policy*) and political will (*politics*), thereby reaching realistic utopianism. A case in point is EcoPeace Middle East, an initiative that brings together Israeli, Jordanian, and Palestinian environmentalists that use the environment as a way to reach peace.

To reach such goals, one would need greater transnational cooperation, whether inter- or intra-regional, as well as a clear roadmap with a vision on cross-issue cooperation. That will require setting long-term rules, not just short-term economic gains. It will also mean putting an end to economic nationalism and populism (whose main elements are anti-elitism, anti-internationalism, conspiracy theories, as well as hostility to taxes and to foreigners; both nationalism and populism are currently driving the Western world). Ideally, this enhanced transnational cooperation should involve the WTO, the International Energy Agency, the Global Environment Facility, the World Bank, the International Monetary Fund, the G20, the MEF, and the UN Environment Program working together in the context of ecumenism and cosmopolitanism, away from balkanization.

Today subnational/local politics and businesses are a major part of climate action, making it a classic example of multilevel governance. Mayors/governors are much more involved in mitigating climate change than prime ministers and presidents of nations. We have witnessed the case of a bottom-up American network determined to fight climate change called *We Are Still In*, which comprises US state, local and business leaders. Similar is the role of the *United States Climate Alliance* or *America's Pledge*, both platforms committed to fight climate change from the bottom up.

227. Lockwood and Whally, "Carbon-Motivated Border Tax Adjustments: Old Wine in Green Bottles?," *The World Economy* 812 (2010).
228. The words by Immanuel Kant were: "Thoughts without content are empty, intuitions without concepts are blind." *See* Immanuel Kant, *Critique of Pure Reason*, p. 51 in the version published in 1781; p. 75 in the version published in1787.
229. Baron de Montesquieu, 2 *Oeuvres Completes: De L'esprit des Lois*. Gallimard/Pleiade, 1951.

We should reduce the demand of energy, especially fossil-fuel consumption, but that is not the prognosis, which justifies a transition to clean energy (i.e., green gold) to help the world grow cleanly. For some countries, the gradual shift to clean energy is also justified by the fact that they wish to minimize geostrategic risks. For climate deniers, the concept of clean energy is easier to accept than that of climate change. If power and transportation are clean, most of the climate-change issue is solved. Solar and wind energy are on their way to becoming very important sources of clean energy. Clean energy will also help diversify the economy by creating green jobs and more trade in renewable energy. The easiest way to introduce renewable energy into the market is via regional markets, as opposed to international/global markets. With the help of simple, clear rules, the market would need to continue to be liberalized and a gradual shift would need to be made from subsidizing fossil fuels to subsidies for renewables.

Moreover, thanks to technology, smart meters and smart grids will continue to rise as part of the digitalization process and will become more adaptive. But there are dangers to technology: cybersecurity has been an issue due to hacking when it comes to energy security.[230] Investing in innovation and technology in renewable energy will remain indispensable elements for success in climate-change mitigation.

In terms of a new architecture for effective climate change mitigation, I have offered the idea of a climate club linked to the international trading system, where the benefits of being a member of this club outweigh the obligations. This new design of transnational governance is justified due to how difficult it is to reach unanimity (which requires the lowest common denominator) among so many countries in the world that have very diverse agendas. Instead of "waiting for Godot,"[231] I propose an alternative legally binding and enforceable regime, where a small group of big countries (and big corporations) that are responsible for most of the global GHG emissions can work effectively with each other toward clean and sustainable economic growth. There is momentum as many companies and major world economies are going sustainable in their ambitions.

One should also add that young people are changing the climate agenda and actively participating to shape it. If they become more politically engaged by voting at elections in Western liberal democracies, a great transformation toward sustainability may be on its way. Today's youth are much better educated about climate change than any previous generation. Initiatives such as compulsory climate education one hour per week in Italian schools are welcome. Educating people on sustainability is key. A generational transformation is under way, with great hope for a sustainable future.

230. K. Zetter, "Inside the Cunning, Unprecedented Hack of Ukraine's Power Grid," Wired, March 3, 2016, available at https://www.wired.com/2016/03/inside-cunning-unprecedented-hack-ukraines-power-grid/.
231. I use the title of the famous play of the 1940s by Samuel Beckett ("*Waiting for Godot*") as a metaphor for a legally binding and enforceable global/universal agreement on climate change that will never happen. In *Waiting for Godot*, two characters were waiting for the arrival of Godot who never arrived. Equally, in the climate field, the international community has been waiting for a legally binding and enforceable universal agreement on climate change that has not yet arrived.

Nevertheless, getting people to shift to renewable energy is not easy because it is intermittent and more expensive than fossil fuels, which means that energy bills will be more expensive. Which is why it is imperative to show the economic dimension to convince people of the momentum. Competition, technology, and innovation will bring the cost down for all of us. So the future is bright.

The biggest challenge of our time is to mitigate climate change. The energy transition has social implications such as what to do with people currently hired in the carbon-intensive industries based on the "leave no one behind" principle. All countries need to find a way to eliminate all GHG emissions by mid-century if they are serious about reducing the risks posed by climate change. International environmental agreements have not thus far produced concrete results and, therefore, energy trade policy must be utilized to decrease fossil-fuel consumption. A solution is to make climate change obligations enforceable. If the benefits are tangible, policy is easier to get through Parliament in Western liberal democracies. Moreover, local individual actions matter because ordinary citizens can talk to their members of Parliament and Parliamentarians can then discuss in Parliament.

In addition, governments can make the biggest difference by taxing CO_2 more widely. The largest emitters of GHG should unilaterally implement upstream carbon taxes, which reinvest the revenues into the development of renewable and sustainable energy alternatives. To overcome the problems of free-riding, these unilateral climate change efforts should be supported with border tax adjustments to incentivize other nations to work toward an international goal of decarbonizing the economy. In an international community made up of states with different priorities and different stages of economic development and industrialization, it has been difficult to reach a consensus to achieve an environmental agreement that effectively tackles countries' dependence and overuse of fossil fuels. In addition, the price of fossil fuel does not reflect the cost of climate change: there are, among others, health-related issues that come at a heavy price to society; alternative energy sources to fossil fuels have the potential to be cheaper than fossil fuels through proper public policies; and the digital revolution is already consuming much less carbon than the normal economy does.

The Paris Agreement on Climate Change may have been hailed a victory by environmentalists, and it did mark a significant achievement, especially in light of the failings of the Kyoto Protocol and the impasse at the Copenhagen Conference. However, while setting inspiring goals for global GHG emissions reduction, it did not create enforcement mechanisms and therefore fails to tackle the important issue of free-riding. It is, therefore, necessary to turn to trade law and policy to effectively tackle climate change.

There are many carbon pricing mechanisms in place across the globe, but they are not effectively contributing to carbon abatement. Carbon pricing is only part of the solution to inducing global carbon abatement, but, if enacted by the major GHG emitters, it would have a substantial impact. We compared two models of carbon pricing: cap-and-trade and carbon taxation.

Cap-and-trade may ensure emissions certainty, but it is undermined by price volatility. The hybrid cap-and-trade with a price floor and ceiling mitigates this volatility to a certain extent, but has the drawback of creating greater emissions uncertainty. Furthermore, the cap-and-trade system is complex to administer.

In contrast, a carbon tax is far simpler to graft onto an existing national taxation system. Ideally, this unilateral effort could be extended to plurilateral action. An upstream carbon tax with reinvestment of its revenues into renewable and sustainable energy alternatives is the carbon pricing model which the largest GHG-emitting nations should unilaterally enact. The reinvestment of revenues directly into green spending ensures that the carbon tax is part of a broader climate change policy and is therefore in line with the FASTER Principles of Successful Carbon Pricing.

In order for such a carbon tax to be effective, a border tax adjustment on carbon needs to be implemented alongside. It is this mechanism in particular that will incentivize other nations to fully decarbonize their economies because the BTA reduces carbon leakage, ensures the competitiveness of carbon-abating nations vis-à-vis their negligent neighbors, and will induce a complete decarbonization of the international economy. If the US, EU and China were to unilaterally enact an upstream carbon tax with reinvestment, alongside complementary BTAs, the climate change mitigation effort would be significantly advanced, and the international trading system would therefore successfully reduce fossil-fuel consumption. For climate change mitigation to be effective in the future by using a carbon tax, taxing planes would be a good place to start, given that the number of flights in the future will only increase and, with it, also with emissions of GHGs from aviation.[232]

In the next chapter, we will explore the WTO and carbon market clubs, looking more specifically on how to put a price on carbon.

232. That said, there is thinking and discussion about electric flights in hybrid form as a way to minimize the impact of aviation on climate change. This idea could start with small hybrid planes carrying 50-100 passengers on regional routes and then becoming more ambitious as technology improves. In fact, Rolls-Royce, Airbus, and Siemens have teamed up to electrify a 100-seater regional aircraft called Bae146 to produce a cleaner and more efficient plane by combining combustion engines with electrical power. *See* The Economist, "Electrifying Flight: Hybrid Vigour," June 29, 2019, p. 78.

CHAPTER 3
World Trade Organization and Carbon Market Clubs

This chapter explores emission units as a linkage in climate clubs in the hope of making a remarkable difference in climate change mitigation. It analyzes emission units trading in the context of regional trade agreements as a novel, promising, and effective way to mitigate climate change. It then examines Article XX of the General Agreement on Tariffs and Trade (GATT) as a potential and encouraging remedy for the development of climate clubs. Specifically, it explores the scope of application of Article XX and investigates to what extent it can be applied to climate concerns. It then provides an analysis of climate-club measures according to the jurisprudence of the World Trade Organization on GATT Article XX.

I INTRODUCTION[1]

While global attention is focused on fighting the COVID-19 pandemic, climate change remains a defining, long-term challenge that requires policies to create sustainable economies. This challenging task is increasingly gaining attention among policymakers worldwide, since climate change is rapidly becoming a top priority in national politics and international trade remains an essential part of the global economy. To address climate change, the international community created the United Nations Framework Convention on Climate Change (UNFCCC) in 1992. It became the main forum of climate change negotiations, and in 1997 the Kyoto Protocol was adopted and set legally binding greenhouse gas emissions (GHG) reduction targets for industrialized countries. In 2019, many governments began the push for net-zero emissions with national targets and new policy strategies.[2]

---

1. The research input of Ellis Malle, Marion Kerestedjian, and Gulce Budak is acknowledged while writing this chapter.
2. ICAP, Emissions Trading Worldwide: Status Report 2020. Berlin: International Carbon Action Partnership, 2020. Available at https://icapcarbonaction.com/en/?option=com_attach&task=download&id=677.

If we look back at the development of emissions trading, in 2005, parties with commitments under the Kyoto Protocol[3] accepted targets for limiting or reducing emissions. To meet their emission-reduction targets with minimum adverse effects on their economies, the parties to the UNFCCC[4] adopted the Paris Agreement on Climate Change. This international treaty establishes the parameters of a new climate regime applicable to all states based on a system of nationally determined contributions (NDCs).[5] Similarly, the UNFCCC[6] introduced emission-reduction policies that require foreign products to mirror the climate costs of their production processes or to favor domestic climate-friendly producers over foreign ones.[7] Accordingly, emissions trading systems (ETS) have set a limit on the amount of emissions permissible for a company,[8] thereby putting a price on emissions and stimulating emissions reductions.

As the cornerstone of the European Union's (EU) climate change policies, the ETS was designed to ensure that states fulfil their commitment to reducing GHGs[9] cost-effectively. Through creating transferable units, the ETS established a system where parties had an economic incentive to buy and sell emissions allowances and creating the first international trading system for GHGs. However, these targets have created questions on how international trade rules may apply to the emissions trading markets. This is because emissions trading varies from country to country and has a range of different design elements.

Furthermore, issues related to emissions trading have not been raised in inter-governmental trade disputes at the World Trade Organization (WTO) to date.[10] The reason for that is that there is an absence of clear understanding on how environmental issues should be dealt with under the current WTO provisions as there is a void of clear understanding between contracting members on how to manage the issues at the interface of environment and trade. The existing rules of the General Agreement on Tariffs and Trade (GATT) and the WTO do not address climate-change problems and policies.

---

3. Kyoto Protocol to the United Nations Framework Conversion on Climate Change, 1998 (Annex B Parties).
4. Piggot, Georgia, Peter Erickson, Michael Lazarus, and Harro Van Asselt, *Addressing Fossil Fuel Production under the UNFCCC*. Paris and Beyond, Stockholm Environment Institute, 2017.
5. NDCs are national climate plans in which the parties to the Paris Agreement must indicate, among other things, the mitigation pledges they intend to achieve. Such pledges are non-legally binding, but parties are obliged to implement domestic mitigation measures "with the aim of achieving" them.
6. The Kyoto Protocol is a protocol to the UNFCCC. It imposes legally binding commitments on certain parties to the UNFCCC to achieve GHG reduction targets. The US is the only country that did not ratify the Kyoto Protocol.
7. Christina Voigt, "WTO Law and International Emissions Trading: Is There Potential for Conflict?," 2:1 *Carbon & Climate Law Review* 54-66 (2008). Available at https://papers.ssrn.com/sol3/papers.cfm?abstract_id=1145242.
8. Niels B. Bekkhus and Karel Van Hecke. "The European Union Emission Trading Scheme," 61:1 *Studia Diplomatica* 115-147 (2008).
9. Richard Youngs, *Climate Change and EU Security Policy: An Unmet Challenge*. Carnegie Endowment for International Peace, 2014.
10. Kateryna Holzer, "Proposals on Carbon-related Border Adjustments: Prospects for WTO Compliance," 4:1 *Carbon & Climate Law Review* 51-64 (2010).

In addition, it is clear that dealing with global climate change is an issue far more complex than any of the issues previously dealt with in WTO disputes. Despite that, rules for the allocation and trade of emission units may also constrain the import of energy products generated by fossil fuels' combustion.[11] Thereby, such climate measures could be confronting WTO rules that seek to guarantee nondiscrimination and market access between the WTO's member states. Similarly, emission units will be subject to WTO provisions, should they be classified as goods. However, classifying emission units is a difficult task, as none of the WTO agreements explicitly addresses their trade.[12]

While there are some differences of opinion regarding the legality of WTO law on environmental issues, such as emissions trading units, there are, nevertheless, exceptions in the general structure of WTO law that allow outside, non-trade interests to be assessed and balanced against trade imperatives. For instance, GATT Article XX(b) and XX(g) provides space for shielding environmental measures from the strictures of GATT law, provided certain conditions are met. GATT Article XIV and the Enabling Clause both constitute an acknowledgement from the WTO members of the necessity to address other legitimate objectives within the organization, while departing from certain established principles. Further, the General Agreement on Trade in Services (GATS) also contains environmental exception in Article XIV, and so does the plurilateral Agreement on Government Procurement in Article XXIII. Similarly, most modern regional trade agreements (RTAs)[13] contain provisions that echo the GATT's general exceptions on the environment and conservation of exhaustible natural resources.

With a view to bring least contributing states to climate change into the global climate effort and catalyze national pledges, the Paris Agreement on Climate Change has been built and elaborated around a bottom-up structure, relying upon NDCs[14] set up according to particular national circumstances and priorities.[15] On this premise, state parties ought to be held liable for their pledged contributions only, despite potentially damaging climate change impacts on global environment, and extensively vulnerable state parties. Accordingly, although the Paris Agreement has gained in popularity compared to previous initiatives, its flexibility and lack of enforcement measures have failed to target effective GHG emissions. It appears that current NDCs are covering approximately one-third of the deep cuts required to meet the 2°C

11. Christina Voigt, "WTO Law and International Emissions Trading: Is There Potential for Conflict?," 2:1 *Carbon & Climate Law Review* 54-66 (2008). Available at https://papers.ssrn.com/sol3/papers.cfm?abstract_id = 1145242.
12. Marisa Martin "Trade Law Implications of Restricting Participation in the European Union Emissions Trading Schemes," 19 *Georgetown International Environmental Law Review* 437 (2006-2007).
13. RTAs, which are reciprocal preferential trade agreements between two or more partners, constitute one of the exemptions and are authorized under the WTO, subject to a set of rules. Available at, WTO | Regional Trade Agreements gateway, https://www.wto.org/english//tratop_e/region_e/region_e.htm.
14. UNFCCC, Decision 1/CP.21, Adoption of the Paris Agreement (December 12, 2015), U.N. Doc. FCCC/CP/2015/L.9/Rev/1, Article 3.
15. *Ibid.* Article 4.3.

temperature goal.[16] What is more, its support from major emitters has recently started plummeting with the withdrawal decision from the US, leaving remaining parties to consider less ambitious reductions and free-riding. In such context, strengthening national pledges' ambition is absolutely inevitable to have a slight chance of reaching net-zero by mid-century.

Going further, however, consideration should be given to how additional ambition could be achieved by any other means than strictly relying upon fallible UNFCCC agreements. The key challenge lying ahead would entail determining a balanced and suitable combination between existing climate treaties and complementary institutional modalities such as club-like arrangements.[17]

Despite its voluntary nature, according to William Nordhaus, the climate-club approach relies on a top-down structure setting from the beginning ambitious standards to reach important levels of abatement.[18] Following this approach, only states that have strong climate commitments could embark on a climate club and benefit from its exclusive goods such as preferential terms of trade or investment.[19]

The practical sticking point between the general bottom-up approach to climate commitments and the voluntary club approach lies within the induced penalties that are being imposed to non-members. There is no denying that climate clubs would embody a larger dimension, either in terms of participation or emissions reductions, when small trade penalties exist,[20] intended as an economic impetus for joining a club structure. Doubts may arise as to whether the current climate regime would allow to proceed with such incentive. Indeed, the framework for climate-change mitigation is already strictly regulated under the Paris Agreement, standing firmly for a facilitative and non-punitive process for implementation and compliance.[21] Therefore, NDCs are not subjected to any material sanctions for noncompliance, except for a peer-review mechanism.

By contrast, although one could query whether such framework could achieve substantial emission cuts at a sufficient pace to meet long-term temperature goals, it shall not be for state parties to decide, through reinforced mitigation in the form of clubs, if each contribution is satisfactory, following their own standards, and impose trade barriers accordingly.

Hence, now is the time to determine whether the carbon-club approach could be conciliated with the bottom-up international climate regime. Trade law could be touted as the solution to legitimize the existence of climate clubs and their direct implications on trade rules. In this sense, climate change law and trade law can be mutually

16. UNEP (2017), The Emissions Gap Report 2017 https://wedocs.unep.org/bitstream/handle/20.500.11822/22070/EGR_2017.pdf.
17. World Bank Group (2016), "Carbon Market Clubs and the New Paris Regime," v.
18. William Nordhaus, "Climate Clubs: Overcoming Free-riding in International Climate Policy," 105:4 *American Economic Review* 1344.
19. Jon Hovi, Detlef F. Sprinz, Hakon Saelen, and Arild Underdal, "The Club Approach: A Gateway to Effective Climate Co-operation?," 49 *British Journal of Political Science* 1076.
20. William Nordhaus, "Climate Clubs: Overcoming Free-Riding in International Climate Policy," 105:4 *American Economic Review*.
21. UNFCCC, Decision 1/CP.21, Adoption of the Paris Agreement (December 12, 2015), U.N. Doc. FCCC/CP/2015/L.9/Rev/1, Article 15.

supportive. Indeed, Article 3.5 of the UNFCCC is not only providing for states cooperation in fighting climate change, but also averts from potential restrictions on international trade, leaving a window open for unilateral measures.[22] With international cooperation falling short on providing satisfying results, states tend to resort to unilateral action, frequently in the form of trade measures.[23]

However, Article XX of the GATT could pave the way for the acceptation of concerted climate-related trade restrictions, among its general exceptions, so as to justify the process the existence of climate clubs. It is henceforth essential to query whether relying on Article XX to legitimize climate clubs would remain nothing but a pipe dream, or rather a true way out of the climate stalemate.

After this introduction, section II explores emission units as a linkage in climate clubs. Section III analyzes emission units trading and regional trade agreements. Section IV examines GATT Article XX as a potential remedy for the development of climate clubs. Specifically, section IV.1 explores the scope of application of Article XX and investigates to what extent it can be applied to climate concerns,[24] while section IV.2 provides an analysis of climate-club measures according to the jurisprudence of the WTO on GATT Article XX.

II EMISSION UNITS AS A LINKAGE IN CLIMATE CLUBS

Section II provides an exploration of emission units as a linkage in climate clubs.

1 Climate Clubs

1.1 The Concept

As decision-making under the UNFCCC meetings and agreements has a limited ability to react to emergency and evolving conditions of climate change, the need for a new institutional arrangement has a high priority. Hereby, the concept of a climate club is offered as an alternative forum to the UNFCCC platform for the fight against climate change.[25] A climate club serves the purpose of forming a global carbon market with the integration and harmonization of climate-mitigation policies of various jurisdictions.

22. UN Framework Convention on Climate Change (adopted May 9, 1992), 1771 U.N.T.S. 107 (UNFCCC), Article 3.5.
23. Cottier and Payosova, "Common Concern and the Legitimacy of the WTO in Dealing with Climate Change" [2016] ELECD 1460 in Panagiotis Delimatsis (ed.) in *Research Handbook on Climate Change and Trade Law*. Edward Elgar Publishing, 2016, 22.
24. For further analysis on common concerns more broadly, *see* R. Leal-Arcas, "Sustainability, Common Concern and Public Goods," 49:4 *The George Washington International Law Review* 801-877 (2017).
25. William Nordhaus, "Climate Clubs: Overcoming Free-Riding in International Climate Policy," 105 *American Economic Review* 1339 (2015). William Nordhaus, *The Climate Casino: Risk, Uncertainty, and Economics for a Warming World*, Yale University Press, 2013; Robert Falkner, *A Minilateral Solution for Global Climate Change? On Bargaining Efficiency, Club Benefits and International Legitimacy*, 2015. Robert Gampfer, "Minilateralism or the UNFCCC? The Political Feasibility of Climate Clubs," 16 *Global Environmental Politics*, MIT Press 62 (2016). Kenneth W.

Participant jurisdictions to club scheme will set the same level of targets and ambition in the climate negotiations and consequently, they can more easily integrate their carbon markets.[26] A climate club can provide clearer certainty regarding the conditions and rules governing the mitigation efforts.[27] In order to incentivize the participation of countries in the club arrangement, the combination of trade-restrictive and liberalizing instruments can be employed for climate-mitigation purposes, especially market-based tools, which can efficiently coordinate ambitious countries in their climate commitments and encourage other countries' mitigation efforts to enjoy the benefits of club membership. Under this new mechanism, emission reduction will be an incidental co-benefit to the economic benefits.[28]

Moreover, climate club will reduce the economic costs of abatement efforts of emissions and volatility of the carbon market compared to non-club prices.[29] Most importantly, climate club by having a harmonized price for the emissions will eliminate the competitiveness concerns and carbon leakage of the industry to the jurisdiction having less stringent environmental protection. A climate club will signal promising multilateral commitment for climate change, which can elevate international cooperation. The abatement activities as a result of club cooperation will improve environmental integrity, air quality, reduce fossil fuel dependence, and encourage low-carbon production facilities within the club members.[30]

1.2 Emission Units as a Linking Instrument

In order to incentivize membership, benefits of emissions reductions should be linked to excludable economic benefits[31] and one of them would be a common emissions trading system (ETS). A comprehensive and enforceable climate club can be designed with direct linkages of different ETSs. Herein, the linkage in the climate club will be emission units. Under the climate club, all Members will be bonded with absolute emissions caps and have the option to trade the emission units among each other. With this arrangement, the emissions caps, flexibility, the structure of the ETS can be negotiated among a small number of like-minded countries and facilitate an easier negotiation platform. Within the common ETS, the trading of emission units among

Abbott, "Strengthening the Transnational Regime Complex for Climate Change," 3 *Transnational Environmental L* 57 (2014).

26. Currently, cooperation among countries is limited to informal dialog, information, and experience exchange. For example, the International Carbon Action Partnership (ICAP) creates a forum for sharing and evaluating best practices for mandatory cap-and-trade systems.
27. M. Mehling and E. Haites, "Mechanisms for Linking Emissions Trading Schemes," 9 *Climate Policy* 169 (2009).
28. William Nordhaus, *The Climate Casino: Risk, Uncertainty, and Economics for a Warming World.* Yale University Press, 2013, at 255.
29. W. Blyth and M. Bosi, *Linking Non-EU Domestic Emissions Trading Schemes with the EU Emissions Trading Scheme*, OECD/IEA, Paris, www.oecd.org/dataoecd/38/7/32181382. pdf (2004).
30. H.Westskog, "Why Should Emissions Trading Be Restricted?," 2 *Climate Policy* 97 (2002).
31. M.R. Zefferman, "Cultural Multilevel Selection Suggests Neither Large or Small Cooperative Agreements Are Likely to Solve Climate Change Without Changing the Game," 13 *Sustain Sci* 118 (2018), https://doi.org/10.1007/s11625-017-0488-3 at p. 114.

members are subjected to the same rules, freely traded among different systems, and can be valid for the compliance of mitigation obligations in each jurisdiction.[32] The key feature in designing a climate club is the coordination of legislation and rules governing each system through an international treaty, which is binding on the domestic ETS legislation in each member country.[33]

Emission units, issued in line with an abatement, target the already-enabled cost-effective compliance with climate club obligations through the allocation of units among members.[34] The limited number of units realizes higher ambition for mitigation of climate change effects and supports low-carbon technologies in production processes. It is offered as an intermediary step in the creation of a global carbon market, which has significant coverage of economic sectors and countries. The size of the ETS market can expand and as a result market liquidity will increase, and carbon prices become more stable compare to individual national systems.[35] Furthermore, the possible objections to unilateral decisions on carbon-pricing can be avoided with an ETS of club countries, as unilateral measures might have market distortions and unfair competition effects, especially on developing countries.

The linking with the ETS requires the agreement on the definitions of common characteristics of the system; design of the allocation scheme of the units; monitoring, reporting, and verification method; all kinds of technical designs and procedures for cooperation.[36] Non-club members will be discriminated from the system as their climate commitments, the stringency of targets, and design of linked ETS will have considerable differences compared to a climate club. As a result, the climate club system creates a differentiated treatment among trade partners based on their membership status that violates the nondiscrimination principle of WTO law, namely the most-favored-nation (MFN) principle. WTO law prescribes a carbon pricing policy, which is nondiscriminatory and does not have extra-jurisdiction effects. However, the club scheme is based on exclusiveness and its success can only be achieved if the non-members are discriminated against the advantages of ETS linkages.

In order to decide on the violation of the MFN principle of WTO, we must first analyze whether the emission units are regulated under the WTO Agreements. The three main subjects that are forming the structure of the WTO Agreements are goods, services, and intellectual property. The design and characteristics of the emission units

32. E. Haites and F. Mullins, Linking Domestic and Industry Greenhouse Gas Emission Trading Systems, EPRI, International Energy Agency (IEA) and International Emissions Trading Association (IETA), Paris (2001).
33. M. Mehling, "Bridging the Transatlantic Divide: Legal Aspects of a Link Between Regional Carbon Markets in Europe and the United States," 2 *Sustainable Development Law and Policy* 46 (2007).
34. Ulrike Will, *Climate Border Adjustments and WTO Law Extending the EU Emissions Trading System to Imported Goods and Services*. Brill Nijhoff, 2019, at 4.
35. Richard B. Stewart et al., "Building Blocks for Global Climate Protection," 32 *Stan Envtl LJ* 34 (2013), at 374.
36. R.N. Stavins and J. Jaffe, Linking Tradable Permit Systems for Greenhouse Gas Emissions: Opportunities, Implications, and Challenges, International Emissions Trading Association (IETA) and Electric Power Research Institute (EPRI), Geneva (2007).

are likely to be fall under the subjects of goods or services and, therefore, their governing agreements are GATT and GATS, respectively.

2 Emission Units as a Good

2.1 The Characterization of the Units

Products are defined as commodities produced by labor, intellectual effort, or natural process that can be transported from place to place.[37] They can be immaterial if they are still performing the role parallel to products.[38] Greenhouse gas (GHG) emissions are produced by labor. Afterward, ETS gives them a price and make them tradable. Emission units fulfill the purposes of accounting in the ETS, in which they are tracked and recorded through a registry.[39] Units represent a certain amount of energy or GHG consumed. They are transferable among facilities. Emission units are movable, digital, and have a unique property feature.[40] In this regard, they can be considered an intangible good. The registry appoints them a unique serial number and that is how they can be traced among different holders.[41] Therefore, the number of units is limited and distributed to the businesses within the jurisdiction that issues them. However, contrary to a regular product, the emissions system does not aim to make units valuable; instead, the savings of units are rewarded. Furthermore, the club issues units to facilitate the compliance of international obligations deriving from a climate club. Therefore, they function more as a means of payment instead of an investment good.

2.2 MFN Principle

The different treatment to emission units from a non-climate-club jurisdiction is a restriction to the MFN principle under the Article I:1 of the GATT. The principle requires that "any advantage, favor, privilege or immunity granted ... shall be accorded immediately and unconditionally to like the product of all contracting parties."[42] Article I:1 covers all types of trade restrictions on goods, such as tariffs, taxes, charges, or standards, and requires the extension of all possible advantages accorded to one

37. Appellate Body Report, United States—Final Countervailing Duty Determination with Respect to Certain Softwood Lumber from Canada (adopted February 17, 2004) WT/DS257/AB/R, paras. 58-59.
38. E. Vranes, "Climate Change and the WTO: EU Emission Trading and the WTO Disciplines on Trade in Goods, Services and Investment Protection," *Journal of World Trade* 707 (2009) at 717. The ECJ ruled in the Almelo v. Energiebedrif ljsselmij case that electricity is good; therefore, the requirement of tangibility is less relevant now. NV Case C-393/92, Municipality of Almelo v. NV Energiebedrif ljsselmij (1994) ECR 1-1477.
39. Charlotte Streck and Moritz von Unger, "Creating, Regulating and Allocating Rights to Offset and Pollute: Carbon Rights in Practice 2016," *CCLR* 178 (2016), at 179.
40. E. Vranes, "Climate Change and the WTO: EU Emission Trading and the WTO Disciplines on Trade in Goods, Services and Investment Protection," *Journal of World Trade* 707 (2009), at 717.
41. Charlotte Streck and Moritz von Unger, "Creating, Regulating and Allocating Rights to Offset and Pollute: Carbon Rights in Practice 2016," CCLR 178 (2016), at 181.
42. Appellate Body Report, Canada—Certain Measures Affecting the Automotive Industry (adopted June 19, 2000) WT/DS139/AB/R, WT/DS142/AB/R, para. 79.

member to all WTO member states. The differentiated treatment to emission units derives from the climate change policy and membership to a climate club, instead of the national origin of the emission units. However, the Appellate Body prohibited any de jure or de facto discrimination under Article I(1) of the GATT.[43] Targeting a specific country based on climate policy could give rise to allegations of discrimination as it might be considered as de facto discrimination on the grounds of national origin.[44] Similarly, any exemptions for the countries which has climate measures would imply advantages within the meaning of Article I:1 and should be extended to other WTO Members. Therefore, the country's membership in a climate club or carbon abatement system would not be a justifiable reason to apply differentiated treatment to emission units under the MFN principle.

The principle is applicable if three criteria are fulfilled cumulatively. These are: (1) an advantage, favor, privilege, or immunity accorded to the product of one country; (2) the advantage should be granted immediately and unconditionally; and (3) the like products of another country.

2.2.1 *Advantage*

The club members will make a regulatory distinction according to the origin of emission units. The climate club members' emission units are linked with each other; therefore, this has implied advantages to emission units of club members, whereas it discriminates against non-club members. This regulation has an extraterritorial effect since it is a unilateral measure to control emissions outside the jurisdiction of climate club members.

2.2.2 *Advantage Accorded Immediately and Unconditionally*

Any advantage, favor, privilege, or immunity given to a product of one state must extend to like products of all WTO Members. This means the application of any possible benefit, without a delay and without any further requirement that affects competitiveness.[45] If the emission units from the club and non-club jurisdiction are found to be like but are treated differently just because the WTO member is not a part of the climate club, the club membership will be a conditional advantage within the scope of Article I:1 GATT and restricts the application of MFN principle.

43. *Id*, para. 178.
44. Christine Kaufmann & Rolf H. Weber, "Carbon-Related Border Tax Adjustments: Mitigating Climate Change or Restricting International Trade?," 10 *World Trade Review* 497 (2011), at 503.
45. Appellate Body Reports, European Communities—Measures Prohibiting the Importation and Marketing of Seal Products (adopted June 18, 2014) WT/DS400/AB/R, WT/DS401/AB/R, paras. 5.88-5.90, 5.93.

2.2.3 *Like Products*

The like products analysis derives from the Appellate Body decisions regarding the national treatment principle under Article III of the GATT. The four general criteria adopted in the likeness test are the following: "(i) the features, nature, and quality of the products (ii) the end-uses of the products, the extent products able to serve the same or similar end-uses, (iii) consumers "tastes and habits, the extent consumers perceive and treat the products as an alternative mean of performing particular functions or satisfy demand, and (iv) international tariff classification of the products."[46]

2.2.3.1 The Product Characteristics

In order to compare the likeness of emission units that are deriving from different countries, several features come under scrutiny. Each ETS, in which the units are issued, is based on different key design features depending on the local context and government policy.[47] The key point of the likeness analysis lies in the comparison between the club's unit system and non-club systems structural effect on the characteristics of the units.

Herein, the systems should be based on the same method, which could be cap-and-trade[48] or baseline-and-credit system.[49] The former approach imposes compliance obligations, whereas the latter one only supports voluntary action to create and reduce the price for emissions.[50] The international agreement that forms the climate club and creates the ETS, sets a target for the maximum amount of emissions permitted within the club. This differentiates the product characteristics of units since without sharing the same target with the club; non-club members could issue an excessive number of units and the effectiveness of emissions reduction will be limited.

The emission units should be issued to represent the same amount of GHGs or energy per unit. The covered amount of emissions and sectors will vary among different emission-units systems. For example, New Zealand ETS covers land-based sectors, whereas the EU does not.[51] Moreover, which entities are required to hold units, what standards and conditions will govern the issuance of units become relevant in the analysis of the unit's product characteristics.

The climate club is likely to have strict penalty mechanisms and an enforcement system in case of noncompliance. Therefore, if the units from a non-club jurisdiction

46. Peter van den Bossche, *The Law and Policy of the World Trade Organization: Text, Cases and Materials*. Cambridge University Press, 2008, at 345.
47. Peter Sopher, "Emissions Trading around the World: Dynamics Progress in Developed and Developing Countries 2012," *CCLR* 306 (2012), at 306.
48. Under a cap-and-trade system, an overall emissions cap is set to achieve emissions reductions.
49. Under a baseline-and-credit system, carbon credits are issued for reductions of emissions against a business-as-usual baseline.
50. Charlotte Streck and Moritz von Unger, "Creating, Regulating and Allocating Rights to Offset and Pollute: Carbon Rights in Practice 2016," *CCLR* 178 (2016), at 178.
51. Peter Sopher, "Emissions Trading around the World: Dynamics Progress in Developed and Developing Countries 2012," *CCLR* 306 (2012), at 307.

are found to be like products, its lax enforcement regulation could damage the clubs' system. For example, a penalty system in which the companies are exempted from further liability for their emissions above the cap if the penalty is paid will harm the stringency of the club scheme.[52] The facilities will prefer to hold units from a system that is governed with more flexible rules for the purpose of compliance with the club's ETS. This situation will lower the achieved carbon abatement goal as causing a carbon leakage.

Furthermore, the consistency among the registries of the units is important.[53] Club members will have one registry that governs all the administrative aspects of the unit issuance, whereas non-club members registries will work independently and each will have separate rules of conduct leading to differentiation in the characteristics of the units. One another aspect is the option of banking.[54] Banking means that participants can hold unused allowances for using in future trading periods. Banking provides advantages, if the emissions prices are expected to increase, or as a hedge against future price increases. The consumers might prefer the unit that has a banking option, and this may result in an inflow from club jurisdiction to non-club ones.[55]

Product characteristics are also affected by the way of the distribution of emissions allowances. For example, if the units are distributed based on the grandfathering[56] or benchmarking[57] instead of an auction, the firms that participated in that ETS can generate profits. Free allocation of units is a core regulatory aspect to address competitiveness concerns.[58] The different allocation methods damage the comparability of the emissions control regulations and complicate the likeness analysis of units from different jurisdictions. Free allocation can be perceived as an implicit subsidy in a system where the units are auctioned.[59] Whereas the club members who have an auction system will be at a competitive disadvantage.[60] Consequently, we could conclude that the characteristics of each unit are one of its kind, as they are subjected to a different set of rules and derive from varying political arrangements.

52. Angelica P. Rutherford, "Linking Emissions Trading Schemes: Lessons from the EU-Swiss ETSs 2014," *CCLR* 282 (2014) at 287.
53. *Ibid.*
54. Christian de Perthuis and Raphael Trotignon, Governance of CO_2 Markets: Lessons from the EU ETS, 2013 Les Cahiers de la Chaire Economie du Climat Working Paper (2013), at 6.
55. Erik Haites, Harmonisation between National and International Tradable Permit Schemes, CATEP Synthesis Paper, OECD (2003), at 5.
56. In grandfathering allowances are given out to operators for free on the basis of historic emissions. This approach is criticized as it is rewarding the higher emitters.
57. Benchmarking is used for the free allocation of the emission units. Benchmark is determined by the product-related GHG emission benchmarks and product performance. In this way, GHG intensive installations will receive fewer free units, compared to low-carbon installations. Moreover, the calculation of the benchmark can vary among different ETS jurisdictions. For example, in EU ETS, benchmarks are selected from the 10% most efficient installations in each sector. Consequently, inefficient installations need to purchase more units or reduce their emissions. European Commission, *EU ETS Handbook*, https://ec.europa.eu/clima/sites/clima/files/docs/ets_handbook_en.pdf, at 40.
58. Tobias Hausotter et al., *Competitiveness and Linking of Emission Trading Systems*. Adelphi, 2010, at 46.
59. *Ibid.* at 47.
60. Angelica P. Rutherford, "Linking Emissions Trading Schemes: Lessons from the EU-Swiss ETSs 2014," *CCLR* 282 (2014), at 288.

2.2.3.2 The End-Use

Even though the design features are differentiating in each unit system, ultimately all units serve the purpose of emission abatement with a quantity-based market mechanism. However, since the emission reduction will occur at different rates due to varying characteristic features of the units, it would be a more correct approach to evaluate the end-uses as similar, not exactly the same.

2.2.3.3 Consumer Preferences

Consumers of the emission units are businesses, financial institutions, and other regulated entities. The consumers in the club jurisdictions will prefer the club units to ensure compliance with the climate club scheme.[61] Otherwise, they face a risk of economic loss by way of penalties or repurchase of units for ensuring compliance.[62] This shows us that the consumers do not regard the units from different jurisdictions as like. Moreover, the club's emission units are one part of larger cooperation among countries and they can address the climate change concerns more effectively. Therefore, the consumers sharing the ethical concerns of protecting natural resources and lowering GHG levels can direct their free choice to club emission units. In this way, they signal that the club and non-club units are not like in terms of consumer preferences.

2.2.3.4 Tariff Classification

The emission units currently do not have a tariff classification and Harmonized System (HS) code.[63] HS Code, a six-digit harmonized code, is used by customs authorities around the world to identify products when assessing duties and taxes, which are called tariffs for imports. However, the effect of tariff classification in the likeness analysis is limited. This is because countries do not always use the same code to define products. Furthermore, the tariff lines are based on a very detailed analysis of each product and might assign different codes to products that are considered as like from the perspective of the other three criteria in the likeness analysis.[64] On the other hand, even though two products have the same tariff classification, they could still be

61. Christina Voigt, "WTO Law and International Emissions Trading: Is There Potential for Conflict 2008," 2:1 *Carbon & Climate Law Review* 54 (2008), at 56.
62. Sikina Jinnah, Emissions Trading under the Kyoto Protocol: NAFTA and WTO Concerns, Graduate Student Theses, Dissertations, & Professional Papers (2002) at 77.
63. The Harmonized System is a standardized numerical method of classifying traded products.
64. For example, the alloyed and non-alloyed iron have very similar physical characteristics, end-use, consumer preferences and they are likely to substitute each other. However, they have been assigned different tariff lines. Ulrike Will, *Climate Border Adjustments and WTO Law Extending the EU Emissions Trading System to Imported Goods and Services*, Brill Nijhoff, 2019, at 140.

considered as unlike products, if their product characteristics differ to a great extent.[65] Different characteristics of the emission units from the club and non-club jurisdictions affect the value of carbon they represent.

2.2.3.5 Competitiveness and Substitutability

The Appellate Body can analyze their direct competitiveness and substitutability with the economic technique of calculating the elasticity of substitution.[66] What will be scrutinized is the effect of the measure that modifies the conditions of competition in the relevant market to the detriment of products from a non-climate club member.[67] The physical product characteristics will also shed light on the competitiveness analysis.[68] The primary market, which the emission units created, is the exchange of units among consumers.[69] As the club members harmonize their emission units in the primary market and refuse the recognition of units from non-members, the market access is strictly restricted and competitiveness is affected.

3 Emission Units as a Service

3.1 The Characterization of the Units

The GATS covers all sectors except services supplied in the exercise of governmental authority. Services are non-physical, intangible, non-storable, and invisible.[70] The emission-units certificates are also intangible and benefit their holder by implying the right to emit and to trade emission allowances. Units function as a decarbonization service for which investors need to pay.[71] The emission units as certificates could have an effect on the trade of foreign environmental or financial services.

---

65. *Ibid.* at 140. An example of it can be found in Appellate Body Reports, Philippines—Taxes on Distilled Spirits (adopted January 20, 2012) WT/DS396/AB/R, WT/DS403/AB/R, paras. 163-164.
66. Appellate Body Report, European Communities—Measures Affecting Asbestos and Asbestos-Containing Products (EC-Asbestos) (adopted April 5, 2001) WT/DS135/AB/R, at para. 99.
67. The Appellate Body Report, Korea-Measures Affecting Imports of Fresh, Chilled and Frozen Beef (Korea-Beef), (adopted January 10, 2001) WT/DS161/AB/R, WT/DS169/AB/R, para.137.
68. Appellate Body Report, European Communities—Measures Affecting Asbestos and Asbestos-Containing Products (EC-Asbestos) (adopted April 5, 2001) WT/DS135/AB/R, at para. 114.
69. Jacob Werksman, "Greenhouse Gas Emissions Trading and the WTO," 8 *Review of European Community and International Environmental Law* 251 (1999), at 255.
70. Ulrike Will, *Climate Border Adjustments and WTO Law Extending the EU Emissions Trading System to Imported Goods and Services*. Brill Nijhoff, 2019, at 282.
71. Glenn M. Wiser, "The Clean Development Mechanism versus the World Trade Organization: Can Free-Market Greenhouse Gas Emissions Abatement Survive Free Trade," 11 *Georgetown International Environmental Law Review* 531 (1999), at 558.

3.1.1 *Environmental Services*

Environmental services comprise: (a) sewage disposal services, (b) refuse and disposal services, (c) sanitation and similar services, and (d) other environmental protection services not classified elsewhere.[72] The emission units can fall under the last category (d), as not falling under the specific characteristic listed in the GATS Schedule.

3.1.2 *Financial Services*

Financial services are "any service of financial nature offered by a financial service supplier of a WTO Member."[73] Different types of it are listed in paragraph 5(a)(v)–(xvi) GATS Annex on Financial Services. For the sake of emission-units analysis, the following financial services might become relevant: derivative instruments including futures and options,[74] transferable securities[75] and a general category of other negotiable instruments and financial assets.[76]

Derivatives are distributed on a private law basis.[77] Even though private actors can trade the emission units among each other, units are not distributed on a contractual relationship.[78] For this reason, units cannot fall under the category of the derivative. Emission units represent a guarantee of a right to payment; they are transferable and can be a medium of investment.[79] Therefore, emission units as a service can be qualified under transferable security.[80] Furthermore, they could discharge debts and legacies, and consequently fall under the category of financial assets.[81]

---

72. GATS Schedule of Specific Commitments (April 15, 1994) GATS/SC/31 (EC *Schedule*), at 58-60.
73. GATS Annex on Financial Services, para. 1, https://www.wto.org/english/tratop_e/serv_e/10-anfin_e.htm.
74. Para. 5(a)(xiii)(C) GATS Annex on Financial Services *Ibid.* Derivatives can be defined as an arrangement or instrument (such as a future, option, or warrant) whose value derives from and is dependent on the value of an underlying asset. Panel Report, China—Certain Measures Affecting Electronic Payment Services (July 16, 2012) WT/DS413/R para. 7.152.
75. Para. 5(a)(xiii)(E) GATS Annex on Financial Services *Id.* Transferable security is a document held by a creditor as a guarantee of his or her right to payment; a certificate attesting ownership of stock, shares, etc.; the financial asset represented by such a document, which is issued to investors to finance a business venture or a pledge of a financial or physical property to be surrendered in the event of failure to repay a loan.
76. Para. 5(a)(xiii)(F) GATS Annex on Financial Services *Id.* Assets are "property owned by a person or company regarded as having value and being available to med debts, commitments or legacies." They "can be converted into money" for the owner's benefit and "can be purchased and sold on an open market."
77. Ulrike Will, *Climate Border Adjustments and WTO Law Extending the EU Emissions Trading System to Imported Goods and Services*. Brill Nijhoff, 2019, at 292.
78. James Munro, "Trade in Carbon Units as a Financial Service under International Trade Law: Recent Developments, Future Challenges 2014," *CCLR* 106 (2014), at 108.
79. *Ibid.* at 109.
80. Ulrike Will, *Climate Border Adjustments and WTO Law Extending the EU Emissions Trading System to Imported Goods and Services*. Brill Nijhoff, 2019, at 293.
81. James Munro, "Trade in Carbon Units as a Financial Service under International Trade Law: Recent Developments, Future Challenges 2014," *CCLR* 106 (2014), at 110.

If the emission units do not fit into one of the above categories, they can be analyzed for a general category which is other negotiable instruments.[82] Emission units are easily transferable from one person to another and can be freely negotiated.[83] Units have the signature of the maker, which is government agencies.[84] However, units are registered to a system; hence, they cannot be transferred as easy as cash and they do not state an "unconditional promise to pay a specified sum of money on its face[85]." In that case it difficult to categorize units as a negotiable instrument.[86] In conclusion, units are an intangible asset that pertains to financial matters such as utilization to avoid financial liability and can be traded in an open market system.[87] Thus, emission units are likely to be qualified under the financial services, either as transferable securities or any other financial assets.[88]

3.2 *MFN Principle*

The secondary market of the emission units is where the buyers and sellers trade financial instruments based on units for investment purposes. These financial instruments are subjected to the financial services provisions of the GATS and therefore, any advantages among club members need to extend to all WTO members regardless of their club membership.[89] The climate club aims to put an extra charge on foreign services imported to club jurisdiction or not to recognize foreign emission units as the equivalent of the club emission units. The nonparties to the climate club do not take part in carbon abatement efforts equally with a club member and, therefore, cannot benefit from the advantage of the recognition of their emission units in the club members' system. Exclusion of a WTO member from the ETS of the climate club could result in allegations of market access restriction and violation of the MFN principle. MFN obliges that any advantage accorded to services and service suppliers should be

82. Para. 5(x)(F) first alternative GATS Annex on Financial Services, available at https://www.wto.org/english/tratop_e/serv_e/10-anfin_e.htm. A negotiable instrument is a written instrument "that (1) is signed by the maker or drawer, (2) includes an unconditional promise or order to pay a specified sum of money, (3) is payable on demand or at a definite time, and (4) is payable to order or bearer."
83. Panel Report, China— Certain Measures Affecting Electronic Payment Services (July 16, 2012) WT/DS413/R para. 7.154.
84. F. Deane, "Emissions Trading and the GATS Financial Services Provisions: A Case Study of the Australian Carbon Pricing Mechanism," 13 *Journal of International Trade Law and Policy* 44 (2014), at 5.
85. Sikina Jinnah, Emissions Trading under the Kyoto Protocol: NAFTA and WTO Concerns, Graduate Student Theses, Dissertations, & Professional Papers (2002) at 60.
86. F. Deane, *Emissions Trading, and WTO Law: A Global Analysis*. Edward Elgar Publishing, 2015, at 117-118.
87. James Munro, "Trade in Carbon Units as a Financial Service under International Trade Law: Recent Developments, Future Challenges 2014," *CCLR* 106 (2014), at 113.
88. In the EU, emission allowances have been uniformly classified as financial instruments under the 2014 revised Markets in Financial Instruments Directive (MiFID II), https://eur-lex.europa.eu/legal-content/EN/TXT/HTML/?uri=CELEX:32014L0065&from=EN.
89. Jacob Werksman, "Greenhouse Gas Emissions Trading and the WTO," 8 *Review of European Community and International Environmental Law* 251 (1999), at 257.

extended immediately and unconditionally to the like services and service suppliers of any country. The concepts in this definition will be examined below in detail.

3.2.1 Like Services

Article I of the GATS mentions both services and service suppliers. Therefore, they are both relevant for the likeness analysis. Likeness does not require two services to be identical, but they need to be comparable or in a competitive relationship with each other.[90] The four criteria for the likeness of services are: (1) the characteristics of the service and the service supplier, (2) the end-use, (3) the consumer preferences, and (4) modes of supply.[91]

3.2.1.1 The Service Characteristics

In terms of service characteristics criteria, the structure of the emission units is relevant. Units are issued within a system, which varies in terms of their size, design, features, and geographical scope.[92] This also affects the characterization of the units. Herein the emissions certificates could be of various characteristics:

- Mandatory, voluntary, or a combination of both in which the participation can only be mandatory for the facilities that emit more than the agreed threshold.[93]
- Based on different caps[94] and base years or use flexible caps. The cap stringency determines the carbon price therefore, the scarcity of units is necessary to create price incentive.[95] Additionally, the demand for units by the participants, for whom the cost of reducing emissions is higher than purchasing units, increases the value per unit.[96] The comparable caps are a sign of a comparable level of effort and signal the responsibility of the same amount of mitigation costs for the participant countries. This is the biggest necessity to convince countries in taking part in a club system.
- Some systems employ cost-containment measures to create stable and low carbon prices, by using the tools of offset and borrowing provisions[97] or price

90. Appellate Body Report, Argentina—Measures Relating to Trade in Goods in Services (adopted May 9, 2016) WT/DS453/AB/R (Argentina—Financial Services), paras. 6.23-6.26.
91. *Id* paras. 6.30-6.33.
92. Andreas Tuerk et al., "Linking Carbon Markets: Concepts, Case Studies and Pathways," 9 Climate Policy 341 (2009) DOI: 10.3763/cpol.2009.0621 at 342.
93. Peter Sopher, "Emissions Trading around the World: Dynamics Progress in Developed and Developing Countries 2012," *CCLR* 306 (2012), at 311.
94. Cap is the maximum amount of GHG emissions allowed to be emitted in the system. When it is combined with the trading element, the emissions abatement goal is achieved in a cost-efficient manner.
95. European Commission, *EU ETS Handbook*, available at https://ec.europa.eu/clima/sites/clima/files/docs/ets_handbook_en.pdf, at 23.
96. *Ibid.* at 16.
97. High rates of borrowing from future commitment periods can harm the ambition for carbon emissions reduction targets. Moreover, future abatement costs will rise as the units will be scarce in the future.

caps.[98] The discrepancy among systems in relation to these tools can affect the price of the units and ultimately their characteristics.

- The target can change from absolute targets to intensity targets in a unit system. In an intensity-based target, facilities are encouraged to increase their emissions, because in the lack of absolute targets they will be given extra units for their increased output.[99]
- Different amounts of emissions reduction target per year can be determined.[100]
- The compliance period of the facilities can vary in the club and non-club jurisdictions.[101]
- The system could foresee differential treatment to new entrants and the qualification of being a new entrant along with the advantages of this status might deviate in each system.[102]
- The allocation method can vary among the auctioning, free allocation, or a mixture of both.[103] The allocation method might have considerable effects on the legitimacy of the ETS in general.[104] The excessive number of free allocated units are considered as a subsidy and minimizing the carbon-reducing effect of the system.
- The share of auctioned certificates[105] might be different in a club and non-club members' system. Moreover, the timing, administration, and auction design could be incompatible between the club and non-club units.[106] A weak auction methodology could result in a collusive behavior among bidders and consequently, a noncompetitive carbon price in the market.
- The method of determining the free allocation and the timing of the allocation can diverge in different ETS.[107]

98. Andreas Tuerk et al., "Linking Carbon Markets: Concepts, Case Studies and Pathways," 9 Climate Policy 341 (2009) DOI: 10.3763/cpol.2009.0621 at, 348.
99. *Ibid.* at 348.
100. European Commission, *EU ETS Handbook*, available at https://ec.europa.eu/clima/sites/clima/files/docs/ets_handbook_en.pdf, at 14.
101. Andreas Tuerk et al., "Linking Carbon Markets: Concepts, Case Studies and Pathways," 9 Climate Policy 341 (2009) DOI: 10.3763/cpol.2009.0621 at 347.
102. The distortion of rules for the new entrants between the club and non-club systems will incentivize companies to start up or expand their new facilities in countries, which provides more advantages to new entrants. Such; as free allocation of units. This will affect the overall cap effectiveness. European Commission, *EU ETS Handbook*, https://ec.europa.eu/clima/sites/clima/files/docs/ets_handbook_en.pdf, at 59.
103. Andreas Tuerk et al., "Linking Carbon Markets: Concepts, Case Studies and Pathways, 9 Climate Policy 341 (2009) DOI: 10.3763/cpol.2009.0621 at 347.
104. M.J. Mace et al., "Analysis of the Legal and Organisational Issues Arising in Linking the EU Emissions Trading Scheme to Other Existing and Emerging Emissions Trading Schemes," Final Report, FIELD/IEEP/WRI, London (2008).
105. Auctioning is a transparent allocation method that enables participants to purchase emission units at the market price by bidding. European Commission, *EU ETS Handbook*, https://ec.europa.eu/clima/sites/clima/files/docs/ets_handbook_en.pdf, at 28.
106. Auctioning will ensure an open, transparent, harmonized, and the nondiscriminatory market for the purchase of the units. *Ibid* at 28.
107. The two methods of free allocation are grandfathering and benchmarking. Free allocation lowers the cost of compliance and protects the global competitiveness of the sector. Therefore, it provides an advantage to the system in which it is employed. *Ibid.* at 41.

- The allocation of units for free or via auctioning will vary according to sectors.[108]
- The revenue from the sale of the emission units could be directed to different aims and institutions.[109] As being a novel source of income for the governments, it is important to divert the income to aims supporting further enhancement in climate change mitigation.
- Different GHGs could be covered. Covered gases can include CO_2, CH_4, N_2O, HFC, PFC, and SF_6.[110] Alternatively, units might represent the energy consumed.
- Covered and exempted sectors into the ETS may be mismatched.
- The type and source of offset credits can change in the club and non-club system. This difference has the potential to affect the supply of emission units and as a result, the prices are influenced.[111]
- The emission units from the different systems might cover a non-identical time period.
- Compliance with the cap is ensured through penalty and enforcement structures. Sanctions might range among different systems from fines to criminal penalties according to the national legal systems in the lack of harmonized club ETS.[112] The stringency of the compliance ensures the efforts of the club jurisdiction is achievable.
- Monitoring, reporting, and verification method[113] will be different in each system.
- Unit providers could be state or private entities. Moreover, the registry of the units[114] and the operator of the ETS will differ and could use different approaches to monitor the system.
- The units are not linked to a multilateral treaty that forms the climate club, therefore their effectiveness can be limited. The treaty defines a common policy goal, abatement burdens, uniform standard for emission units, and all aspects of the ETS. Within the club system standardized rules, procedures, and methodology will be used to ensure transparency and equal treatment to all market participants. Most importantly burden-sharing will be equally distributed among the club members. Climate club will have strict regulations to minimize uncertainties; and to incentivize compliance, as well as mechanisms

108. *Ibid.* at 24.
109. *Ibid.* at 35.
110. Peter Sopher, "Emissions Trading around the World: Dynamics Progress in Developed and Developing Countries 2012," CCLR 306 (2012), at 312.
111. Andreas Tuerk et al., "Linking Carbon Markets: Concepts, Case Studies and Pathways," 9 *Climate Policy* 341 (2009) DOI: 10.3763/cpol.2009.0621 at 348.
112. European Commission, *EU ETS Handbook*, https://ec.europa.eu/clima/sites/clima/files/docs/ets_handbook_en.pdf, at 16.
113. A credible method is essential in forming a linked, consistent, accurate, and transparent ETS. *Ibid.* at 82.
114. The registry of the units is an electronic accounting system that provides the calculation of the units issued, then the owner of the units and their transaction; and the relevant information about them. *Ibid.* at 129.

to hold members accountable in noncompliance.[115] National systems are generally found to be time-consuming, complicated, and not adequately transparent. This created uncertainty for companies who take part in the trade of units.[116]

These different characteristics of the emission units as services lead to the result that the characteristics of the club and non-club units are unlike.

3.2.1.2 End-Use

In terms of the end-uses, club and non-club units serve the purpose of lowering the emissions with a market-based approach. However, the different structures of the unit system can affect the extent and scope of contribution to the aim. Therefore, end-uses of the club and non-club certificates are not the same, but similar.

3.2.1.3 Consumer Preferences

For the consumer preferences, unit traders perceive club and non-club units to be substitutable to each other if they serve the same aim. This is because the flexibility and the lowest price of the system weigh more than the characteristics of the unit from the perspective of the consumers. However, in the case of nonrecognition of the non-club units, the traders are left with no choice but to prefer club units for compliance purposes. Additionally, the environmental consciousness of consumers can direct them to choose units that are subjected to a higher set of criteria for carbon abatement. As a result, the consumer preferences will be directed to perceive units as unlike.

3.2.1.4 Modes of Supply

Trade in services categorizes four different modes of supply of services. Mode 1 is a service that is transboundary without traders moving. Mode 2 is also a transboundary service with the movement of consumers, whereas in mode 3 suppliers of the firm or a natural person need to move across the border to provide services.[117] The emission units are the electronic service, and their service supplier is the relevant authority or registry of the ETS in each country. The service consumers are the facilities that need units to continue their productions, therefore, the units are traded among registry and facilities. Under the climate club arrangement, the units can be traded among suppliers and consumers from different jurisdictions; however, they do not need to cross the

115. Britta Rennkamp and Andrew Marquard, "South Africa's Multiple Faces in Current Climate Clubs," 24 South African Journal of International Affairs 443 (2017) https://doi.org/10.1080/10220461.2017.1421479.
116. European Commission, *EU ETS Handbook*, https://ec.europa.eu/clima/sites/clima/files/docs/ets_handbook_en.pdf, at 43.
117. Y. Wang, "Most-Favoured-Nation Treatment under the General Agreement on Trade in Services-And Its Application in Financial Services," 30 *Journal of World Trade* 91 (1996), at 94.

border to supply service of units. For this reason, the emission units can be categorized under supply mode 1. As both the trade of club and non-club units does not require any of the parties to cross the border, the two types of units belong to mode 1 supply. Therefore, they are like in relation to modes of supply.

3.2.1.5 Competitiveness and Substitutability

Deriving from the explanations above, the characteristics of the units differ to a great extent. In terms of the end-uses, they are similar but not identical. The consumers also consider the services of the unit to be unlike. Whereas in the modes of supply, club and non-club units can be like services. Article II of GATS uses the broader term of likeness. The Appellate Body in the *Argentina—Financial Services* case recalled the likeness analysis under GATT Article I and concluded that the word "like" is applicable to the products that are in a competitive relationship; therefore, the likeness is also a determination of the competitive relationship between emission units of club members and non-members. As they are competing for the same market and substitute each other, units of club members and non-members are found to be in a competitive relationship.

On the other hand, the Appellate Body indicated in the *Argentina—Financial Services* case, paras. 6.25-6.26, that not all services or products in a competitive relationship are like, the case-by-case analysis will be determinative in the likeness. Therefore, if we take the characteristics as a more dominant criterion since the structure of the whole emissions trading system is based on the characteristics and the nature of the units, we can conclude that the club and non-club units would be unlike. The design of ETS has significantly different features, raising from the varying policy priorities, economic and legal structure, and international arrangements. Different coverage of sectors, auction formats, emission-reduction targets, and types of the target such as intensity-based or absolute emissions caps affect the effectiveness of the system and the result of the achievement of the goals. Therefore, it would not be possible to consider units from different jurisdictions as like services. As climate club will be compromised by a group of countries that have clearly defined abatement goals, they are likely to tolerate a higher price of units and design a system that has more strict rules. This creates a huge discrepancy between the units from non-club jurisdictions, as their sensitivity is more directed to economic concerns and competitiveness.

3.2.2 *Advantages Immediately and Unconditionally Provided and Treatment No Less Favorable*

If the emission units from the club and non-club countries are qualified as like services, they should have the same market access,[118] meaning an advantage. The market

118. Article XVI:1 GATS; With respect to market access through the modes of supply identified in Article I, each Member shall accord services and service suppliers of any other Member treatment no less favorable than that provided for under the terms, limitations, and conditions

access advantage should be provided as soon as possible and should not require the non-club member to pay compensation or any other extra charge.[119] However, within the climate club, the unitholder from non-club jurisdictions will not be able to trade the units they have with the club members' jurisdiction as their units are excluded from the club schema. The climate club members will be the exclusive supplier of the emission units within their territories, this contradicts with the Article XVI:2(a) GATS.

III EMISSION UNITS TRADING AND REGIONAL TRADE AGREEMENTS

Section III analyzes emission units trading and regional trade agreements.

1 Introduction

This section aims to analyze the relevance of specific WTO rules on the environment, with specific attention to climate change. In analyzing the issue, this section intends to incorporate RTAs as a useful tool to regulate emissions trading. The latter will be analyzed in line with WTO law. Factors such as border tax adjustments (BTA) will be considered as essential elements for emissions trading.

2 Emissions Trading under International Law: An Overview

In the wake of frustrations on climate change and the slow pace of the UNFCCC and other agreements on climate change, several bilateral and plurilateral technological initiatives have been launched to deal with climate change. However, to date, there has been no agreement on the eminent policy instrument that is regulated by international economic law, the WTO, investments agreements or free trade agreements.

After adopting the Paris Agreement on Climate Change, emissions trading systems (ETSs) became increasingly embraced as a policy to reduce emissions cost-effectively through national and regional agreements.[120] Emissions' trading among

agreed and specified in its Schedule. GATS: General Agreement on Trade in Services, April 15, 1994, Marrakesh Agreement Establishing the World Trade Organization, Annex 1B, 1869 U.N.T.S. 183, 33 I.L.M. 1167 (1994).

119. Y. Wang, "Most-Favored-Nation Treatment under the General Agreement on Trade in Services-And Its Application in Financial Services," 30 *Journal of World Trade* 91 (1996), at 96.

120. There are two types of emissions trading. The first type is regulated markets, which trade carbon credits issued under projects under the Kyoto Protocol. These credits can be used for compliance purposes. In some cases, subject to rules about fungibility, credits from one scheme can be used for compliance in another. The other type of emissions trading is voluntary markets, which trade credits that are generated by activities and projects that are not approved or regulated under legal mechanisms. Carbon credits in this market are called voluntary emission reductions or verified emission reductions (VERs). Usually, VERs cannot be submitted to comply with the surrender obligation of a state's or region's emissions trading scheme.

parties was firmly established by Article 17 of the Kyoto Protocol[121] as a policy tool to reduce emissions of GHGs cost-effectively. The Protocol introduced international emissions trading as a way of countries to meet GHG reduction commitments. Based on the Kyoto Protocol, there is a limited number of carbon units that can be traded under the Protocol that includes assigned amounts of units (AAUs),[122] removal units (RMUs),[123] certified emission reductions (CER)[124] and emission-reduction units (ERUs).[125] After the Kyoto Protocol ratification, the EU set up a climate change policy, a complex set of measures tackling all the six GHGs listed under the UNFCCC.

While most emissions trading systems are national and regional in character, the system has expanded globally. As to date, there are 21 emissions trading schemes in place, nine in actively development and 15 under consideration.[126] The EU ETS system remains the world's most extensive system to combat carbon emissions. It operates in all of the 28 EU countries pre-Brexit[127] plus Liechtenstein and Norway, limiting emissions from more than 11,000 heavy users of energy including power stations and industrial plants and airlines operating between ETS member countries.[128] Research has also shown that the EU emissions trading system has helped to drive innovation in low-carbon technologies such as renewable power sources and energy efficiency, one of the original objectives of the system. Increased use of these technologies also helps to reduce greenhouse gas emissions.[129]

Further to that, between 2008 and 2012, the EU ETS was responsible for ensuring industrial, and power sites with the largest emissions contributed to the EU achieving its commitments[130] under the Kyoto Protocol. Between 2012 and 2020, a single centralized cap covering the whole EU was set. The cap expected emission reductions by at least 1.74% each year up to 2020, and by 2.2% a year from 2021 onwards.[131] In

121. Article 17 of the Kyoto Protocol provides that countries with commitments under the Kyoto Protocol can acquire emission units from other countries with commitments under the Protocol and use them toward meeting a part of their targets.
122. These are quantities of emissions of carbon dioxide-equivalent that countries are allowed to emit under the Kyoto Protocol.
123. These are carbon credits issued in respect of certain land use, land-use change and forestry (LULUCF) activities, such as reforestation.
124. These are emissions generated under Clean Development Mechanism (CDM) projects.
125. These are emissions generated under joint implementation (JI) projects.
126. William Acworth, Christopher Kardish, International Carbon Action Partnership (ICAP), "*Carbon Leakage and Deep Decarbonisation Report*," June 2020. Available at https://icapcarbonaction.com/en/publications.
127. As a result of the Brexit referendum, the UK formally left the EU in January 2020. A UK Emissions Trading Scheme (UK ETS) has replaced UK participation in the EU Emissions Trading System (EU ETS). Operators had still to comply with their obligations under the EU ETS relating to the 2020 scheme year, which ended on 30 April 2021.
128. EU ETS Handbook, 2015, European Union. Available at https://ec.europa.eu/clima/sites/clima/files/docs/ets_handbook_en.pdf.
129. Graham Research "*How do emissions trading systems work?*," 2018 available at https://www.lse.ac.uk/granthaminstitute/explainers/how-do-emissions-trading-systems-work/.
130. The EU ETS covers more than 40% of carbon dioxide emitted in Europe. The region's ETS covers industry sectors that consume large quantities of energy, such as power stations, refineries, iron and steel, cement and lime paper, aviation, ceramics just to name a few.
131. European Commission "*Report from the Commission to the European Parliament and the Council*" COM, 2015, 576.

December 2020, the council transmitted NDC submission on behalf of EU and member states. The submission contains an updated and enhanced target of at least 55% reduction in greenhouse gas by 2030. The NDC submission shows that the EU is strongly committed to the Paris Agreement and its objectives, as the goal of the block is to become climate neutral by 2050.[132]

On the other hand, the World Bank's "State and Trends of Carbon Pricing" reports[133] that there are more than 50 implemented or scheduled carbon-pricing initiatives worldwide, which include ETSs in Switzerland, South Korea, New Zealand and some of states in the United States and Canadian provinces and other national-level emission allowances. In November 2020, the International Carbon Action Partnership (ICAP) and Carbon Market Watch have advocated for increased linking of domestic emissions trading schemes.[134]

Following the United Kingdom's (the UK) decision to leave the EU in 2016, the country maintained the same stand on its role with the EU ETS. The UK set out the design of the new UK ETS, the first phase of which will run from 2021 to 2030,[135] and replaces the operability of the EU ETS in the UK. This could operate as either linked or stand-alone ETS, as the UK remains flexible in considering a link with the EU ETS. Other jurisdictions that have established or are developing ETSs including the United States of America (USA) under the Acidic Rain Program (ARP),[136] Tokyo Metropolitan Government Bureau of Environment: Tokyo Cap and Trade, China's emissions trading scheme,[137] New Zealand Emissions Trading Scheme, Chicago Climate Exchange, a Federal Canadian Scheme, to name a few.

132. Council of European Union *"Submission to the UNFCCC on Behalf of the European Union and its Member States on the update of the nationally determined contribution of the European Union and its Member States,"* 14222/1/20, 2020, available at https://www.consilium.europa.eu/media/47652/st14222-re01-en20.pdf.
133. The World Bank "*State and Trends of Carbon Pricing,*" Washington, May 2018, available at https://openknowledge.worldbank.org/bitstream/handle/10986/29687/9781464812927.pdf?sequence=5&isAllowed=y.
134. Webinar on Emissions trading: International Law and Dispute Resolution, hosted by the British Institute of International and Comparative law(BIICL) on November 18, 2020.
135. A Press release, Meeting climate change requirements from 1st January 2021, available at https://www.gov.uk/government/publications/meeting-climate-change-requirements-if-theres-no-brexit-deal/meeting-climate-change-requirements-if-theres-no-brexit-deal.
136. Established under title IV of 1990 Clean Air Act. The program sets a permanent cap on the total amount of sulfur dioxide (SO2) and nitrogen dioxide (NO_2) in the United States of America. *See* Klaassen, Ger, and Andries Nentjes. Emission Trading for Air Pollution Control in Practice. Report. International Institute for Applied Systems Analysis (IIASA), 1995. 11-19.
137. China launched its national ETS politically in 2017 and was set to start by 2020, but was still under development as of December 2020. China is hopinig that its ETS will become the largest carbon market in the world by volume. With its ETS, China aims to use market mechanisms to help bring emissions to peak by 2030 and get to net zero by 2060. *See* "Explainer: How will China's new national carbon emissions trading scheme (ETS) work?," *Reuters*, July 14, 2021, available at https://www.reuters.com/business/chinas-national-emissions-trading-scheme-ets-2021-07-14/.

3 The Relationship Between Emissions Trading Systems and International Emissions

3.1 Background

Given the uncertainty of the emissions trading schemes including competitiveness of prices, and other associated costs, the prospects of linking the markets have formed interests in linking their schemes with other countries to strengthen a coordinated international market for emissions. The Kyoto Protocol established mechanisms that help industrialized countries to lower the costs of achieving their commitments to limit or reduce GHGs.[138] The mechanisms include the Joint Implementation, the Clean Development Mechanism (CDM) and the International Emissions Trading (IET).[139]

As it stands, emissions trading is a market-based system to reduce emissions of climate-damaging greenhouse gases. The market is based on the principle of "Cap-and-Trade" and Baseline-and-Credit systems. According to the OECD, in a cap-and-trade system, an upper limit on emission is fixed and emission permits are either auctioned out or distributed for free according to a specific criterion. Under a baseline-and-credit system, there is no fixed limit on emissions, but emitters who reduce their emission more can earn credits that they can sell to others who need them to comply with regulations.[140] However, in most cases, most emission allowance permits have not been distributed for free; in circumstances where the emission allowances are being allocated, they often include different allowances brackets.[141]

Nevertheless, emitters with high reduction costs can buy emission allowances and postpone their action, thereby complying with the GHG policy more cheaply than they otherwise would have been able to. One way to equalize emissions costs between domestic and foreign producers competing on the market of a country putting an ETS in place is to include imports in the ETS.[142] That means that importers will be obliged to surrender emissions allowances if they want to sell products from a sector covered by the ETS. Accordingly, governments allocate or auction permits to companies. An emission permit allows flexibility, and regulated companies can trade permits with one another.

On the other hand, the international corporation on climate change has seen various efforts to facilitate emissions trading and linkage between carbon markets. The

138. Article 6 of the Kyoto Protocol.
139. Article 17 of the Kyoto Protocol.
140. OECD "*Interactions Between Emission Trading System and Other Overlapping Policy Instruments*" 2011, General Distribution Document, Environment Doctorate, available at https://www.oecd.org/env/tools-evaluation/Interactions%20between%20Emission%20Trading%20Systems%20and%20Other%20Overlapping%20Policy%20Instruments.pdf.
141. This clearly violates the requirement in Article I:1 of GATT 1994.
142. There is, of course, a difference between ETS and International Emissions Trading (IET). Under IET, Annex I, countries under the Kyoto Protocol have the opportunity to purchase emission units from another country when they risk exceeding their reduction target. While the ETS is not a Kyoto flexible mechanism in itself, it is an independent EU policy to reduce GHG emissions. The main difference between the two is the nature of the participating entities. In the IET, emissions trading occurs between countries, while under the ETS, trading occurs between private companies within the EU.

Kyoto Protocol and the subsequent Marrakesh and Copenhagen Accords have all played an essential role in linking the emission markets. Articles 3, 10, and 17 of the Kyoto Protocol established a foundation for international emissions trading and transfer of tradable permits between countries. Similarly, the Marrakesh Accords have played a vital role in facilitating IET by providing a set of rules and procedures aiming to improve the process of trading emission permits across different jurisdictions.[143]

The development of ETS as an instrument to control GHG emissions has proven to be quite dynamic and somehow successful. Several linking[144] proposals are currently on the table. Several jurisdictions are also considering or have already established links between their systems. Linking enlarges the permit markets by connecting isolated regional emissions trading schemes.[145] In this context, it is argued that linkage between schemes may be unilateral, where permits can only be transferred in only one direction, or bilateral where permits may flow in either direction. The primary goal of linking is to ensure full cost-efficiency, as linking schemes tend to favor the low-cost abetments opportunities. In this respect, deployment of such schemes may be limited by the presence of discretionary quantity control systems such as WTO law.

Nevertheless, under the Paris Agreement, countries could pursue different options to ensure that international linking of ETSs is appropriately reflected in formulating and accounting for NDCs. The Agreement creates a framework for mechanisms that allow nations and subnational actors to trade emissions. In October 2003, the Directive[146] establishing the emissions trading schemes was adopted. The Directive paves the way for linking the EU ETS with other countries ETSs through mutual recognition agreements in order to increase the diversity of low-cost compliance options within community scheme leading to a reduction of the overall costs of compliance within the Kyoto Protocol while improving the liquidity of the community market in the greenhouse gas emission allowances.[147] For instance, in December 2019, an agreement was established between European Union and the Swiss Confederation on the linking of their Green Houses Emissions Trading System.[148] Pursuant to Article

143. Dmitry Fedosov, "Linking Carbon Markets: Development and Implications," 10:4 *Carbon & Climate Law Review* 202-216 (2016).
144. Linking means one system's permit or other offset unit can be used, directly or indirectly, by a participant in another system for compliance. *See* G. Grull and L. Taschini, "Linking Emission Trading Schemes: A Short Note," 1:3 *Economics of Energy & Environmental Policy* 115-122 (2012).
145. George Grull and Luca Taschin "Lining Emission Trading Schemes; A Short Note," 1:3 *Economics of Energy & Environmental Policy* 115-118 (September 2012), International Association for Energy Economics.
146. Directive 2003/87/EC.
147. Proposal for a Directive of the European Parliament and of the Council, of October 2003, establishing a scheme for green house gas emission allowance trading within community and amending Council Directive 96/61/EC, OJ L 275, October 25, 2005, p. 32. Available at European Commission. https://www.europarl.europa.eu/RegData/docs_autres_institutions/commission_europeenne/com/2010/0095/COM_COM(2010)0095_EN.pdf.
148. Directive 2003/87/EC of the European Parliament and of the Council of 13 October 2003 establishing a scheme for greenhouse gas emission allowance trading within the Community and amending Council Directive 96/61/EC. Available at L 275/32 Official Journal of the European Union, 25.10.2003, https://eur-lex.europa.eu/legal-content/EN/TXT/PDF/?uri=CELEX:32003L0087.

25(1.a) of Directive 2003/87/EC, agreements should be concluded with third countries provided that the system in the other country is compatible, mandatory and has an absolute emissions cap. The rule of a thumb of this agreement is to ensure that the emission caps are met, free allocation and methodologies are applied as well as the use of international credits.

Concerning the free allocation of emission allowances, the initial stage of emissions trading is free allocation of allowances, whereby the government distributes emission allowances[149] to firms participating in an ETS. The free allocation has been used to reduce the financial burden of an ETS, mainly for domestic firms. It is also used as a way of addressing competitiveness and carbon leakage concerns under existing national cap-and-trade-systems.[150] For instance, under the EU ETS phase 3 (2013-2020), free allocation took account of product benchmarks, basing allocation on the most efficient operators' performance. This aimed to provide an incentive for efficiency gains (i.e., lower emissions per unit produced) at the operator level, but may still allow an increase in emissions in the industrial sector if production level grows.[151] The incentive negatively affected the state budget and creates the risks of windfall profits. In turn, the consistency of free allocation of allowances raised concerns with the WTO subsidy and anti-dumping rules. However, only free allowances accessible to carbon leakage are subsidies; otherwise, free allowances accessible to all facilities subject to ETS are not subsidies.[152]

In relation to WTO rules provided under the Agreement on Subsidies and Countervailing Measures (SCM Agreement, ASCM), subsidies are defined as a financial contribution by a government or public body to an individual or business. This financial contribution can be in many forms—such as grants, loans, loan guarantees or tax breaks.

In the energy context, subsidies are classified into twofolds: subsidies granted to energy producers and downstream industries,[153] and investment subsidies. Nonetheless, this section focuses only on subsidies granted to energy producers and downstream industries. In respect of the SCM Agreement,[154] there should be two conditions cumulatively applying to a situation in order for a subsidy to exist: (a) a financial contribution by government or any public body within the territory of a WTO member *or* (b) any form of income or price support in the sense of Article XVI of GATT 1994.[155]

149. Ways of distributing emission allowances include free allocation or auctioning.
150. I. Jegou and L. Rubini, "The Allocation of Emission Allowances Free of Charge: Legal and Economic Considerations," Transition to a Low Carbon Future Series' Issue Paper no. 18, 2011, ICTSD, at 3 and 6.
151. Special report, "*the EU's Emission Trading System: Free allocation of allowances need better targeting,*" 2020.
152. The USA Department of Commerce, "*Issues and Decisions Memorandum for the Final Determination in the Countervailing Duty Investigation of Fluids End Blocks from Italy and Germany,*" Federal Register, Vol. 85, No. 239, Friday, December 11, 2020, Notices.
153. This category includes direct payments that support production for instance deficiency payments and operating subsidies to producers as well as consumer subsidies); tax-related subsidies (exemptions from taxation, tax credits, etc.); policies that reduce costs of inputs (budgetary subsidies to energy inputs, price controls for inputs.
154. Article 1.1 of ASCM.
155. *See* discussion on what constitutes income price support in the sense of GATT 1994 Article XVI.

Regarding the first condition, there is a subsidy if there is financial contribution by a government or any public body within the territory of a WTO member where: (i) a government practice involves a direct transfer of funds (e.g. grants, loans, and equity infusion), potential direct transfers of funds or liabilities (e.g. loan guarantees); (ii) government revenue that is otherwise due is foregone or not collected (e.g. fiscal incentives such as tax credits); (iii) a government provides goods or services other than general infrastructure, or purchases goods; (iv) a government makes payments to a funding mechanism, or entrusts or directs a private body to carry out one or more of functions illustrated in (i) to (iii) above which would normally be vested in the government.[156]

It appears from this definition that for a financial contribution to qualify as a "subsidy" for the purpose of SCM agreement it must be provided by the government or public body.[157] Whether there has been a cost to the government is irrelevant in this context; instead, to qualify as a subsidy, a financial contribution or income or price support must confer a benefit.[158] For instance, it has is found that the ETS program that the EU characterizes as an environmental program, provides an additional free allowance to companies that are on a leakage list, automatically the free allowance blocks the listed installations from purchasing additional emission from the government, meaning that the government has already given up its entitlement to collect revenue. Thus, these additional allowances provided to companies on the carbon leakage list constitute financial contributions, as the installation companies were not subject to tax.[159]

In *Canada–Aircraft*,[160] the WTO's Appellate Body confirmed the Panel's finding that a financial contribution had to make the recipient "better off" than it would have been before. Based on that, under the free allocation, when allowances are distributed for free,[161] instead of being exchanged for money, they confer benefit to firms.[162] That alone meets the criteria of Article 1.1(a) of the SCM Agreement. Nevertheless, these

156. Agreement on Subsidies and Countervailing Measures (ASCM). Available at https://www.wto.org/english/docs_e/legal_e/24-scm.pdf.
157. The term government includes the local and regional governments while the term public body does not cover all entities owned and/or controlled by the government by only those who possess, exercise, or are vested with governmental authority. *See* Article 1.1 of GATT 1994.
158. Peter Van Deb Bossche and Denise Prevost, *Essentials of WTO Law*. Cambridge University Press, 2016.
159. International Economic Law and Policy Blog, "*Countervailing the EU's Trading Scheme*," Part 2, December 2020, Available at https://bit.ly/3h0WGBL.
160. Canada—Measures Affecting the Export of Civilian Aircraft (WT/DS70): Report of the Appellate Body. (2001). In World Trade Organization (Ed.), *Dispute Settlement Reports 1999* (World Trade Organization Dispute Settlement Reports, pp. 1377-1439).
161. Emissions allowance, which was received for free, can always be sold in the market if a firm achieves emissions reduction and has no need in the allowance to comply with its emissions quota. *See* Kateryna Holzer, "*WTO law issues of emissions trading*" Working Paper No. 2016/1| April 2016. Available at https://boris.unibe.ch/84032/1/WTO%20law%20issues%20of%20emissions%20trading.pdf.
162. L. Rubini, "Subsidies for Emissions Mitigation under WTO Law," in G. Van Calster and D. Prevost (eds.), Research Handbook on Environment, Health and the WTO, Edward Elgar, 2013, at 575-576.

practices, in most cases, have not been found inconsistent with WTO law, except for when the preferential allocation of allowances is linked to export performance.

Moreover, under the WTO law, subsidies are either prohibited or actionable.[163] Article 3 of SCM explicitly prohibits the export of subsidies meaning subsidies that are contingent upon export performance or is contingent upon the use of domestic over imported products. A prohibited subsidy does not have to be targeted to a specific industry. In relation to free allocation of emission allowances, the fundamental issue is that free allowances are usually available only to certain firms or industries,[164] namely to those under significant risk of carbon leakage; thus, the free allocation is likely to be viewed as a specific subsidy, and as such, could potentially be actionable.

For instance, it has been found that the ETS program also provides for additional emissions allowances to specific company installations as a result of those installations being placed on the carbon leakage list.[165] These additional allowances are countervailable.[166] In addition to that, it has also been discovered that some ETS regulatory systems have established distinct rules for companies on carbon linkage list; these rules are designed for the companies on linkage list to not incur full costs like other companies during the allocation of emission allowances.

Such privilege includes the ability of the carbon linkage companies to claim a higher percentage of rebates compared to other companies. It is, however, very difficult for a country to bring a claim of subsidy if a complaining country itself does not have an ETS in place, as it would be difficult to claim that the free allocation of emissions allowances causes adverse effects to its domestic industries, which bear no emissions costs at all. Only countries with ETS stand a chance of bringing complaints claiming that their domestic industries are adversely impacted by imports from countries where emission allowances are distributed for free.[167]

Free allocation of emission allowances could violate WTO rules on subsidies, but no disputes have been brought forward to the WTO dispute settlement mechanism concerning the matter. Thereby, the chances of disputes brought forward under ASCM against the free allocation of emission allowances appear to be very narrow.

3.2 Linking Emissions Trading Schemes

Under present circumstances, linking of ETSs can be affected where and when emissions are reduced. Allowing allowances from one jurisdiction to be used for compliance in another jurisdiction enables GHG abatement to take place wherever it is

163. Actionable subsidies are subsidies that impose adverse effects upon the interests of other WTO member countries. *See* Article 2 of SCM agreements.
164. Kateryna Holzer, "*WTO Law Issues of Emissions Trading,*" Working Paper No. 2016/1 April 2016.
165. International Economic Law and Policy Blog: Countervailing. Available at https://ielp.worldtradelaw.net/2020/12/countervailing-the-eus-emissions-trading-scheme-part-2-.html.
166. However, the respondent parties state that the operator only receives the order to comply with the rules and obligations set by the system and thus such allowances are not accountable.
167. Currently, there are very few countries in which domestic producers bear emissions costs.

cheapest.[168] Although allowances may be transferred in both directions, a difference in abatement opportunities and costs across jurisdictions implies that there is a net flow of allowances from the jurisdiction with lower abatement costs to the jurisdiction with higher abatement costs.[169] Such differences in abatement options will inevitably affect the price of allowances within these systems, which might negatively affect the economy or likely create a challenge for one of the systems to be linked, essentially for countries with stringent price control measures. Similarly, when there is a change in any of the jurisdictions between any linking parties, the net flow allowance may also change.

When linking ETSs internationally, allowances can flow across international borders.[170] Consequently, this can change the level of emissions in the participating countries. The question is whether and how countries should account for such links. Countries could pursue different options to ensure that international linking of ETS is appropriately reflecting and observing the Paris Agreement rules. First, they could account for the linking of ETS as provided under Article 6.2 of the Paris Agreement. Second, countries with a linking agreement or a joint ETS could communicate a single NDC or communicate two targets in their NDCs (a common ETS target and separate targets for their non-ETS sectors). Third, countries could also decide not to account for the link where the shift in emissions from linking is very small in relation to the country's total emissions.[171]

Further, if a country has decided to link its allowances across international borders, allowances could either flow through two separate ETSs,[172] or allowances could flow through joint ETS.[173] Article 6 of the Paris Agreement offers a voluntary cooperation framework for a mechanism that allows nations and subnation actors to trade emissions.

As long as countries could pursue different options to ensure international linking of ETSs, it is appropriate for those countries to observe the Paris Agreement. Article 6.2

168. L. Schneider, J. Cludius, and S. La Hoz Theuer, *Accounting for the linking of emissions trading systems under Article 6.2 of the Paris Agreement.* International Carbon Action Partnership (ICAP), 2018.
169. International Carbon Action Partnership (ICAP), "*Accounting for the Linking of Emissions Trading Systems Under Article 6.2 of Paris Agreement*," November 2018. Available at https://icapcarbonaction.com/en/?option = com_attach&task = download&id = 598.
170. *Ibid.*
171. International Carbon Action Partnership (ICAP), "*Accounting for the Linking of Emissions Trading Systems Under Article 6.2 of Paris Agreement*," November 2018. Available at https://icapcarbonaction.com/en/?option = com_attach&task = download&id = 598.
172. Two countries, or subnational jurisdictions located in different countries, could establish separate ETSs and link their systems by mutually recognizing allowances from the other jurisdiction. Allowances can flow between accounts of the participating systems and thus across international borders. *See* International Carbon Action Partnership (ICAP), "*Accounting for the Linking of Emissions Trading Sytems Under Article 6.2 of Paris Agreement*," November 2018.
173. A group of countries, or subnational jurisdictions, could participate in a joint ETS. In this case, allowances flow only between registry accounts within the joint ETS. If the ETS covers more than one country, these allowances can flow across international borders. *See ibid.*

of the Agreement[174] establishes a framework for using "internationally transferred mitigation outcomes" (ITMOs) to achieve NDCs. Linking ETSs is a crucial application of this framework. According to Article 6.2, a state could then import mitigation outcomes (MOs) generated on the territory of another state (through its domestic policies or a bilateral agreement) and use these "foreign" MOs to meet its mitigation pledge under the Paris Agreement. The rationale for allowing such international transfers is to help parties to meet their NDCs in a cost-effective way.[175] In other words, the Paris Agreement's Article forms a legal framework to allow the use of market-based climate change mitigation mechanism and offer parties the opportunity to corporate with one another when implementing their NDCs. Besides that, many emerging or planned systems (i.e. in the Republic of Korea, China, or Taiwan, to mention a few), meanwhile, are somewhat hesitant to engage into cross-jurisdictional linking yet, with some waiting for their respective systems to sufficiently mature before any linking, while considering or already have considered intra-national linkages of subnational markets.[176]

3.3 *Purpose of Linking Emissions Trading Schemes*

The methods applied for linking emissions have already demonstrated positive changes in a few regions, such as the EU.[177] The EU ETS illustrates a unilateral approach under which other carbon markets have to adapt to it primarily. The scheme has created a general model for establishing carbon markets, allowing future linking and, at the same time, offering flexibility to take into account the economic circumstances and priorities of other jurisdictions.

Most notably, the EU ETS is committed to achieving net-zero emissions at net-zero costs.[178] A cap-and-trade system, i.e., an ETS with a fixed cap, is progressively leveraged to support this type of goal by achieving an emissions reduction target by a specific time at minimum or zero cost. In December 2019, the European Commission announced the EU Green Deal:[179] a new policy intended to accelerate GHG emissions reduction across the EU. In achieving the European Green Deal, the commission

174. According to Article 6.2 of the Paris Agreement, countries can also engage in "cooperative approaches" to transfer "mitigation outcomes" and use these internally transferred mitigation outcomes toward their NDCs. Nevertheless, the agreement does not define what "cooperative approach" is.
175. Isabella Alloisio "*Can Emissions Trading Work Without Article 6 of Paris Agreement?,*" January 2020. Available at https://energypost.eu/can-emissions-trading-work-without-article-6-of-the-paris-agreement/ and, Geraud de Lassus Saint-Genies, "*Article 6.2 of the Paris,*" CIGI Papers No. 200, November 2018. Available at https://www.cigionline.org/sites/default/files/documents/Paper%20No.200.pdf.
176. Dmitry Fedosov, "Linking Carbon Markets: Development and Implications," 10:4 *Carbon & Climate Law Review 202-216* (2016).
177. *See* https://ec.europa.eu/clima/policies/ets_en.
178. Paolo D'Aprile, Hauke Engel, Godart van Gendt, Stefan Helmcke, Solveigh Hieronimus, Tomas Nauclér, Dickon Pinner, Daan Walter, and Maaike Witteveen, *Net Zero-Europe: Decarbonization Pathways and Socioeconomic Implications*. McKinsey&Company, 2020.
179. European Commission." *The European Green Deal.*" COM, 640 final. Brussels: European Commission.

endorsed the intentions of reaching net-zero emissions by 2050 in line with the objectives of Paris Agreement.

The interim target is to reduce at least 55 per cent of emissions by 2030.[180] According to the European Environment Agency (EEA), the EU is on track to meet GHGs set targets as projected. The EEA report[181] confirmed that the EU has a good tracking record of meeting its decarbonization targets. For instance, when the EU signed the Kyoto Protocol in 1997, it committed to reducing its GHG emissions by 8 per cent by 2012 during the first commitment period from 2008 to 2012.[182] In turn, it reduced emissions by 18 per cent.

Further, the preliminary national estimates indicate that the EU-28 (that is, including the UK pre-Brexit) emissions fell by 3.6 from 2018-2019. Accordingly, the latest EU statistics show that the region's emissions have fallen by 26% below 1990 levels and therefore on track to meet its 20% reductions of GHGs by 2020.[183] This highlights the results of effective climate policies implemented across the EU and the commitments set by the European Council in 2007 for the EU to become highly energy efficient and low carbon economy by 2020.[184]

Furthermore, decarbonization requires additional complementary policies beyond carbon pricing. Funding to support the deployment and development of low-carbon technologies for industries is an example of policies targeted both upstream and downstream. Upstream support focuses on research and development and other inputs to stimulate the supply of new technologies, while downstream support focuses on the diffusion of promising technologies. The market for low-carbon technologies in transport, buildings, and energy is far more advanced than in the emissions-intensive industry, owing to more concerted government policies spanning decades.[185] With all these effective plans and efforts, already the social implications of emissions trading allowances have shown remarkable positive economic results, from capital reallocation, employment to trade and productions and more positive results are projected.

Under present circumstances, linking emissions trading is the only serious candidate for introducing enforcement mechanisms to climate-change agreements, particularly decarbonization. One way of addressing this could be through climate clubs,[186] a proposal that has attracted many policymakers.[187] In achieving a sound

180. European Commission Communication (COM(2019) 640) "The European Green Deal,"2019. Available at https://icapcarbonaction.com/en/?option=com_attach&task=download&id=598.
181. EEA, Trends and Projections in Europe 2020: Tracking progress towards Europe's climate and energy targets, ISSN 1977-8449, November 2020.
182. Kyoto Protocol, MEMO 04/43Bruxelles, 2004, available at https://ec.europa.eu/commission/presscorner/detail/en/MEMO_04 43.
183. Article 6.2 of the Paris Agreement.
184. EC, Brussels European Council 8/9 March 2007—Presidency conclusions (7224/1/07 REV 1).
185. O. Sartor and C. Bataille, C. "Decarbonising Basic Materials in Europe: How Carbon Contracts-for Difference Could Help Bring Breakthrough Technologies to Market," 2019, IDDRI Study No. 06/19, IDDRI, Paris.
186. A club is a voluntary group deriving mutual benefits from sharing the costs of producing an activity that has public-good characteristics. *See* William Nordhaus, "Climate Clubs: Overcoming Free riding in International Climate Policy," 105:4 American Economic Review 1339-1370 (2015). The notion of a climate clubs refers to small groups of countries that share a strong

working system, the clubs should contain attractive incentives such as minimum carbon tariffs for members and punitive measures to countries that would not obey the rules.[188] However, this would breach WTO rules against trade discrimination, unless it can be justified under the WTO's exception rules. Alternatively, another way of forming climate clubs without breaching WTO law is for the climate clubs to be regulated by regional trade agreements, which are permissible under the WTO law.

Based on the above, most modern RTAs[189] contain provisions that echo the GATT's general exceptions on environment and conservation of exhaustible natural resources. Whether RTAs could regulate emissions trading without discrimination or failure to with other WTO rules is an open question that we intend to address.

4 A Club of Emissions Trading

4.1 The Concept

While it is easy to design potential international climate agreements, the reality is that it is challenging to construct one that is effective and stable, as it is difficult to induce countries to join an international agreement with a significant reduction in GHG emissions. The Paris Agreement on Climate Change and the Kyoto Protocol have made significant efforts to construct an international climate-change agreement that harmonizes different countries' policies, thereby limiting or reducing GHG emissions. However, a country would need to use certain conditions or criteria for linking its emissions trading. Such conditions could be established unilaterally through the inclusion of the clause in the ETS legislation specifying the condition for acknowledging other countries' emissions allowances, or bilaterally/plurilaterally through the conclusion of a mutual recognition agreement (MRA) over the ETS-related issues with other countries.[190]

commitment to climate change action and want to work together to achieve a transition to a decarbonized future. The agreement envisioned here centers on an "international target carbon price." The climate club concept was introduced by Professor William Nordhaus, who was awarded the Nobel Prize in Economics in 2018 for his pioneering work on integrating climate change into economics. He put an added spotlight on the idea that carbon clubs might be a policy path worth considering.

187. *See* J. Hovi et al., "Climate Change Mitigation: A Role for Climate Clubs?," 2 *Palgrave Commun* 16020 (2016); David G. Victor, "The Case for Climate Clubs." E15Initiative. Geneva: International Centre for Trade and Sustainable Development (ICTSD) and World Economic Forum, 2015. www.e15initiative.org/.
188. Gabriel Weil, "Incentive Compatible Climate Change Mitigation: Moving Beyond the Pledge and Review Model," 42 *William & Mary Environmental Law and Policy Review* 923 (2018).
189. RTAs, which are reciprocal preferential trade agreements between two or more partners, constitute one of the exemptions and are authorized under the WTO, subject to a set of rules. Available at, WTO | Regional Trade Agreements gateway. https://www.wto.org/english//tratop_e/region_e/region_e.htm.
190. James Chapman, "Linking a United States Greenhouse Gas Cap-and-Trade System and the European Union's Emissions Trading Scheme," 11:1 *Vermont Journal of Environmental Law* 45-106 (2009). Accessed December 17, 2020. doi:10.2307/vermjenvilaw.11.1.45. and A Tuerk et al. "Linking Emissions Trading Schemes," Climate Strategies, 2009 at 4-5.

In the context of linked schemes, a balance needs to be found between leaving each government with sovereignty over its own system while providing linking partners adequate authority to influence those changes in linked systems that would materially affect their own system. Otherwise, when allowance prices are too high or too low, price uncertainty risks undermines the cost-effectiveness. If allowances prices are too low, the economic incentives for emission reduction become less appealing and insufficient in a long run.

Putting a price on emissions gives domestic companies an incentive to work and produce in an increasingly climate-friendly manner, but liking the price control mechanism from different jurisdictions is not an easy task. If one of the ETSs leaves some degree of the discretion to the functioning of its price controlling mechanism, there is a risk of free-riding behavior.[191] Regarding this, delegating the regulation of linked ETS to a supranational level would limit free-riding behaviors among members and, at the same time, it would reduce regulatory uncertainty.[192] On the other hand, the EU Commission believes that different price control mechanisms for emission allowances prices would lead to a competitive disadvantage and "carbon leakage"[193] if the EU's trading partners are not pricing carbon in a similar way.[194]

In the meantime, the Paris Agreement and the Kyoto Protocol have established features designed to cut GHG emissions such as the international cap-and-trade system for emissions.[195] However, the Protocol is projected to accomplish little due to its designed incentive structure.[196] For instance, the Paris Agreement tries to limit free-riding by having all countries as parties, while leaving actual policy design to countries operating under a "pledge and review" arrangement. To get the desired global result, one still has to act through one's national government. So what is needed is a policy that can be adopted by individual nations without plunging them into unproductive economic pain and which can then evolve into a collective global policy

191. Weitzman, "For International Cap-and-Trade in Carbon Permits, Price Stabilization Introduces Secondary Free-Rider-Type Problems," 74:2 *Environmental and Resource Economics* 939-942 (2019).
192. Tuerk, Mehling, Flachsland, and Sterk, "Linking Carbon Markets: Concepts, Case Studies and Pathways," 9:4 Climate Policy 341-357 (2009).
193. The term "carbon leakage" describes emissions occurring when European companies transfer their production sites to countries with less stringent emission-reduction rules or when formerly domestically produced goods are replaced by more carbon-intensive imports. *See* Andrei Marcu, Christian Egenhofer and Sussan Roth, *Carbon Leakage: An Overview*, 2013, CEPS Special Report No 79; Kerstine Appun "Emission Reduction Panacea or Recipe for Trade War?," The EU's Carbon Border Tax Debate" *Clean Energy Wire*, November 2020. Available at https://www.cleanenergywire.org/factsheets/emission-reduction-panacea-or-recipe-trade-war-eus-carbon-border-tax-debate.
194. Kerstin Appum, "Emission Reduction Panacea or Recipe for Trade War? The EU's Carbon Border Tax Debate," November, 2020, available at https://www.cleanenergywire.org/factsheets/emission-reduction-panacea-or-recipe-trade-war-eus-carbon-border-tax-debate.
195. J.M.B urniaux, J. Chateau, and R. Duval, "Is There a Case for Carbon-Based Border Tax Adjustment? An Applied General Equilibrium Analysis," 45:16 *Applied Economics* 2231-2240 (2013).
196. Warwick J. McKibbin and Peter J. Wilcoxen. "The Economic and Environmental Effects of Border Tax Adjustments for Climate Policy," in Brookings Trade Forum 2008/2009, Brookings Institution Press: Washington, DC 1(1), 1-34, 2009.

that provides a consistent worldwide incentive to cut back carbon emissions.[197] All that is required is that a global price floor be agreed and enforced by some coalition or "club" of nations (i.e., a climate club).

Prominent economists have supported this concept.[198] Arguably, all countries that sign up (thereby forming a coalition or "climate club") agree on a price that is to be applied to the carbon emitted within their borders. Ideally, the club would be the entire global community, but smaller coalitions can implement the scheme, and there can even be several different coalitions, each with its own price. Second, each government within the club adopts policy measures to bring their internal carbon price up to that international price. They may employ a domestic carbon tax, or a tradable emission permits scheme with a floor price set at the agreed international price. Lastly, all countries within the club may impose a uniform tariff at their borders on imports from the rest of the world, both to incentivize others to join the club and as a means of restricting carbon leakage.

Out of the three components, the only thing that has to be negotiated and agreed upon is the carbon price. It is easier to negotiate one price than in quantities—especially when the one price can be interpreted as "fair" in terms of equality of marginal effort, rather than negotiating a set of country-by-country quantitative emission targets.[199] This can be done through the form of each potential club member's upfront price commitment, translating the agreed-upon price into domestic terms as well as imposing a meaningful cost or penalty on those who do not join the club, which provides the incentives for them to join. A key part of the club mechanism (and the major difference from all current proposals) is that nonparticipants are penalized. The penalty referred here is uniform *ad valorem* tariffs[200] on the imports of nonparticipants into the club region.

Central to the climate-club proposal is border tax adjustments. In doing so, members of the club would impose a harmonized tariff to apply on all goods imported from nonparticipating countries. Non-membership would then encounter the carbon tariff whenever trading with countries in the club. The tariff would both restrict carbon leakage and provide the incentive for new members to join up.[201]

Border tax adjustments are very important as they reduce the increase in foreign GHG emissions from unilateral emission policies, and the scheme is most likely to pass muster under current WTO law than other alternative forms of emissions trading linkage. However, it is essential to acknowledge that a border tax adjustment partially

197. Geoff Bertram, "William Nordhaus's Climate Club Proposal: Thinking Globally about Climate Change Economics," 12:2 Policy Quarterly 23 (May 2016).
198. Peter Cramton, Axel Ockenfels and Steven Stoft, "An International Carbon-Price Commitment Promotes Cooperation." 4 *The Energy Journal* (2015), available at http://www.iaee.org/en/publications/eeeparticle.aspx?id=92.
199. M. Weitzman, "Internalizing the Climate Change Externality: Can a Uniform Price Commitment Help?," 4:2 *Economics of Energy and Environmental Policy* 37-49 (2015), available at https://www.journals.uchicago.edu/doi/full/10.1086/676039.
200. *See* Howard J. Wall, "The Non-equivalence of Specific and 'Ad *Valorem*' Tariffs with Quality-Differentiated Goods," 9:1 *Journal of Economic Integration* 80-93 (1994).
201. W. Nordhaus, "Climate Clubs: Overcoming Free-Riding in International Climate Policy," 105:4 *American Economic Review* 1339-1370 (2015).

shifts emissions regulation from domestic production to domestic consumption. If all imports are charged a border fee that matches the taxes or allowance fees that would have been paid if produced domestically, then all domestic consumption is on equal footing.[202]

Moreover, a carbon-pricing club should have an inclusionary, rather than an exclusionary, aim and should be pursuing the global good, rather than just the self-interest of members. Nevertheless, it should use the same essential defensive tool of a common external tariff or other barriers against non-members to ensure there are a benefit of membership and a cost of defection from the club.[203] For instance, the Comprehensive and Progressive Agreement for Trans-Pacific Partnership (CPTPP), a group of countries forming an exclusive club with various market barriers to be overcome by non-members wishing to trade with the club. This emphasizes the role of club benefits as incentives for participation and compliance. In order for the benefits to be effective between members, they must be shareable among complying participants and excludable to nonparticipants and noncomplying participants. Climate clubs' key issue is mutual agreements between smaller groups with more interests and common objectives achieving reductions in GHGs.

4.2 *Are Emission Units Services or Goods?*

Whether emission allowances can be defined as goods or services under WTO rules, the *US—Softwood Timber Lumber* case[204] sheds light on this question. The case was brought by Canada, which complained that the USA had imposed countervailing duties regarding certain softwood lumber imports from Canada. The USA considered that Canada's granting of tenure and licensing agreements to companies that included the right to harvest publicly owned timber was to be considered a provision of goods by the government, as per regulated by Article 1(a)1 of the SMC Agreement. Canada argued that the government was not providing timber to companies, but merely a right to harvest timber.

On this dispute, the main question was whether standing timber could be classified as "goods." The panel found that the standing timber was "goods" as trees are tangible objects and are capable of being owned. The Appellate Body upheld these findings, and another question arose as to whether an allowance representing one ton of carbon dioxide can be considered as goods, using similar reasoning. On this occasion, a ton of carbon dioxide was not considered goods, but was considered products. Arguably, carbon dioxide is a by-product without market value parse; rather,

202. Paul-Erik Veel, "Carbon Tariffs and the WTO: An Evaluation of Feasible Policies," 12 *Journal of International Economic Law* 749, 751-752 (2009).
203. P. Cramton, A. Ockenfels, and S. Stoft, "An International Carbon Price Commitment Promotes Cooperation," 4:2 *Energy and Environmental Policy* 51-64 (2015).
204. *United States—Measures Affecting Imports of Softwood Lumber from Canada*, Report of the Panel adopted on October 27, 1993, SCM/162, *Basic Instruments and Selected Documents*, 40th Supp., Geneva: GATT, pp. 358-518. Available at https://www.cambridge.org/core/journals/world-trade-review/article/abs/uscanada-softwood-lumber-dispute-and-the-wto-definition-of-subsidy/4BA35822084B4D6B71CE30EA59CD23B1.

what can have the market value is the tradable rights such as allowance. On the other hand, allowances cannot be considered as "goods" as they are a by-product of the production process.

Thus, it has been argued that allowances neither fall within a "goods" category, product nor financial instruments. Therefore, they cannot be regulated by the WTO, unless the WTO provides additional agreements on trade in products. There is also strong support in the academic literature for the view that rules for the allocation and trade of emission units may also constrain the import of energy products generated by fossil-fuels combustion. Such climate measures could be confronting WTO rules that seek to guarantee nondiscrimination and market access among the WTO's member states. Emission units will be subject to WTO provisions, should they be classified as goods. However, classifying emission units is a difficult task, as none of the WTO agreements explicitly addresses their trade.

4.3 Emissions Trading Measures in Line with WTO Rules

When we consider possible mechanisms to combat carbon emissions, it is generally assumed that the most effective approach will be to impose trade sanctions to the highest polluters. However, potential obstacles with that proposal are the fact that most countries are constrained by their obligations under the GATT/WTO system such as fair competition and predictability through binding and transparency. GATT Articles also set maximum tariff schedules (GATT Article II) and establishes the principle of national treatment (GATT Article III), prohibiting discrimination between domestic and imported goods with respect to internal taxation and regulation. Any measures that violate the MFN or national treatment principles or that impose tariffs that exceed their GATT Article II schedules must be justified under an express exception. Thus, the most straightforward way to trade carbon allowances that will comply with WTO law, and avoid the violation of Articles I, II, and III of the GATT, would be to design rules or to trade under a carbon or energy tax.

Depending on the issue, there are exceptions on WTO law provided under Article XX of the GATT. For instance, a general and permanent exception to the (MFN) principle under WTO law that permits trade benefits under climate clubs might be a policy option worth exploring by WTO members. This exception could constitute an incentives-based system that serves as a first step for countries to address climate change. However, it is unlikely that the initiatives proposed will be explored or negotiated at once. If the amendment related to the MFN provisions under the GATT is accepted, it would then require a unanimous decision by all members.

Regarding climate clubs, Nordhaus and Yang raised fundamental issues with the current international climate-mitigation models and identified the formation of climate clubs on the basis of introducing carbon border tax adjustments on emissions.[205] The proposal is ambitious and backed by detailed modelling. However, it is not without

205. William D. Nordhaus and Zili Yang "A Regional Dynamic General Equilibrium Model of Alternative Climate Change Strategies," 86:4 *The American Economic Review*741-765 (1996).

questions. The main question is whether a border tax adjustment addressing carbon emissions could satisfy requirement provided under Articles I, II, and III of the GATT 1994, and whether it would be deemed a border duty on imports or an internal tax.[206]

Similarly, several WTO members have questioned the legality of carbon border adjustment tariffs/mechanism (CBAM). The members have argued that the CBAM should be designed and implemented in a manner that will recognize carbon-pricing systems that are in place in other countries (including at the subnational level) while aligning with international obligations and standards.[207]

Under Article II of the GATT, each WTO Member bound itself to a certain maximum ceiling of tariffs on a product-by-product basis and in exchange for similar tariff reductions by its trading partners. According to WTO law, a carbon tax or emission allowance requirement on imports could be permitted if such border adjustment does not discriminate imports against domestic products (GATT Article III) and does not discriminate against imports from other countries (GATT Article I). This type of competitiveness provision could pass WTO scrutiny without any reference to the environmental exceptions in Article XX of the GATT 1994.

There are two trade measures that could be applicable against high-carbon imports in the combat against climate change. These include import restrictions in respect of locally emitted carbon and import restrictions regarding foreign emitted carbon.[208] Restrictions regarding locally emitted carbon usually bring imported products into the fold of domestic regulations on climate change, targeting the carbon they emit within the importing country. For as long as such restrictions do not discriminate imports against domestic products, nor between imports of different origins, these kinds of restrictions are accepted under WTO rules.[209]

On the other hand, import restrictions regarding foreign emitted carbon such as import bans, punitive tariffs, and anti-dumping duties are countervailing. These restrictions mostly violate WTO law: the prohibition in Article XI of the GATT, which imposes the general elimination of all quantitative restrictions.[210] Once border adjustment of domestic climate legislation is permitted, and it is applied on a nondiscriminatory basis, there would not even be a need to go to the exceptions of GATT Article XX. As clarified by the WTO Appellate Body in *China—Auto Parts*,[211] border duties apply by virtue of the event of importation, whereas internal taxes are triggered by an internal factor, taking place within the customs territory. To be considered an internal

206. Joost Pauwelyn, "Carbon Leakage Measures and Border Tax Adjustments under WTO," in G. Van Calster and D. Prévost (eds), *Res. Handbook on Env't, Health & the WTO* (2013), 16.
207. WTO, "EU Carbon Border Adjustment Mechanism take centre stage at Market Access Committee," November 2020, available at https://www.wto.org/english/news_e/news20_e/mark_16nov20_e.htm.
208. Trade restrictions such as tariffs, taxes or emission regulations in respect of carbon emitted by imported products in their country (or countries) of production and/or during international transportation outside the importing country.
209. Geert Van Calster and Danis Prevost, *Research Handbook in Environment, Health and the WTO*. Edward Elger Publishing Inc, 2013.
210. For more analysis on border tax adjustment on carbon markets in line with the WTO laws, *see* section 4.4.
211. Appellate Body Report, China—Measures Affecting Imports of Automobile Parts, 158, WTO Doc. WT/DS339,340,342/AB/R (adopted August 31, 2009).

tax, a carbon tax would need to be structured as an act of sale distribution, or use of the imported product after clearing customs.[212] GATT Article III does not place any quantitative limits on "internal charges," but contains the basic obligation that countries cannot treat imports less favorably than they treat their own domestic products.

Depending on whether a country were to impose a tariff on imports or rather frames the adjustment in the form of a tax, anti-dumping duty, technical regulation or carbon label, the WTO consistency of competitiveness provisions can vary substantially.[213] In *United States—Taxes on Petroleum and Certain Imported Substances,*[214] Canada and the European Economic Community (EEC) challenged a U.S. tax on certain imported substances that were manufactured using feedstock chemicals that would have been taxed if sold in the United States. The Panel asserted that a border tax adjustment on certain feedstock chemicals was inconsistent with the GATT because the pollution created in the production of the imported substances did not occur in the United States. The Panel concluded that the tax adjustment rules of the GATT distinguish between taxes on products and taxes not directly levied on products; they do not distinguish between taxes with different policy purposes.[215] Whether a sales tax is levied on a product for general revenue purposes or to encourage the rational use of environmental resources is, therefore, not relevant for determining the eligibility of a tax for a border tax adjustment.

In a similar vein, alternatives, such as environmental dumping, have been suggested to frame additional custom duties on imports from countries that do not observe carbon restrictions. Arguably, an importing country that adopts binding carbon cuts should then have the right to impose anti-dumping duties, i.e., extra tariffs to offset the dumping up to the margin of dumping that would include the amount of the social cost of the carbon.[216] Doing so would correct the failure of governments to internalize the full cost of carbon-intensive products.[217] Provided the fact that dumping duties take the form of tariffs, they are, however, explicitly permitted under WTO law, regardless of whether the resulting tariff might exceed a country's maximum ceiling.[218] Usually, an import is considered to be dumped in a situation where the sale of product's price is cheaper in the foreign market as opposed to the price in the exported country/market. Similarly, in carbon markets, for dumping purposes, export prices are not compared to carbon-restricted domestic prices or to an ideal market price that

212. Gabriel Weil, "Incentive Compatible Climate Change Mitigation: Moving Beyond the Pledge and Review Model." 42 *William & Mary Environmental Law and Policy Review* 923 (2018).
213. P. Cramton, A. Ockenfels, and S. Stoft "An International Carbon Price Commitment Promotes Cooperation," 4:2 *Energy and Environmental Policy* (2015).
214. Report of the Panel, United States—Taxes on Petroleum and Certain Imported Substances, WTO Doc. L/6175—34S/136 (June 17, 1987).
215. *United States—Taxes on Petroleum and Certain Imported Substances,* para. 5.2.4, available at https://www.wto.org/english/tratop_e/dispu_e/gatt_e/87superf.pdf.
216. C. Satapathy. "WTO Agreement on Anti-Dumping: Misuse and Case for Review," 32 *Economic and Political Weekly* 2210-2212 (1999).
217. Joost Pauwelyn, "Carbon Leakage Measures and Border Tax Adjustments under WTO Law," in D. Prevost and G. Van Calster (eds.), *Research Handbook on Environment, Health and the WTO,* Edward Elgar, 2012.
218. GATT Article II:2(b).

internalizes the social cost of carbon but they are compared to normal prices in the country of export itself.

Further, countervailing duties have been assessed as another alternative form of additional tariffs on imports (border tax adjustments). With anti-dumping, the WTO explicitly permits the imposition of extra tariffs to offset a foreign subsidy,[219] even if the resulting tariff exceeds a country's maximum ceiling. In relation to carbon markets, this happens when a government fails to impose and collect a carbon tax or when a government forces local producers to internalize the full cost of carbon emitted. Under WTO rules, countervailing duties to offset subsidies by foreign governments can only be levied in case the subsidy is specific to an enterprise or industry or group of enterprises or industries. Not imposing a carbon tax is a country-wide policy and not likely to meet the specificity requirement. A country only has to convince the other WTO members that a competitiveness provision is only the extension of domestic climate policy applied on an equal footing to imports.[220]

4.4 *Border Carbon Adjustments in Line with WTO Law*

The relevant general rules of BCAs can be found in Articles I,[221] II,[222] III,[223] XI,[224] and XVI[225] of GATT. However, Article II of GATT regulates tariffs applied on importations and is one of the few pathways that would allow introducing a specific WTO-compatible BCA design. The key WTO provision on border tax adjustments (BTAs) is Article II:2(a) of the GATT, which provides that "nothing in this Article shall prevent any contracting party from imposing at any time on the importation of any product: (a) a charge equivalent to an internal tax imposed consistently with the provisions of paragraph 2 of Article III in respect of the like domestic products or in respect of an article from which the imported product has been manufactured or produced in whole or in part." Under this rule, members are permitted to impose an additional charge on imported products on entry into their customs territory as long as it corresponds to an internal tax or charge imposed on like domestic products.[226] The same applies to border carbon adjustments (BCAs).

Similarly, more provisions on BTAs are covered under Article III:2 GATT; the Article states that "imported products shall not be subject to any internal taxes or other internal charges in excess of those applied directly or indirectly, to like domestic products." On reading Article III:2, it applies only on taxes on products. In a

219. A subsidy under WTO rules is defined as a financial contribution by the government or public body that confers a benefit. Article 1.1 of Agreement on Subsidies and Countervailing Measures.
220. J. Frankel and J. Aldy, "Addressing the Leakage/Competitiveness Issue in Climate Change Policy Proposals." Brookings Trade Forum, 69-91 (2008).
221. Most-Favoured-Nation Principle.
222. Schedules of Concession.
223. National Treatment Principle.
224. Quantitative Restriction Principle.
225. GATT rules on Subsidies.
226. Article II Schedules of Concessions. Available at https://www.wto.org/english/res_e/publications_e/ai17_e/gatt1994_art2_gatt47.pdf.

combination, the GATT border tax rules creates a scheme in which the WTO members are permitted to balance their internal tax burdens on imports to ensure that domestic goods remain price-competitive in international markets. Put it differently, under the WTO's border tax adjustment rules, a carbon or energy tax could be rebated on exports and imposed in an equivalent amount on like imports, ensuring a competitive advantage.[227]

Further, if a carbon tax applied to imports were judged to be triggered by virtue of the event of importation, GATT Article II:2(a) allows a WTO member to impose a charge equivalent to an internal tax in respect of the like domestic product or in respect of an article from which the imported product has been manufactured or produced. To be border-adjustable, a tax must be applied to products, rather than producers as provided under Article XI of GATT.[228]

By definition, BCAs are trade-related policy instruments used to offset differences in the stringency of climate policies between trade partners.[229] They do so by imposing a tax or other regulatory measures on imports based on their carbon content and/or by exempting exports from domestic carbon constraints. However, only indirect taxes can be adjusted at the border.[230] Because only a few GATT and WTO disputes have dealt with BTAs, the main guidance of BTAs operations is the Report of the GATT Working Party on Border Tax Adjustments of 1970[231] working party. The working party offers significant guidance usage of GATT rules on BTA if it does not meet consensus on everything.

A starting point in the analysis of the legality of BCAs in the context of WTO law is the determination of whether the requirement to surrender emissions allowances is a tax at all and, if it is a tax, whether it is an indirect tax, and whether it is compatible with WTO rules. These questions are essential since rules on border tax adjustments vary with the type of measure. When the ETS requirement is a tax and adjusted on the importation, it falls under GATT Article III:2. Accordingly, the tax burden for imports must be the same as for the like domestic products.[232] On the other hand, when it is a domestic regulation when applied to imports, it falls under the provisions of GATT

227. Warren H. Maruyama, "Climate Change and the WTO: Cap and Trade Versus Carbon Tax." 45:4 *Journal of World Trade* 679-726 (2011).
228. No prohibitions or restrictions other than duties, taxes or other charges, whether made effective through quotas, import or export licences or other measures, shall be instituted or maintained by any contracting party on the importation of any product of the territory of any other contracting party or on the exportation or sale for export of any product destined for the territory of any other contracting part. *See Jurisprudence on the Scope and Meaning Obligation GATT*. Available at https://www.wto.org/english/tratop_e/markacc_e/03_gabrielle_marceau_and_julia_kuelzow.pdf.
229. Holzer, Kateryna, and Nashina Shariff, "The Inclusion of Border Carbon Adjustments in Preferential Trade Agreements: Policy Implications." 6:3 *Carbon & Climate Law Review* 246-260 (2012).
230. GATT Working Party on Border Tax Adjustments (L73464, December 2, 1970, BISD 18S797, para. 14).
231. GATT Article II:2(b).
232. World Trade Organization (ed.), "Japan-Taxes on Alcoholic Beverages, Complaints by the European Communities (WT/DS11): Report of the Appellate Body)." Chapter. In Dispute Settlement Reports 1996-124.World Trade Organization Dispute Settlement Reports.

Article III:4 and, hence, the treatment of like imported products may sometimes be different, but never less favorable.[233]

Furthermore, studies[234] have considered emission allowances to be a tax eligible for adjustment at the border. Arguably, emissions allowance requirements can qualify as a tax[235] even if emissions allowances are distributed for free, given that allowances always have an opportunity cost. The key feature is the existence of payment to the government. For instance, in an auctioning, the government converts trade permits to taxation and decides to take all of the revenues generated and recycle them to the economy.[236]

In relation to emissions trading, it means that importers will be obliged to surrender emissions allowances, if they want to sell products from a sector covered by the ETS. This follows the principle of import-side border tax adjustment. Instead of taxes, however, adjusted will be costs of emissions allowances. A core question is whether WTO rules permit border tax adjustment for carbon taxes. The answer to the question is yes, provided that the carbon tax does not discriminate in favor of domestic producers or to favor imports from certain countries over others.

Further, WTO rules permit countries with carbon taxes to adopt nondiscriminatory harmonizing tariffs as these protects energy-intensive trade-exposed industries by eliminating the competitive advantage enjoyed by exports from countries that do not tax carbon emissions. These tariffs also create incentives for non-carbon-taxing countries to adopt carbon taxes, since harmonizing tariffs represent revenue that the exporting country could collect by imposing its own carbon tax. With border tax adjustments, import fees are levied by carbon-taxing countries on goods manufactured in non-carbon-taxing countries.

Regarding the assessment of the above, opinions are divided on whether WTO law permits border tax adjustments for taxable inputs that are not physically incorporated into the final traded product. It is not clear whether an import tax could vary based on the amount of carbon dioxide emitted during a good's production.[237] WTO

233. World Trade Organization (ed.), "Korea—Measures Affecting Imports of Fresh, Chilled and Frozen Beef (WT/DS161, WT/DS169): Report of the Appellate Body." Chapter. In Dispute Settlement Reports 2001, 1:5-58, para. 137, World Trade Organization Dispute Settlement Reports.

234. Joost Pauwelyn, "US Federal Climate Policy and Competitiveness Concerns: The Limits and Options of International Trade Law." Nicholas Institute for Environmental Policy Solutions Working Paper 702 (2007) and J. de Cendra, "Can Emissions Trading Schemes be Coupled with Border Tax Adjustments: An Analysis Vis-à-Vis WTO Law," 15:2 *RECIEL* 136 (2006). Available at https://www.cambridge.org/core/services/aop-cambridge-core/content/view/645C28F035EA5FE80EF1210E307F35CB/S1474745611000292a.pdf/carbonrelated_border_tax_adjustment_mitigating_climate_change_or_restricting_international_trade.pdf.

235. According OECD, tax is defined as an "unrequited payment to the government" or "a compulsory contribution imposed by the government for which taxpayers receive nothing identifiable in return." *See* https://www.oecd.org/tax/tax-policy/oecd-classification-taxes-interpretative-guide.pdf.

236. A. Petsonk "The Kyoto Protocol and the WTO: Integrating Greenhouse Gas Emissions Allowance trading into Global Market Place," 10:1 *Duke Environmental Law and Policy Forum* 185, at 211 (1999).

237. Border Adjustments—Carbon Tax Center. Available at, https://www.carbontax.org/issues/border-adjustments.

rules would have to be interpreted in a way that considers products not to be "like" each other based on their carbon footprints.

Additionally, under WTO rules, rebates for exports and taxes on imports are equivalent,[238] and such rebates are not considered prohibited subsidies. Put it differently, rebating or exempting indirect taxes such as the VAT on exports is permitted under GATT Articles VI and XVI and the SCM Agreement and is, therefore, not treated as an export subsidy, as long as the rebate/exemption rate is not greater than the rate at which the tax is levied domestically.[239]

Regarding indirect taxes, not every internal tax can be adjusted and imposed on the imports. Only indirect taxes (taxes applied to products) can be adjusted at the border,[240] which is permitted by WTO rules.[241] The permissibility of border tax adjustments was first addressed by the working party, which concluded that, their convergence of views to the effect that taxes levied directly on products were eligible for tax adjustment.[242] This concept of taxes applied to products (i.e., indirect taxes) was adopted in *US—Tuna II (Mexico)*,[243] in which the Panel concluded that, under the national-treatment principle of Article III GATT 1994, contracting parties may apply border tax adjustments with regard to those taxes that are borne by products, but not for domestic taxes not directly levied on products. The principle also applies to WTO subsidy rules. GATT Article VI:4 prohibits the application of countervailing (anti-subsidy) duties for rebates or exemptions upon exportation of taxes "borne by the like product."[244]

Since only indirect taxes can be adjusted at the border, the following assesses the legality of border tax adjustment under the WTO law, and whether the ETS requirement to surrender emissions allowances can qualify as an indirect tax. According to the definition of direct versus an indirect tax in the SCM Agreement,[245] in more abstract terms, direct taxes are those imposed directly on producers, who are also responsible

238. GATT Article VI:4 and Ad Note to GATT Article XVI. *See also* Article 3.1 of SCM Agreement and, footnote 1 and paras. (g) and (h) of Annex I to the SCM Agreement. The Working Party on Border Tax Adjustments.
239. Rchard Egling "Border-Adjustable Taxes under the WTO Agreements," Available at https://www.whitecase.com/publications/alert/border-adjustable-taxes-under-wto-agreements.
240. The availability of border adjustment only for indirect taxes follows from the text of the legal provisions of GATT Article II:2(a), Article VI:4 and Ad Article XVI. It is also confirmed by the GATT Working Party on Border Tax Adjustments (L73464, December 2, 1970, BISD 18S797, para. 14).
241. The WTO Agreement on Subsidies and Countervailing Measures ("SCM Agreement") defines indirect taxes as "sales, excise, turnover, value added, franchise, stamp, transfer, inventory and equipment taxes, border taxes and all taxes other than direct taxes and import charges." Available at https://www.wto.org/english/res_e/booksp_e/discussion_papers9_e.pdf.
242. 1970 Working Paper Party on Border Adjustments, available at https://www.worldtradelaw.net/reports/gattpanels/bordertax.pdf.download.
243. *United States—Measures Concerning the Importation, Marketing and Sale of Tuna and Tuna Products*—Appellate Body Report and Panel Reports pursuant to Article 21.5 of the DSU—Action by the Dispute Settlement Body—WT/DS381/49/Rev.1.
244. *United States—Measures Concerning the Importation, Marketing and Sale of Tuna and Tuna Products*—Appellate Body Report and Panel Reports pursuant to Article 21.5 of the DSU—Action by the Dispute Settlement Body—WT/DS381/49/Rev.1.
245. The term "direct taxes" means taxes on wages, profits, interests, rents, royalties, and all other forms of income, and taxes on the ownership of real property." The term "indirect taxes"

for paying them.[246] At the same time, it is largely recognized that these types of taxes indirectly affect products. In contrast, indirect taxes are imposed directly on products, also known as "consumption taxes." Whereas producers are responsible for paying direct taxes, consumers are responsible for paying indirect taxes (with the producer typically collecting the tax from the consumer at the point of sale and passing it onto the government).[247]

The logic behind the distinction between the two categories lies in the "destination principle," which states that products should be taxed in the country of consumption.[248] Indirect taxes, by being attached to a product, can only be paid in the country of consumption. Based on this, carbon taxes can be classified as indirect taxes and thus, in principle, be adjustable.[249] In 1987, the trade body ruling in the *U.S.—Superfund* case[250] argued that taxes or charges intended to curb pollution were no different from general revenue-generating taxes.

On this view, if the ETS requirement is a tax when adjusted on the importation, it thus falls under GATT Article III:2 and, accordingly, the tax burden for imports must be the same as for the like domestic products. If it is a domestic regulation, when applied to imports, it falls under the provisions of GATT Article III:4 and, accordingly, the treatment of like imported products may sometimes be different but never less favorable. Needless to say, there is no consensus on whether the ETS requirement to surrender emissions allowances can qualify as a tax.

The key issue here is how broadly Article III: 2 of GATT can be interpreted, and more precisely the terms "internal taxes applied indirectly to products." The main concept of the carbon tax is to internalize the social cost of carbon in the ultimate price of products to give both producers and consumers an incentive to limit the use of carbon-intensive products and shift to greener energy. Based on that, a carbon tax is a tax applied at least indirectly to products. Nevertheless, as is the case of taxes and permissible border tax adjustments, not all domestic regulations can be applied to imports at the border.[251] Process measures, such as carbon regulations, that fall outside

---

means sales, excise, turnover, value added, franchise, stamp, transfer, inventory and equipment taxes, border taxes and all taxes other than direct taxes and import charges. Available at https://works.bepress.com/tristan_brown/1/download/.

246. Tristan R Brown, "Rebates, Subsidies, and Carbon Regulation: The ASCM and Climate Policy" Available at https://works.bepress.com/tristan_brown/1/download/.

247. *Ibid.*

248. *Ibid.*

249. *See* footnote 58 of the SCM Agreement. A carbon tax imposed on products is arguably a specific excise tax and thus explicitly covered as an adjustable indirect tax. A carbon tax imposed on producers does not fall under any of the types listed under direct taxes; hence, even a carbon tax on producers would seem to be an indirect tax as it is other than direct taxes. *See* generally Bacchus, James, *The Case for a WTO Climate Waiver*. Report. Centre for International Governance Innovation, 2017. 6-11, at p. 10. Available at http://www.jstor.org/stable/resrep 17314.8; and Tristian Brown "Rebates Subsidies and Carbon Regulation: The ASCM and Climate policy" Biobased Industry Centre, 2010.

250. GATT Panel Report, United States-Taxes on Petroleum and Certain Imported Substances (*US—Superfund*), para. 5.2.8.

251. The members recognize that internal taxes and other internal charges, laws, regulations, and requirements affect the internal sale, purchase, transportation, distribution or use of products, and internal quantitative regulations requiring the mixture, processing or use of products in

the scope of GATT Article III are as well presumed to be prohibited under GATT Article XI.[252] Article XI prohibits any kind of quantitative restrictions on imports. Under this Article, a member cannot have quotas or total bans on imported or exported products.

Carbon regulations could fall under the WTO agreement on Technical Barriers to Trade (TBT),[253] since regulations that address "terminology, symbols, packaging, marking or labelling requirements" are covered by the TBT Agreement. These requirements are covered as soon as they apply to a product, process or production method;[254] however, the rules are still the same: carbon regulations on imports must be nondiscriminatory and not create an unnecessary obstacle to international trade.[255] At the same time, to ensure that a border tax adjustment does not discriminate against imports, it should incorporate two elements, namely (1) the possibility of foreign producers showing the actual emission released during the production process[256] and (2) when an importing industry does not want and cannot show its production process emission, the border tax adjusted country should assume that the imported product is using the best available technology.[257]

In *EC—Asbestos*,[258] the Appellate Body considered whether France's ban on asbestos and products containing asbestos violated the TBT Agreement as a technical regulation that was more restrictive of trade than necessary. The Panel decided it was not necessary to address whether the measure constituted "arbitrary or unjustifiable discrimination." Besides, the Appellate Body chose not to rely on GATT Article XX. The Appellate Body relied, instead, on a judicious view that the substantive obligations of Article III:4 of the GATT were not violated.

Also, the Panel looks at whether the measure can be justified under the chapeau of GATT Article XX. The chapeau analysis is applicable as a last measure to ensure that

---

specified amounts or proportion, should not be applied to the imported or domestic products to afford protection to domestic production. Article III National Treatment on International Taxation. Available at https://www.wto.org/english/res_e/booksp_e/gatt_ai_e/art3_e.pdf.

252. In the *Tuna—Dolphin* dispute, the US ban on certain tuna captured in a way that risks killing dolphins was found to violate GATT Article XI and not justified under the environmental exceptions in GATT Article XX. According to the panel, regulation for taxes can only be adjusted at the border if they apply to the product as such, and not if they regulate the producer. In conclusion, the panel found that the regulation could not be adjusted at the border for imported tuna. Hence, GATT Article III did not cover the US tuna ban; instead fell under the violation of GATT Article XI. Even though the panel seem to almost decided against border adjustment for carbon regulations, nevertheless the panel reports were never adopted.
253. "Technical regulation" is defined by Annex 1.1 of the TBT Agreement as a "document which lays down product characteristics or their related processes and production methods, including the applicable administrative provisions, with which compliance is mandatory." *See also* Appellate Body Report, *European Communities—Measures Affecting Asbestos and Asbestos-Containing Products*, WT/DS135/AB/R, adopted April 5, 2001, DSR 2001:VII, 3243, para. 67).
254. Agreement on Technical Barriers to Trade, Article 2.2.
255. Charles Owen Verril, "*Maximum Carbon Intensity Limitations and the Agreement on Technical Barriers to Trade*," CCLR, 2008.
256. WTO AB April 29, 1998, United States-Standards for Reformulated and Conventional Gasoline, WT/DS2/AB/R.
257. R. Ismer and K. Neuhoff, "*Border Tax Adjustment: A Feasible Way to Address Non-Participation in Emission Trading*," Cambridge Working Papers in Economics (CWPE 0409, CMI Working paper, 36, 2004, at 11.
258. Appellate Body Report, European Communities—Measures Affecting Asbestos and Asbestos-Containing Products, WT/DS135/AB/R, adopted April 5, 2001.

the GATT exceptions are not misused.[259] The chapeau analysis is not a test that evaluates the measure itself; it examines how the measure is applied.

Moreover, the GATT makes it "illegal" for members to enact laws that limit free trade based on environmental concerns, unless a GATT exception applies. For instance, the invocation of a GATT Article XX(b) exception has never been successful because most measures fail the "necessity" test.[260] The Appellate Body has not yet found a measure to fall within the human health exception. We argue that emissions trading could be expressed as WTO-permissible, crucially if the trading schemes do not discriminate some imports against other products. This type of competitiveness provision could pass WTO scrutiny without any reference to the environmental exceptions in GATT Article XX.

Even if emissions trading may violate GATT rules, the violation may still be justified by the environmental exceptions in GATT Article XX. Such justification would then most likely center on whether, under the introductory phrase of GATT Article XX, emission credit requirement or other regulations on imports were applied on a variable scale that considers local conditions in foreign countries. The overall analysis on the carbon market is that ETSs are likely to violate several GATT and GATS obligations, but that virtually all violations can be justified on environmental grounds under general exceptions agreements.

The US— Superfund case[261] has promoted a series of still-unanswered questions that still perplex the trading system, including a lack of clarity on exactly what kinds of taxes are eligible for border adjustment under WTO law as well as under what conditions rebates are applied to environmental, energy or climate taxes.[262] Besides, there has been no definitive Appellate Body ruling adopted by the Dispute Settlement Body on the general permissibility of border tax adjustments. Not only that but also GATT Panel reports do not establish when a tax is "borne by" a product, nor do they provide a definitive list of the taxes that are expressly eligible or ineligible for border adjustment. Put it differently, the underlying philosophy behind the GATT rules governing border tax adjustments was the ensuring of trade neutrality between imported and domestically produced goods.[263] It is this neutrality, not the identity of the taxpayer (or type of tax), that WTO rules on border tax adjustments seek to protect.

259. *See* WTO Appellate Body Report on United States-Standards for Reformulated and Conventional Gasoline, WT/DS2/AB/R, at 20-21 (April 29, 1996).
260. *See* WTO Appellate Body Report on EC Measures Concerning Meat and Meat Products (Hormones), WT/DS26/AB/R, WTDS/48/AB/R (January 16, 1998).
261. United States-Taxes on Petroleum and Certain Imported Substances (*US—Superfund*), para. 3.1.8.
262. Scott Vaugan "Carbon Without Borders: Can Trade Policy Support Ambitious Climate Action?," Available at https://www.iisd.org/articles/carbon-without-borders-can-trade-policy-support-ambitious-climate-action.
263. Richard Eglin, "*Border-Adjustable Taxes under the WTO Agreements,*" 2017; *see also* GATT (1970) Report on Border Tax Adjustments, adopted on December 2 (L/3464), para. 14, available at https://www.wto.org/gatt_docs/English/SULPDF/90840088.pdf.

5 WTO Rules Applicable to Environmental Policies and Measures

5.1 *Scope and Coverage of the Rules on Environmental Issues*

The protection and preservation of the environment are fundamental goals of the WTO. They are enshrined in the Marrakesh Agreement, which established the WTO and complements the WTO's objective to reduce trade barriers and eliminate discriminatory treatment in international trade relations.[264] However, there is an absence of clear understanding on how environmental issues should be dealt under the current WTO provisions as there is a void of clear understanding between contracting members on how to manage the environmental issues and trade interface. The existing GATT and WTO rules were not drafted to address climate change problems and policies, and up to date, there is no dispute regarding environmental issues that has been raised within the WTO dispute settlement mechanism yet.

Hence, WTO rules do not restrict the set of taxes and regulations that a nation may impose domestically on products to protect the environment, but they do require nondiscrimination[265] in applying such policies to domestic and imported products. In certain circumstances, the contracting members can adopt trade-related measures to protect the environment on a condition that misused such measures are avoided with the fundamental principles of transparency, and accountability.

Another possible option for overcoming this constraint is through the GATT's Article XX(g), which permits exceptions to the MFN principle concerning natural resource conservation. Second, by adding the preferential trade agreement to existing WTO provisions in accordance with Article XXIV, and third, is by WTO members creating a new MFN exception that explicitly permits discriminatory climate club trade benefits. The latter is less likely to happen, as it will undermine the WTO rules on nondiscrimination rule. This could also set off a series of potential violations to the World Trade Organization (WTO) fundamental principles.[266]

Under GATT general regime, various articles must be taken into account to avoid WTO rules violations. For instance, GATT Articles I, II,[267] and III are particularly important with respect to imports, and GATT Article XVI, as well as the 1994 Agreement on Subsidies and Countervailing Measures (SCM), to exports.[268] Expressly,

---

264. WTO Trade and environment. Available at https://www.wto.org/english/tratop_e/envir_e/envir_e.htm.
265. Under WTO law, nondiscrimination is expressed in two principles in the GATT: first, the Most Favored Nation (MFN) principle, embodied in Article I of GATT, which requires contracting members to accord the same treatments to all contracting members; and second, Article III of GATT covers the National Treatment rule (NT) that demands WTO members to treat imported products not less favorably than like domestic products. Both articles are closely related to environmental issues, essentially to matters that concern emission-trading units.
266. Tariff bindings, national treatment, prohibition on Quantitative Restrictions-QRs, and Most Favored Nation-MFN).
267. Article II permits contracting parties to incorporate into their Schedules acts yielding rights under the General Agreement but not acts diminishing obligations under that Agreement; Panel Report, US—Sugar, para. 5.2.
268. Stephanie Monjon and Philippe Quirion, *A Border Adjustment for the EU ETS: Reconciling WTO Rules and Capacity to Tackle Carbon Leakage* (2011).

the rules require WTO members to treat foreign goods no less favorably than comparable domestic goods and are not based on quantitative restrictions.

Moreover, WTO rules, including specialized agreements such as the Agreement on Technical Barriers to Trade (TBT)—which deals with products regulations—prohibit countries from maintaining technical regulations that restrict trade if their concerns can be addressed in a less trade-restrictive manner. TBT Agreement Article 2.2 recognizes the protection of the environment as a legitimate objective. Further, Article 2.5 of the TBT permits the application of technical regulations if applied for environmental reasons per international standards.

In addition, Article 27.2 Agreement on Trade-Related Aspects of Intellectual Property Rights (TRIPS) contains an environmental exception concerning patents. Members may exclude an invention from patentability when the prevention of domestic and commercial exploitation is necessary to protect human, animal or plant life or health or to avoid serious prejudice to the environment. Furthermore, Article 6 of the Agreement on Sanitary and Phytosanitary Measures (SPS) (which concerns food safety and animal and plant health), provides scope for environmental objectives to be followed and for necessary trade-related measures to be adopted. Article XI of GATT establishes a prohibition on quantitative restrictions and seeks to prohibit such trade actions as quotas, embargoes, and licensing schemes on imported or exported products.

The GATS also contains exceptions on environmental matters in Article XIV, as does the plurilateral Agreement on Government Procurement in Article XXIII. The intension of these provisions is to allow the departure from a strict application of the trade principles such as MFN and National Treatment clauses to pursue the overriding policy goals listed in Article XX of GATT.

5.2 *General Exceptions According to GATT Article XX*

In so far as it applies, if contracting members are challenged with violating any of the above principles/agreements, they have recourse to the GATT General Exceptions covered under Article XX of GATT. The Article is a limited and conditional exception from obligations under other provisions of the general agreement. But the party invokes the exception that must demonstrate that restrictive measures are "necessary."[269] According to Article XX(b), the measures excepted to be "necessary" are those related to protecting human, animal, plant life or health.

Article XX(g) allows contracting parties to take measures only to matters relating to protecting the conservation of exhaustible natural resources, and if such measures are made effective in conjunction with domestic production or consumption restrictions. This implies that Article XX(g) covers measures that are "necessary" for conservation of exhaustible national resources but covers a wide range of measures that impact the conservation of natural resources such as greenhouse gas emissions. In

269. The Panel Report on "*Canada—Administration of the Foreign Investment Review Act*" (L/5504, adopted on February 7, 1984, 30S/140, 64, para.5.20).

United States—Restrictions on Imports of Tuna[270] the Panel considered that GATT both Article XX(b) and (g) were not limited to policies related to living things or resources located within the territory of the country introducing the policy, but was nevertheless limited by the jurisdiction in general.

In order to assess the necessity element of a specific trade measure such as protection of human, animal, plant life or health protection; it has to be considered if such objectives could be obtained through a nondiscriminatory basis between countries where the same conditions prevail, or a disguised restriction on international trade as provided under the chapeau of Article XX. The provision seeks, among other things, to ensure that environmental measures are not applied arbitrarily and not framed as disguised protectionism.[271]

However, in order to determine "necessity," various concepts are usually used to make sure the measure is not trade-restrictive than necessary, and it is proportional to the objective pursued. In other words, when deciding on Article XX of GATT, the Panel usually follows three steps. The first step considers whether the policy protects humans, animals, and plants' life and health. This is an easy test to satisfy because the panels do not examine the necessity of the measure, but protecting the life and health of humans, animals or plants. When concluded that the policy goal falls within the subsection (b) of GATT Article XX, the necessity test review will be conducted. The "necessity" test is not a test of whether the policy itself is necessary, but whether the measure is necessary to achieve a policy goal. The test examines whether an alternative measure is consistent with the GATT or, at least, less inconsistent with the GATT.

Finally, the Panel looks at whether the measure can be justified under the chapeau of Article XX. This was affirmed in the in *US—Gasoline*.[272] Here the U.S. Environmental Protection Agency (EPA) issued the Gasoline Rule to reduce vehicle emissions of toxic air pollutants and ozone-forming volatile organic compounds. The Rule mandated that all gasoline sold in the United States needed to conform to minimum "cleanliness" requirements. Most American refineries were allowed to use the individualized baseline, while foreign producers were required to meet the general statutory baseline based on the quality of U.S. gasoline. Developing countries protested this particular act as discriminatory.

The Panel ruled in favor of the developing countries, despite differences in their process and production methods. The adjudicators found that the products were "like" goods under the WTO and in breach of the nondiscrimination obligation of GATT Articles I and III. In this specific case, the U.S. tried to challenge the panel's decision

270. Panel Report on "United States—Restrictions on Imports of Tuna" (DS29/R of June 16, 1994, not yet adopted, paras. 5.15 to 5.20 and 5.31 to 5.32).

271. *See* Appellate Body Report, United States—Import Prohibition of Certain Shrimp and Shrimp Products—Recourse to Article 21.5 of the DSU by Malaysia, WT/DS58/AB/RW, adopted November 21, 2001, DSR 2001:XIII, 6481.

272. Panel Report, United States—Standards for Reformulated and Conventional Gasoline, 2.5-2.7, WT/DS2/R (January 29, 1996); Appellate Body Report, United States—Standards for Reformulated and Conventional Gasoline, § I.B, WT/DS2/AB/R (April 29, 1996) [hereinafter U.S.-Gasoline AB Report].

based on GATT Article XX(b) that provides an exception for trade restrictions necessary to protect human, animal, plant life, or health. In the light of the findings, the Panel concluded that that the U.S. did not qualify for Article XX(g) deference; instead, it was ruled that the U.S. was unnecessarily discriminatory as the baseline established by the EPA was not consistent with Article III:4 of the GATT and cannot be justified under paragraph (b), (d) and (g) of Article XX of GATT.

In addition, the Panel concluded that the chapeau of Article XX should not be so applied as to frustrate or defeat the legal obligations of the holder of the right under the substantive rules of the General Agreement. If those exceptions are not to be abused or misused, in other words, the measures falling within the particular exceptions must be applied reasonably, with due regard both to the legal duties of the party claiming the exception and the legal rights of the other parties concerned.[273]

Despite these exemption provisions on the environment and natural resources, only a few articles are significantly considered in emissions trading schemes such as National Treatment and Most-Favored-Nation rules (Articles III and I of the GATT, respectively). Similarly, the same is true for GATT Article XI on quantitative restrictions and Articles 2.2 and 2.5 of the TBT Agreement. Lastly, GATT Article XX and its chapeau are significant to emissions trading, as the provisions on general exemptions lay out several specific instances in which WTO members may be exempted from GATT rules.

6 Regional Integration and Emissions Trading: The Case of RTAs in the WTO Framework

6.1 WTO Legal Framework

The lack of a global treaty that adequately addresses climate change has spurred countries worldwide to codify climate objectives through new and existing bilateral efforts. One effort that is gaining traction is the addition of climate-related provisions in RTAs,[274] wherein most countries are using the agreements to put forward their citizens' well-being. RTAs, which comprise bilateral trade agreements, free trade agreements, economic partnerships, and other arrangements aiming at trade liberalization between countries, have grown substantially over the past decade. These agreements are negotiated, implemented, and resolved according to WTO rules.

In recent years, RTAs have been advocated as one way of securing trade liberalization between trading blocs as they are considered to be less cumbersome to negotiate than multilateral trade agreements such as the WTO. Usually, RTAs happen to be between two or more governments and are reciprocal by nature. Examples of

273. WTO, Committee on Trade and Environment, "*GATT Dispute Settlement Practice Relating to GATT Article XX, paragraphs (b), (d) and (g),*" WT/CTE/W/203, March 8, 2002. Available at https://docsonline.wto.org/dol2fe/Pages/FormerScriptedSearch/directdoc.aspx?DDFDocuments/t/wt/cte/w203.doc.

274. Sam Baker, "Leveraging Trade Agreements to Reduce Greenhouse Gas Emission in Accordance with the Paris Agreement," 2016, CMC Senior Theses. 2154.

RTAs include the United States-Mexico-Canada Agreement (UMSA), the Central American-Dominican Republic Trade Agreement (CAFTA-DR), the European Union (EU) and the Asian-Pacific Economic Corporation. In addition, there are new trade agreements that have been concluded or are being negotiated, including the Comprehensive and Progressive Agreement for a Trans-Pacific Partnership (CPTPP), the European Union–Mercosur trade agreement, the Regional Comprehensive Economic Partnership, and in Africa the Tripartite Agreement between parties to the Common Market for Eastern and Southern Africa (COMESA), the East African Community (EAC), the Southern African Development Community (SADC), and the African Continental Free Trade Agreement (AfCFTA).[275] What aids the negotiations process in RTAs is that fewer parties share common values and interests compared to the multilateral trading system. In fact, as of 2016, all WTO members had at least one RTA in force.[276]

Under WTO law, RTAs mean any reciprocal trade agreement between two or more partners, not necessarily belonging to the same region. The WTO has allowed countries to negotiate rules and commitments beyond the organization's multilateral system. Among the issues that have been allowed under RTAs include services, intellectual property, environmental standards, investment and competition. In an alignment with WTO rules, RTAs provide an additional layer of discipline in three notable ways, namely: (1) reaffirming WTO rules to make disciplines available to alternative mechanism provided by RTAs; (2) agreeing on provisions that deepen or expand multilateral commitments; and (3) agreeing on provisions that define the scope of non-actionable subsidies between parties to the agreement.

Based on the WTO principle of nondiscrimination, setting up a customs union or free trade area, in theory, would violate this principle. However, Article XXIV of the GATT, Article V of GATS and paragraph (2c) of the Enabling Clause for developing countries permit WTO members to conclude RTAs on the condition that specific criteria are met. In this regard, GATT Article XXIV is most commonly viewed as an exception to the most favored nation rule contained in GATT Article I, allowing a subset of members to liberalize trade between them without extending such liberalization to all other WTO Members. In *Turkey—Textiles* report,[277] the Appellate Body confirmed the conditional right of WTO Members to form RTAs. In particular, the Appellate Body focused on the words "shall not prevent" as proof of the fact that "Article XXIV may, under certain conditions, justify the adoption of a measure which is inconsistent with certain other GATT provisions, and may be invoked as a possible 'defense' to a finding of inconsistency."[278] This applies in particular to emissions trading units as set out in Article 17 of the Kyoto Protocol, that allows countries that

275. The World Bank, Regional Trade Agreements, 2018. Available at https://www.worldbank.org/en/topic/regional-integration/brief/regional-trade-agreements.
276. The World Trade Organization "Regional Trade Agreement" 2021. Available at https://www.wto.org/english/tratop_e/region_e/region_e.htm.
277. Appellate Body Report, Turkey—Restrictions on Imports of Textile and Clothing Products, adopted October 22, 1999, WT/DS34/ABR ("Turkey—Textiles, Appellate Body Report."
278. Appellate Body Report, *Turkey—Textiles* report para. 45.

have emission units to spare, acquire emission units from other countries with commitments under the Protocol and use them toward meeting a part of their targets.

Certainly, RTAs and the WTO have become relevant tools to govern issues related to the environment and climate change. The two share the common objective of trade liberalization.[279] After all, RTAs operate as an exception in the WTO system of rights and obligations. In relation to environmental issue, such as climate change, many RTAs contain chapters and articles that are environmentally specific.[280]

However, parties can elect to more broadly incorporate environmental objectives in their RTAs to promote an integrated approach for addressing environmental concerns in such agreements.[281] A more collaborative approach to trade exists within RTAs because of a smaller number of trading partners who can easily collaborate on adopting stringent policies and rules, including addressing grey areas in the WTO legal framework and inadequate coverage of environment agreements. While the desire to increase market access certainly plays a crucial role in the establishment of RTAs, another factor influencing the formation of RTAs is the opportunity to make progress on issues that are not sufficiently addressed by the WTO agreements, such as the environment. Nevertheless, environmental provisions in RTAs have become far-reaching.

Early RTAs were merely replicating environmental provisions in WTO agreements. Among the first agreements that were enforced to address environmental issues include the NAFTA[282] (now replaced by the US-Mexico-Canada Agreement). Its negotiation coincided with a growing controversy about the relationship between trade and environmental policies. As a result, NAFTA contains unprecedented provisions to maintain and enhance health, safety, and environmental protection in nations that are party to the pact. In response to pressure for enhanced environmental protection, the parties negotiated the North American Agreement on Environmental Cooperation (the Environmental Side Agreement), which strikes a balance between the need to respect the rights of governments to exercise discretion in enforcing their laws and to provide

279. In 2006, WTO members agreed to implement a provisional mechanism to enhance the transparency of RTAs and understand their effects on the multilateral system. Under this process, members notify the WTO about their RTAs and these are discussed by the wider WTO membership based on a factual presentation prepared by the WTO Secretariat. *See* WTO | Regional Trade Agreements gateway. Available at https://www.wto.org/english//tratop_e/region_e/region_e.htm.
280. Peter Draper, Nkululeko Khumalo, and Faith Tigere "Sustainability Provisions in Regional Trade Agreements: Can they be Multilateralised?" RTA Exchange. "2017" Geneva: International Centre for Trade and Sustainable Development (ICTSD) and the Inter-American Development Bank (IDB). Available at www.rtaexchange.org/.
281. Shunta Yamaguchi, "Greening Regional Trade Agreements: Subsidies Related to Energy and Environmental Goods," OECD Trade and Environment Working Paper, 2020/01. Available at https://www.oecd-ilibrary.org/docserver/7e1fe8ed-en.pdf?expires=1608502977&id=id&accname=guest&checksum=A74F176078EFE44F71C385BB720C51D6.
282. Richard Baron and Justine Garrett, "Trade and Environment Interactions: Governance Issues," Background paper for the 35th Round Table on Sustainable Development, OECD, June 28-29, 2017.

a mechanism for addressing a Party's persistent pattern of failure to enforce its environmental law.[283]

In the 1990s, the EU incorporated environmental section with developing countries into its RTAs.[284] Chile and Singapore in 2014 followed the same trend.[285] In 2006, the European Union adopted the Global Europe Strategy,[286] which solidifies the EU's environmental policies. The main concern for the EU to negotiate included the need to address climate change as unfair competitive behavior within environmental trade. After 2006, all EU trade agreements incorporated an environmental section to address climate change, and more specifically decarbonization. For instance, the revision of the Cotonou Agreement establishes cooperation in developing and participating in a carbon market, integrating domestic climate change and development policy, and providing financial and technical support for mitigation and adaptation technologies.[287]

The following considerations inform the rationale for including climate change provisions into RTAs:

- Economic interests for forming emissions trading allowances, reducing the risks of emission leakage, as well as maintaining trade competitiveness among trading partner.
- Becoming cost-effective in the process of reducing GHG emissions. This could be done by liberalizing trade of environmental goods and services, mutual recognition and harmonization of regulatory standards, as well as disseminating technical information.
- It helps countries reduce emissions efficiently as they work as a team. Even though RTAs have incorporated provisions that deal with the environment, alternatives to the bilateral approaches need more actions and willpower by those who need to respond thoughtfully to climate change.

Regarding the above, several analysts have agreed that RTAs can potentially contribute to climate governance due to how they operate. The scope of applicable provisions under RTAs offers an opportunity for policy experimentation to craft and test climate provisions at a limited scale with like-minded countries. In addition, RTAs are uniquely positioned to address various measures at the intersection of trade and climate change

283. John J. Kim and James P. Cargas "The Environmental Side Agreement to North American Free Trade Agreement: Background and Analysis," 23 ELR 10720 (1993). Available at https://elr.info/sites/default/files/articles/23.10720.htm.
284. Appellate Body Report, United States—Import Prohibition of Certain Shrimp and Shrimp Products—Recourse to Article 21.5 of the DSU by Malaysia, WT/DS58/AB/RW, adopted November 21, 2001, DSR 2001:XIII, 6481.
285. Ana Cristina Molina and Vira Khoroshavina, *How Regional Trade Agreements Deal with Disputes Concerning Their TBT Provisions?* WTO Staff Working Paper, No. ERSD-2018-09, 2018, World Trade Organization (WTO), Geneva.
286. Stephen Woolcock, European Union Trade Policy. In *The New Palgrave Dictionary of Economics Online*. Palgrave Macmillan, 2011, reproduced with permission of Palgrave Macmillan.
287. The Cotonou Agreement, 2000. Available at https://eeas.europa.eu/archives/delegations/namibia/documents/eu_namibia/cotonou-agreement_en.pdf.

such as emissions trading, border tax adjustments, etc. In practice, RTAs are considered as a tool that can help in setting common rules for trade-related climate measures by aligning standards and regulations.[288]

6.2 RTAs and Climate Clubs

Following the EU's pioneering initiatives in advancing the climate-change objectives of the Paris Agreement through RTAs,[289] the bloc could intensify the review process of the existing RTAs. The initiative could check the extent to which existing RTAs can support the implementation of NDCs, followed by forming a new cooperation, or renegotiating on the preexisting rules with the contracting partners, such as initiating the formation of the climate club. While a review or a formation of climate club within the existing RTAs may be a plausible proposition for actors seeking to take the lead on climate action such as the EU, climate clubs could, however, build compromises that would be impossible in multilateral or bilateral forums where many countries with diverse interests participate.[290]

In this view, we argue that the review of preexisting RTAs rules may reconsider their climate dimensions and take corrective actions accordingly.[291] Overall, this option appears to be unlikely, at least in the short term. Instead, forming a side corporation-climate club could potentially work better, as the rules will not interfere with the preexisting RTAs provisions, but becoming a part of an overall commitment to cooperate and strengthen enforcement of environmental laws in general under the RTAs. Complementary forms of international collaboration, such as climate clubs, are probably necessary to foster and mainstream the process of gradual and voluntary increase in nationally determined contributions.[292]

One way for countries to show that they take climate-change mitigation seriously is to treat it as a first-tier geopolitical issue. Thereby, countries could start by forming small clubs that would make it easier to negotiate among a small group of large

288. Morin, J.F. and Jinnah, S. "The untapped Potential of Preferential Trade Agreements for Climate Governance, Environmental Politics," 27: 541-565.
289. Kasturi Das, Harro van Asselt, Susanne Droege, and Michael Mehling, "Making the International Trade System Work for Climate Change: Assessing the Options," *Climate Strategies* (2018).
290. Beatriz Leycegui Gardoqui and Imanol Ramirez, "Identifying a WTO Exception to incorporate Climate Clubs" 2015. Available at https://ictsd.iisd.org/bridges-news/biores/news/identifying-a-wto-exception-to-incorporate-climate-clubs.
291. For instance, when Democrats regained control of Congress in 2006, they renegotiated RTAs with Peru, Colombia, South Korea, and Panama to include provisions codifying the parties' commitment to a list of multilateral environmental agreements and creating dispute settlement mechanism that could be used if one or more parties was not fulfilling its commitments. *See* Sakina Jinnah and Elisa Morgera, "Environmental Provisions in American and EU Free Trade Agreements: A preliminary Comparison and Research Agenda" 22:3 *Review of European Comparative & International Environmental Law* 324-339, Available at 324-339. http://dx.doi.org/doi:10.1111/reel.12042.
292. Leonidas Paroussos, Antoine Mandel, Kostas Fragkiadakis, Panagiotis Fragkos, Jochen Hinkel and Zoi Vrontisi, "Climate Clubs and the Macro-economic Benefits of International Corporation on Climate Policy." Available at http://www.coalitiontheory.net/content/climate-clubs-and-macro-economic-benefits-international-cooperation-climate-policy.

countries (and major corporations) than among a large group of small countries. The gains from a successful club are sufficiently large that members will pay dues and adhere to club rules to gain the benefits of membership.[293]

Nevertheless, the significant conditions for a successful club must include the following: (i) a public-good-type resource that can be shared; (ii) the cooperative arrangement, including the dues, that is beneficial for each of the members; (iii) non-members can be excluded or penalized at relatively low cost to members; and (iv) membership is so stable that no one wants to leave.[294] The same should apply to climate clubs. The economic benefits generated from environmental commitments should incentivize countries to subscribe to climate clubs. In contrast to that, there should be punitive measures, such as trade sanctions, for members who do not comply with the club's rules.

IV GATT ARTICLE XX AS A POTENTIAL REMEDY FOR THE DEVELOPMENT OF CLIMATE CLUBS

Section IV.1 will explore the scope of application of Article XX, investigating to what extent it can be applied to climate concerns,[295] while section IV.2 will be dedicated to an analysis of climate-club measures according to the WTO jurisprudence on GATT Article XX.

1 The Scope of GATT Article XX and Its Application to Climate Clubs

As a starting place, it is worth mentioning that the GATT was initially enacted in the aftermath of World War II with a view to encourage free international trade through the abolition of tariffs. Its underlying objective lied in the improvement of living conditions, the promotion of full employment, and the optimum use of natural resources. Hence, the Agreement was soon considered as the early "jus commune" of globalization. Bearing this in mind, a number of exceptions had to be included to shield against abusive trade measures.

293. William Nordhaus, "Climate Clubs to Overcome Free-Riding." 31:4 *Issues in Science and Technology* 27 (July 2015).
294. Michael Finus, Juan Carlos Altamirano-Cabrera and Ekko C. Van Ierland, "The Effect of Membership Rules and Voting Schemes on the Success of International Climate Agreement," 125:1-2 *Public Choice* 95-127 (2005).
295. For further analysis on common concerns more broadly, *see* R. Leal-Arcas, "Sustainability, Common Concern and Public Goods," 49:4 *The George Washington International Law Review* 801-877 (2017).

1.1 Material Scope: A Larger Umbrella of Exceptions

1.1.1 A Trade Measure Infringing GATT's Substantive Obligations

First and foremost, a prerequisite was expected in order to proceed with the specific study of the exceptions found in GATT Article XX. This approach entailed spotting primarily a breach to GATT's substantive obligations.[296] To carry out this assessment, the first task implies looking at the different types of trade restrictions that climate clubs involve and then the principles embedded in WTO Agreements, particularly within GATT's provisions, to prevent any such restriction from impeding international trade.

The club theory presupposes the adoption of a plurilateral agreement, the purpose of which will be to achieve harmonized GHG emissions reductions. The favored channel to undertake such reductions would seek an "international target carbon price" or a quantity reduction target. Members that comply with reciprocal environmental standards set by the climate club (frequently linked to the carbon content of traded products) would benefit from exclusive trade advantages, such as the mutual recognition of members' emissions credits, assistance in building institutional capacity, or promotion of cross-border investment in low-carbon infrastructure.[297] What is more, the club functioning relies on the implementation of a range of climate policies, from cap-and-trade to carbon border taxes.

On the other hand, non-membership will have stiff consequences, nonparticipants being penalized either in the form of carbon duties taxing imports at the border according to their carbon content, or penalty tariffs, likened to a uniform tariff percentage of a sanctioning nature for emitting GHGs in the production country.[298] From now onwards, implications on international trade are no longer being ignored. Substantive rights of other World Trade Organization (WTO) members can easily be challenged, among which is the prohibition of tariffs on process and production methods (PPMs), but also the principle of nondiscrimination, consisting of the most-favored-nation and national treatment principles. Exporting states' rights to market access must be properly weighed against the right of an importing country to pursue legitimate policy objectives,[299] such as climate-change mitigation. It is therefore essential to investigate in the following: what are the rationales for GATT's cornerstone provisions, namely Articles I (Most-Favored-Nation Treatment), III (National treatment), XI (quantitative restrictions regulations)? And how can club restrictions relate to the underlying principle of nondiscrimination?

The contentious issue here is that climate-related trade restrictions imply revisiting enduring arguments on the legitimacy of the adjustment of taxes linked to PPMs (processes and production methods) in light of the principle of nondiscrimination. It

296. WTO AB Report, EC-Seal Products, para. 5.185.
297. Nathaniel Keohane, Annie Petsonk and Alex Hanafi "Toward a Club of Carbon Markets," 144:1 *Climatic Change* (Springer 2015).
298. William Nordhaus, "Climate Clubs: Overcoming Free-riding in International Climate Policy," 105:4 *American Economic Review*.
299. Steve Charnovitz, "The WTO's Environmental Progress," 10:3 *Journal of International Economic Law* 685 (2007).

has repeatedly been pictured as a widespread violation of all three principles embedded in Articles I, III, and XI.[300]

Border taxes are no stranger to international trade law. Yet, an important nuance has always existed in the WTO border taxation scheme, namely only indirect taxes are eligible for adjustment, according to the principle of destination.[301] On that basis, carbon-related border adjustment mechanisms (BAMs) imposed in connection with PPMs (also called non-related-product PPMs) pose severe challenges. In this scenario, importing states would put a price on the carbon content of the product, therefore adjusting a direct tax on PPMs. From this perspective, tariffs on PPMs could easily be found in violation of GATT substantive obligations.

But how do PPMs operate in the context of climate clubs? Border tax adjustment would certainly continue along the same trend line in the specific climate context. Generally speaking, PPMs requirements based on the energy efficiency or GHG emissions would be found to violate GATT's provisions just as much.[302] It is a serious thorn in the side of climate clubs, since their underpinning goal is to create a trade incentive to climate-change mitigation, through targeting GHG emissions released in foreign countries. As a result, this first restriction would be a significant hurdle to targeting carbon leakage. However, climate clubs' destiny is not sealed, since members applying carbon border adjustments could seek support under the general exceptions to GATT rules.[303]

1.1.2 An Exhaustive List of Exceptions: Detailing Article XX(b) and (g)

To examine whether WTO-inconsistent measures, such as carbon border adjustment mechanisms, could be justified requires a thorough examination of GATT Article XX, the only provision so far supporting legitimate public-policy objectives, including environmental protection.[304] In the face of climate-change mitigation, Article XX offers useful hints, providing that:

> nothing in this Agreement shall be construed to prevent the adoption or enforcement by any contracting party of measures:
>
> ...
>
> (b) necessary to protect human, animal or plant life or health;
>
> ...

300. GATT Panel Report, *United-States—Restrictions in imports of Tuna*, August 16, 1991, 30 ILM 1594.
301. GATT Report by the Working Party on Border Tax Adjustments (November 20, 1970), L/3464.
302. Gary Sampson, "WTO Rules and Climate Change: The Need for Policy Coherence" in *Global Climate Governance*, in W. Chambers (ed.) *Inter-Linkages: The Kyoto Protocol and the International Trade and Investment Regimes*. United Nations University Press, 2001.
303. Kateryna Holzer, "Summary of Main Findings," in *Carbon-Related Border Adjustment and WTO Law*. Edward Elgar Publishing, 2014.
304. Kateryna Holzer, "The Possibility of Defense under General Exceptions of GATT Article XX," in *Carbon-Related Border Adjustment and WTO Law*. Edward Elgar Publishing, 2014.

(g) relating to the conservation of exhaustible natural resources if such measures are made effective in conjunction with restrictions on domestic production or consumption; ...

Clearly, environmental protection features among the legitimate purposes that should be safeguarded against international trade. However, the decisive question here is whether Article XX was initially conceived as to address climate-mitigation objectives. In fact, although climate concerns are not expressly mentioned in Article XX, they might fit into the broad and ***evolutionary*** terms[305] of Article XX. Indeed, climate change has not always been the "tragedy of the commons," and GATT negotiators had surely not anticipated the consequences of climate change back in 1947, where weather imbalances remained fairly minor. Nevertheless, being aware of the establishment of the UNFCCC in 1992, it is surprising that trade negotiators had excluded any explicit reference to climate change in the text of GATT 1994.

1.1.3 Stretching the Definition of Article XX(b) and (g) to Include Climate-Related Trade Restrictions

1.1.3.1 The Integration of Climate Concerns in the Definition of Article XX(b) and (g)

Article XX(b) strictly mentions trade measures necessary for the protection of human, animal, and plant life and health. The question of the causality between climate change and the depletion of the natural environment no longer arises, since heat waves, floods, the emergence of tropical diseases etc., have kept impacting irreversibly our global environment and human life.[306] Yet, never has the WTO Appellate Body applied Article XX(b) to translational environmental concerns,[307] such as potential club-related restrictions. At this time, breaking the deadlock of Article XX(b) would probably require proving that the measure addresses domestic climate-change concerns[308] and targets a localized protection of human, animal and plant life.

Nevertheless, Article XX(g) is a whole different kettle of fish. Indeed, from the specific wording of the Article, several questions already emerge in relation to climate-related trade measures: first, is the climate an "exhaustible natural resource"? How does the measure relate to climate protection? And are climate measures "made effective"? Although most questions would require careful examination of the WTO jurisprudence (below), climate measures seem, prima facie, to fall within the scope of Article XX(g). In fact, the Appellate Body and GATT panels have found "clean air" to

305. WTO AB Report, US—Shrimp, paras. 129-130.
306. Kateryna Holzer, "The Possibility of Defense under General Exceptions of GATT Article XX," in *Carbon-Related Border Adjustment and WTO Law*. Edward Elgar Publishing, 2014.
307. Bradly J. Condon, "Climate Change and Unresolved Issues in WTO Law," 12:4 *Journal of International Economic Law* 919 (2009).
308. *Ibid.*

count among exhaustible natural resources.[309] Consequently, climate change being caused by the GHG effect, it could easily be seen as a clean-air issue.[310] Besides, clean air being a borderless environmental issue, Article XX(g) might be relevant to transboundary relations,[311] which raises the question of the extraterritorial application of Article XX.

1.1.3.2 The Specific Treatment of Emissions Tradable Permits

Our climate-club structure will explore ways to provide an exclusive market access to members' emissions trading schemes. Under these circumstances, to determine whether carbon trading would be subjected to WTO restrictions entails considering whether emissions credits enter the definition of a "good" under international trade law. Could emission units be treated as imported products? Goods being usually considered as "tangible or movable personal property other than money; (especially) articles or items of merchandise,"[312] emissions allowances would appear quite unconventional, being regarded substantially as a right to emit a certain amount of GHG. Due to their peculiar characteristics, the status of traded emissions remains debatable[313] so that there is a rather large grey area in respect to its direct effect on trade law.

In fact, restrictions on the admittance of emissions allowances issued in other jurisdictions could constitute a direct breach of WTO nondiscrimination rules. In any event, Emissions Trading Systems (ETSs) comprise a number of unique design features that could incidentally affect the fairly broad scope of international trade and would in the end have an import-restrictive impact on "goods" and "services" per se, either in the form of WTO-illegal subsidies, inappropriate dumping, or border carbon adjustment.[314]

With regard to subsidies and anti-dumping constraints, the focus can easily be shifted by phasing out the free allocation of emission units handed over by governments. However, the issue is not clear cut for carbon border adjustment mechanisms, even when relying on the surrender of allowances instead of a proper tax levy. The heated question is whether those mechanisms can still be envisaged as a tax falling under Article III:2? And positively if they are portrayed as indirect taxes (adjustable)? Although some authors assimilated emissions allowances requirements to an indirect

309. WTO AB Report, United States—Standards for Reformulated and Conventional Gasoline (US—Gasoline), WT/DS2/AB/R, adopted May 20, 1996.
310. John M. Broder, "EPA Clears Way for Greenhouse Gas Rules," (*New York Times*, 2009) https://www.nytimes.com/2009/04/18/science/earth/18endanger.html accessed November 24, 2020.
311. Bradly J. Condon, "Climate Change and Unresolved Issues in WTO Law," 12:4 *Journal of International Economic Law* 918 (2009).
312. Dispute Settlement Reports 2002: Volume 9, pp. 3595-4077.
313. Kateryna Holzer, "WTO Law Issues of Emissions Tradingm" *World Trade Institute Working Paper* No. 2016/1 (April 2016).
314. *Ibid.*

tax eligible for border adjustment,[315] in a more probable scenario, it is assumed that club members would find a "pass-through" in the letter and spirit of Article XX.[316]

Objective criteria such as the stringency of emissions caps,[317] or resilient use of governments revenues (fund low-carbon projects)[318] would certainly legitimize trade restrictions under Article XX.

Nonetheless, a final point remains unresolved with respect to Article XX's scope of application: the question of its application to a cross-border club of carbon markets.

1.2 Geographical Scope: The Extraterritorial Application of Article XX to a Club of Carbon Markets

GATT's provisions have been formulated to block out abusive unilateral trade measures, enacted in the form of harmful trade protectionism. Unilateral initiatives have already succeeded at imposing national climate policy standards to foreign countries, having them borne the consequences of those unilateral restrictions. Frequently, unilateral trade measures have intervened in response to proven transboundary environmental harm, on the one hand (Principle 21 Stockholm), and in accordance to Article XX, on the other. The same thinking pattern can be applied to international environmental law and WTO law, the right to sovereignty assuming, in either case, a balancing role.

1.2.1 The Environmental Law Approach Backs up Climate-Clubs Initiatives

The first occurrences of extraterritoriality concerns were highlighted in the context of international environmental law, before the adoption of GATT 1994.

The *Trail Smelter* case appeared as a benchmark decision, opening up the possibility of triggering states' responsibility for transboundary environmental harm, in the present case air pollution caused by industrial fumes.[319] Later, 'Principle 21 of the 1972 Stockholm Declaration, which is a pioneer text in the preservation and enhancement of human interaction with the environment, struck a balance between states' sovereign right to exploit their own resources and their responsibility to not cause transboundary damage to the environment.[320]

The question of states' liability for transboundary environmental harm raises relevant questions in terms of noncompliance consequences. In fact, according to the

---

315. J de Cendra "Can Emissions Trading Schemes Be Coupled with Border Tax Adjustments? An Analysis Vis-à-Vis WTO Law," 15:2 *RECIEL* 136 (2006).
316. *Brazil-Retreaded Tyres*, WTO AB report, para. 227; *EC-Seal Products*, WTO AB report, paras. 5.299-5.300.
317. Werksman and Lefevere "WTO Issues Raised by the Design of an EC Emissions Trading System," (1999) 17-19.
318. Kateryna Holzer, "WTO Law Issues of Emissions Trading," *World Trade Institute Working Paper* No. 2016/1 (April 2016).
319. Reports of International Arbitral Awards: Trail smelter case (United States, Canada) vol. III, 2006, 1965.
320. Declaration of the United Nations Conference on the Human Environment, Stockholm, June 16, 1972, Principle 21.

international environmental jurisprudence, unilateral sanctions even enacted as trade-restrictive measures would conceivably find legal justification.[321]

However and surprisingly, such approach has not been replicated in international climate change governance. Despite being acknowledged as the greatest threat to humankind and the global environment, UNFCCC negotiating parties continue to exclude any reference to such "no-harm" principle,[322] leaving it out of the UNFCCC agreements.[323]

First of all, climate treaties have chosen to translate climate-change mitigation actions into strictly "quantified emission limitation and reduction commitments"[324] which must be understood as substantially differing from states' responsibility to remedy the environmental harm caused, or cease its wrongful conduct[325] under international environmental law.

Moreover, although the Paris Agreement outlines the importance of loss and damages in the promotion of global climate action,[326] it has thereafter specified that under no circumstances should this mention provide a "basis for any liability or compensation."[327]

Nevertheless, there is no doubt that the no-harm principle is relevant to climate-change discussions, since its triggering is dependent upon the fulfillment of conditions such as cross-border environmental damage, causal link with national activities, and a failure to take reasonable measures to prevent such harm to occur.[328] Yet, the path taken by climate negotiators may be understandable regarding the difficulty that proving the satisfaction of previous conditions may represent. Climate change being by definition a global phenomenon, giving effect to transboundary damage in this context would highlight the concomitant responsibility of numerous states, but also and the indirect nature of the climate impact, as well as a very remote causality between GHG emissions and environmental harms.[329]

321. Kateryna Holzer, "The Possibility of Defense under General Exceptions of GATT Article XX," in *Carbon-Related Border Adjustment and WTO Law*. Edward Elgar Publishing, 2014, 158.
322. The general trend in international climate-change governance is to give more weight to principles such as Common but Differentiated Responsibilities and Respective Capabilities (CBDR-RC), instead of relying on customary principles of environmental law. In fact, very rarely has states' liability been retained on the basis of the no-harm principle in climate litigation (due to difficulties of its application).
323. Alexander Zahar, "Mediated Versus Cumulative Environmental Damage and the International Law Association's Legal Principles on Climate Change," 4 Climate Law 217 (2014).
324. Kyoto Protocol to the United Nations Framework Convention on Climate Change (adopted December 11, 1997, entered into force February 16, 2005), 2303 UNTS 148, Article 3(1).
325. Benoit Mayer, "The Relevance of the No-Harm principle to Climate Change Law and Politics," 19 *Asia-Pacific Journal of Environmental Law* 79-104 (2016).
326. UNFCCC, Decision 1/CP.21, Adoption of the Paris Agreement (December 12, 2015), U.N. Doc. FCCC/CP/2015/L.9/Rev/1, Article 8.
327. UNFCCC, Decision 1/CP.21, Adoption of the Paris Agreement (January 29, 2016) UN Doc. FCCC/CP/2015/10/Add.1 para. 52.
328. Benoit Mayer, "State Responsibility and Climate Change Governance: A Light through the Storm," 13 Chinese JIL 539 (2014).
329. Benoit Mayer, "The Relevance of the No-Harm Principle to Climate Change Law and Politics," 19 *Asia-Pacific Journal of Environmental Law* 79-104 (2016).

The irreversible and borderless impacts of climate change necessarily imply setting a large transboundary framework for climate-change mitigation. In response, this context calls to consider concerted climate-clubs initiatives, enacting trade-restrictive measures, designed as to legitimize the protection of global environmental interests.

In addition, for matters of overall consistency between environmental, climate, and trade regimes, there is an absolute necessity to safeguard mitigation actions extraterritorially, for instance within the structure of a climate-club. However, legal foundation for compliance is lacking in international climate law, and only time will tell whether any occurrence of the no-harm principle would ever be included in climate treaties. One thing is for sure: climate change is an area of environmental law in its own right in which the traditional no-harm principle would require adaptation so as to be flexible, while allowing state parties to take preventive measures, why not in the form of trade barriers.

Either way, climate conventions are not the only texts remaining silent with regard to transboundary environmental impact. Even though pre-GATT international trade law already considered extraterritorial reach, the potential application of general exceptions of GATT Article XX to cross-border initiatives remains to be scrutinized.

1.2.2 *The Potential for a Cross-Border Approach to Climate-Change Mitigation under Article XX*

Non-product-related PPM measures were long regarded as compliant with states' rights to sovereignty, and were not therefore per se incompatible with WTO law.[330]

While it is unclear whether a climate club setting trade restrictions jointly would be protected under Article XX GATT, it might reasonably be expected that, since a unilateral "one-off approach" to trade measures would be relevant in the case of transboundary harm, the scope of Article XX could equally be extended to climate-clubs initiatives.

To support this position, public international law long regarded consensual multilateral measures exercising an extraterritorial jurisdiction, through a state-to-state ETS or carbon border adjustment mechanisms, as more desirable than mere unilateral measures, often arbitrary.[331] There is an absolute necessity for club members to lay down their common standards in a multilateral agreement, instead of relying upon unilateral measures, consisting of isolated undertakings. It must be acknowledged, however, that the broader the participation and acceptance from major emitters, the more the guarantee of legitimacy for existing trade measures.[332] Hence, climate club members have every interest in making their club attractive in terms of exclusive advantages granted to participants.

330. Christine Conrad, *Processes and Production Methods (PPMs) in WTO law: Interfacing Grade and Social Goals*. CUP, 2011, 275.
331. Kateryna Holzer "The Possibility of Defense under General Exceptions of GATT Article XX," in *Carbon-Related Border Adjustment and WTO Law*. Edward Elgar Publishing, 2014, 161.
332. *Ibid.*

Additionally, club-structures for the implementation of climate-change mitigation have at the moment no competition in terms of multilateralism. Considering the continuing lack of framework for a global carbon market (previous COPs failing to provide a comprehensive rulebook for the implementation of climate commitments, especially with regard to "Internationally Transferred Mitigation Outcomes" under Article 6(2) of the Paris Agreement) in the context of the UNFCCC, regional club initiatives are the most efficient cooperation existing on the international scene, and could become the norm.

Scrupulous attention should now be paid to the substantive content of Article XX, especially how it has applied to trade restrictions, with a view to legitimizing climate-change mitigation measures under the auspices of a climate club.

2 Climate-Related Trade Measures Subject to the "Two-Tier Test" of Article XX

The "two-tier test" contained in Article XX requires the defendant to prove first that the trade measure fits one of the subparagraphs under Article XX, and second, by reference to the introductory clauses, that it does not constitute "a means of arbitrary or unjustifiable discrimination between countries where the same conditions prevail, or a disguised restriction on international trade."[333]

2.1 Providing a Degree of Connection Between Trade Measures and Climate Policy Objectives: Specific Objectives under Paragraphs B and G of Article XX

It is no surprise that a climate-related trade measure would seek legitimacy under Article XX(b) or (g). The application of those paragraphs is not subjected to the same conditions though. Indeed, while Article XX(b) would rely on a "*necessity test*" similar to other legitimate objectives, Article XX(g) is built on a "*relating to test,*" in pursuance of the letter of the Article.

2.1.1 The Necessity Test of Article XX(b)

In order to complete successfully the assessment of "necessity," a number of distinct aspects must submit to a "weighing and balancing test,"[334] assessing in the first instance various relevant factors: the extent of the contribution of the policy to the achievement of the objective (effectiveness), and its trade-restrictiveness effect, "in the light of the importance of the interests or values at stake" (proportionality).[335] The WTO's Appellate Body has recalled that the more crucial or important the interests and

333. WTO AB Report, United States—Standards for Reformulated and Conventional Gasoline (US—Gasoline), WT/DS2/AB/R, adopted May 20, 1996, paras. 6.20 and 6.35.
334. WTO AB Report, China—Publications and Audiovisual Products.
335. WTO AB Report, *Brazil-Retreaded Tyres*, para. 156.

values at stake, the easier it is to acknowledge the contribution of the measure to achieve those ends.[336]

Thereupon, the interpretation of the Panel in *Brazil—Taxation* has been a real boon for climate clubs initiatives. Indeed, the reduction of CO2 emissions was scrutinized as an issue of high importance. The Panel has reproduced the argumentation of the WTO's Appellate Body in *Brazil—Retreaded Tyres*, to consider that "few interests are more 'vital' and 'important' than protecting human beings (...) and protecting the environment is no less important."[337] With the release of the 5th Annual Report of the IPCC and the adoption of the Paris Agreement on Climate Change, never has the climate threat been acknowledged as more critical. In all likelihood, such context would certainly pave the way for the introduction of climate-related trade restrictions that would make a contribution to climate-change mitigation and fight carbon leakage, regardless of their trade-restrictive effect.

Whether climate-change mitigation achievements are attributed to domestic ETSs could be called into questions, considering the existence of competing policies such as the development of renewable energy, energy efficiency targets, etc. As credible evidence, economic surveys using EU Sectoral Emissions Data make a very strong case that the EU ETS has had an important net impact on GHG emission cuts, suggesting that it "saved about 1.2 billion tons of CO_2 from 2008 to 2016, roughly 3.8% relative to total emissions over this period."[338] Therefore, in most cases, doubts regarding the contribution of ETSs to climate-change mitigation have been allayed.

The question is more sensitive for border carbon adjustments. Indeed, hardly any state has ever experienced those mechanisms, usually appearing in the form of distant long-term policies. In such context, the WTO's Appellate Body has been very clear and specifically considers that, on account of the fact that measures adopted for climate-change mitigation "can only be evaluated with the benefit of time," then the defendant could bring future "quantitative projections" into the debate.[339] Consequently, a club-structure resorting to unexperienced climate policies could easily provide evidence of contribution to climate-change mitigation.

Aside from the "necessary" justification assessing the effectiveness and proportionality of the measure,[340] a review of "possible alternative measures that may be reasonably available to the responding member" should be carried out.[341] In fact, even though the environmental aspiration is often considered to outweigh the potentially trade-restrictive effect of the climate measure, the applicant could still contribute elements to the discussion, seeking a "reasonable alternative" to the restriction. Hence,

336. WTO AB Report, EC-Asbestos, paras. 170-172.
337. GATT Panel Report, *Brazil—Taxation*, paras. 7.913-7.916.
338. Patrick Bayer and Michaël Aklin, "The European Union Emissions Trading System Reduced CO2 Emissions Despite Low Prices," 117:16 Proceedings of the National Academy of Sciences (2020).
339. WTO AB Report *Brazil-Retreaded Tyres*, para. 146.
340. Kateryna Holzer, "The Possibility of Defense under General Exceptions of GATT Article XX," in *Carbon-Related Border Adjustment and WTO Law*. Edward Elgar Publishing, 2014.
341. WTO AB Report, China—Publications and Audiovisual Products, paras. 239 and 242.

the assessment will be finally layered with a suitability test[342] comparing the measure with less trade-restrictive alternatives, "providing an equivalent contribution to the achievement of the objective."[343] There is no question of lowering the climate standards embedded in the measure, as the Paris Agreement requires parties to have the "highest possible ambition."[344]

Nevertheless, why not revise a trade-restrictive measure whose impact on climate-change mitigation could be achieved otherwise? For instance, a climate club establishing a complete import ban on any good produced by non-members having lower climate standards might sound quite extreme, when importers can simply put in place border taxes or request the surrender of allowances. These are corrective mechanisms that are much less invasive, while still a disincentive for producers.

The stringency of the necessity test can be both a blessing and a curse for club members. Indeed, although Article XX(b) would secure the interests of non-members, preserving them from abusive penalties, climate club members could experience a relative difficulty of proving the necessary character of their policies,[345] which is why other avenues should be contemplated as means of defense.

2.1.2 The "Relating to" Test under Article XX(g)

There is a widespread belief that Article XX(g) would be more easily triggered to justify PPM-related measures or even carbon-related border adjustment mechanisms (BAMs) its scrutiny being relatively moderate.[346]

Climate concerns being henceforth integrated in the definition of natural resources, the assessment of "relating to" would explore whether trade measures are "primarily aimed at" the conservation of clean air.[347] This wording has been clarified to require that a "close and genuine relationship of ends and means"[348] to the conservation of natural resources be established. However, the nature of this relationship has not been spelled out. Therefore, proving that climate concerns are not merely a smokescreen, the real purpose of which will be to restrict international trade, appears sufficient for Article XX(g) to kick in.

Nonetheless, analyzing the "genuine relationship" would require focusing on the design and structure of the measure, concurrently with the conditions of the relevant market, among which are the natural resource at stake, the market structure, the

342. Kateryna Holzer, "The Possibility of Defense under General Exceptions of GATT Article XX," in *Carbon-Related Border Adjustment and WTO Law*. Edward Elgar Publishing, 2014.
343. WTO AB Report, *Brazil-Retreaded Tyres.*
344. UNFCCC, Decision 1/CP.21, Adoption of the Paris Agreement (December 12, 2015), U.N. Doc. FCCC/CP/2015/L.9/Rev/1, Article 4.3.
345. Joost Pauwelyn "U.S. Federal Climate Policy and Competitiveness Concerns" (April 2007) Working Paper prepared by the Nicholas Institute for Environmental Policy Solutions, Duke University, NI WP 07-02, 34.
346. Saito, "Yardsticks for ‚Trade and Environment': Economic Analysis of the WTO Panel and the Appellate Body Reports regarding Environment-oriented Trade Measures" *Jean Monnet Working Paper* 14/01.
347. WTO AB Report, *US—Gasoline*, p. 16.
348. WTO AB Reports, *China—Rare Earths*, para. 5.94.

geographical scope of the market, and the interactions between foreign- and domestic-market players.[349] A climate club structure is based on solid trade rules (climate standards, ETS) and would usually bring together regional partners toward joint climate action. Along with the interpretation of Article XX(b), the Appellate Body equally gave weight to "predictable effects" arising out from the careful evaluation of design, structure, and market conditions surrounding the measure.[350] Based on these circumstances, climate clubs, whose ultimate goals are climate-change mitigation and the elimination of carbon leakage, would conceivably fall within the definition of Article XX(g).

What is more, another important condition can be found in the second limb of the Article, requiring measures to be "made effective in conjunction with" domestic restrictions. In this sense, trade measures applicable to foreign countries must operate jointly with the restrictions on domestic production or consumption.[351] There can be no doubt that participants to a climate club are enduring high levels of regulation, since they agreed to comply with a number of stringent climate standards. The very objective of climate clubs being to safeguard industries' competitive interests against carbon leakage, members would have necessarily imposed restrictions on their domestic production or consumption through carbon taxation schemes, or national ETSs. While the Appellate Body has not required that the burden of climate conservation be evenly distributed,[352] domestic restrictions must be "real," all the more so when domestic consumption "accounts for a major part of the exhaustible natural resources,"[353] meaning in the climate change discourse if club members feature among major GHG emitters.

Henceforward, looking at the introductory clause of Article XX would entail examining the manner in which the measure is applied, specifically whether it arbitrarily or unjustifiably discriminates WTO members, or restricts international trade.

2.2 *The "Chapeau" as Anti-abuse Check*

The chapeau recalls that trade measures shall not be "applied in a manner which would constitute a means of arbitrary or unjustifiable discrimination between countries where the same conditions prevail, or a disguised restriction on international trade [...]." Such inclusion early in Article XX reflects a restrictive approach to GATT's exceptions designed to prevent any abuse of the right to invoke exceptions and thus defeat substantive rules.[354] Therefore, not only is the manner in which the measure is

349. *Ibid.* para. 5.9.
350. *Ibid.* paras. 5.99-5.100.
351. WTO AB Reports, *China—Raw Materials*, para. 356.
352. WTO AB Reports, *China—Rare Earth*, paras. 5.133-5.134.
353. *Ibid.* para. 5.136.
354. Donald M. McRae, "GATT Article XX and the WTO Appellate Body," in Bronckers and Quick (eds), *New Directions in International Economic Law: Essays in Honour of John H. Jackson*. Kluwer Law International, 2000, 232.

applied important, but also its design, content, and nature will be relevant for the close examination under the chapeau.[355]

2.2.1 *Sanctioning an "Arbitrary or Unjustifiable Discrimination"*

The requirement for nondiscrimination is already mentioned under the GATT's legal obligations (Articles I and III). Yet the exception clauses of the GATT are based upon identifying a discriminatory intent,[356] compared to the mere discriminatory effect set under substantive rules. Determining the rationale of the discrimination necessarily entails searching for the actual cause of the discriminatory treatment and establishing whether it bears relevant connection to the Article XX-related objective.[357] The immediate response challenge was to query to what extent circumstances existing in different countries had to be taken into account.

Are exporting countries required to fully copy climate policies implemented under a climate-club structure? Are independent climate actions given any effect at the importing states' borders? To what extent shall circumstances existing in different countries be taken into account? Most of these concerns have been dispelled following the *US—Shrimp* case. While the US had initially required that exporting countries "adopt essentially the same policy," excluding the consideration of any "other specific policies (...) adopted for the protection and conservation of sea turtles,"[358] the import ban was eventually revised to allow for the adoption of a "program comparable in effectiveness."[359] This last requirement was regarded as satisfying the Appellate Body's test, leaving sufficient flexibility for exporting states in the application of the measure.[360] Applying such strategy in the context of PPMs or carbon BTAs is absolutely vital, first to avoid at all costs double taxation, but also to take heed of differentiation.

On the one hand, it would be rather inappropriate to impose a carbon BTA when producers are already paying emissions costs under an ETS, an export tax, or even an improved energy intensity target.[361] But performing a comparison between different emission reductions policies would obviously lead to substantial technical difficulties, especially when comparing price-based (carbon taxes, cap-and-trade systems) and non-price-based climate measures (renewable targets, bans, etc.).[362] Besides, drawing lessons from the Appellate Body's reasoning in *US—Shrimp*, the US exempted exporting countries from surrendering emissions allowances when they had already taken

355. WTO AB Report, US—Shrimp, para. 140.
356. Reinhard Quick and Christian Lau, "Environmentally Motivated Tax Distinctions and WTO Law. The European Commission's Green Paper on Integrated Product Policy in Light of the 'Like Product' and 'PPM-' Debates," 6:2 *Journal of International Economic Law* 439 (2003).
357. WTO AB Report *Brazil-Retreaded Tyres,* para. 225-7.
358. WTO AB Report, US—Shrimp, paras. 161 and 163.
359. *Ibid.* para. 144.
360. *Ibid.*
361. Kateryna Holzer, "The Possibility of Defense under General Exceptions of GATT Article XX", in *Carbon-Related Border Adjustment and WTO Law*. Edward Elgar Publishing 2014, 170.
362. Patrick Low, Gabrielle Marceau and Julia Reinaud, "The Interface Between the Trade and Climate Change Regimes: Scoping the Issue", 46:3 *Journal of World Trade* 485 (2012).

comparable action nationally.[363] Therefore, participants to a climate club should adopt this same comprehensive approach, not only for the sake of compliance with the chapeau of GATT Article XX, but equally for international diplomatic reasons.

However, one may wonder how the effectiveness of measures to achieve climate-change mitigation should be assessed. Should it be by reference to the percentage of GHG emissions reduction desired? Or by looking at countries' NDCs? There is a need to build coherence around the climate-change and trade regimes. In fact, in climate-change law particular attention has been attached to the guiding principle of "common but differentiated responsibilities and respective capabilities" (CBDR-RC), which takes on a new dimension in the context of unilateral trade measures.

Indeed, WTO Agreements have included differentiation concerns to some extent, calling for the preservation of the environment in a manner consistent with members' "respective needs and concerns at different levels of economic development."[364] Although the chapeau of GATT Article XX usually targets measures where the same conditions prevail, the Appellate Body has also considered that the discrimination may take place "when the application of the measure at issue does not allow for any inquiry into the appropriateness of the regulatory program."[365] Inherently, domestic climate contributions cannot be the same according to the principle of equity, minding "specific needs and special circumstances of developing countries,"[366] which is why a unilateral measure excluding any account of "national circumstances" would constitute an unjustifiable and arbitrary discrimination. For this reason, solely relying on the publicity of states' NDCs would not be sufficient to decide whether to trigger a border tax adjustment. The preferred avenue should look at reaching a negotiated agreement between club members and their potential trading partners, agreeing on a definition of comparable efforts, mindful of the level of development as well as historical and current GHG emissions.[367]

2.2.2 Restrictions on International Trade: A Critical Assessment of the Trade-Restrictive Effects of Climate Clubs

The formulation of the chapeau of GATT Article XX accurately depicts the relative confrontation between trade and climate change, as it perfectly echoes Article 3.5 of the UNFCCC. As James Bacchus states, "*the truth is the economy and environment are one and the same, and they must be treated as one and the same in all international rules for global governance. The economic future cannot be separated from the environmental*

363. Kateryna Holzer, "The Possibility of Defense under General Exceptions of GATT Article XX", in *Carbon-Related Border Adjustment and WTO Law*, Edward Elgar Publishing 2014, 170.
364. 1994 Agreement Establishing the World Trade Organization, 1867 UNTS 154; 33 ILM 1144 (1994).
365. WTO AB Report, US—Shrimp, para. 165.
366. UNFCCC, Decision 1/CP.21, Adoption of the Paris Agreement (December 12, 2015), U.N. Doc. FCCC/CP/2015/L.9/Rev/1, preamble.
367. Zhang, "Encouraging Developing Country Involvement in a Post-2012 Climate Change Regime: Carrots, Sticks, or Both?," MPRA (2009).

future, and vice versa."[368] In fact, it would be misleading to regard climate and trade merely as two clashing features, when they have such an intimate and inextricable relationship. Doubts remain though as to whether both features could ever be reconciled.

There is a genuine opportunity for trade law to contribute to climate-change mitigation and the energy transition at large. Yet, it has stood as a major obstacle to mitigation progress. The WTO jurisprudence of Article XX resting on a case-by-case basis appears today somewhat inappropriate to effectively tackle the climate crisis. Different multilaterally negotiated pathways have been suggested in order for trade rules to give a push to climate-change mitigation: from considering the free trade of green products, the application of carbon-related BAMs, and the clarification of environment-related WTO rules (Article XX), to fostering green technologies transfer and encouraging green subsidies.[369] However, neither UN climate negotiations nor WTO trade discussions have considered remedying this clashing situation, although it has long been acknowledged that trade liberalization has been considerably increasing GHG emissions, as a result of increased economic activity.[370]

Pending a long-term solution balancing trade and climate interests, the development of a climate club cooperative approach could be part of the response. Different trade policies options have nonetheless been highlighted to clear the way for prosperous club development, starting with the opportunity for an explicit reference to the climate challenge in trade rules.[371] Surely the preferred solution would require WTO members to assimilate climate measures taken pursuant to the Paris Agreement, to legitimate objectives protected under GATT Article XX, so far as consistent with Article 3.5 of the UNFCCC and the chapeau of Article XX.[372] Considering that Article 6 of the Paris Agreement urges "voluntary cooperation" in the pursuit of "higher ambition" through complementary "mitigation and adaptation actions,"[373] climate clubs would, beyond doubts, fall within the definition of a "climate measure" taken for the implementation of climate-change mitigation goals.

It is assumed, however, that the amendment of WTO obligations and provisions would be quite unfeasible under current circumstances, the decision-making process being generally practiced by consensus.[374] Another solution would hence need consideration: indeed, the content of WTO waivers being rather flexible in the alteration of trade rules, the adoption of a climate waiver would allow the realignment of trade rules

368. Bacchus, "The Content of a WTO Climate Waiver", *CIGI Papers* No. 204 (December 2018) .
369. Beatriz Leycegui and Imanol Ramirez, "Addressing Climate Change: A WTO Exception to Incorporate Climate Clubs" (E15 Expert Group on Measures to Address Climate Change and the Trade System, May 2015).
370. Report by UNEP and WTO, "Trade and Climate Change" https://www.wto.org/english/res_e/booksp_e/abstract_trade_climate_change_e.pdf.
371. James Bacchus, "Global Rules for Mutually Supportive and Reinforcing Trade and Climate Regimes" (E15 Expert Group on Measures to Address Climate Change and the Trade System, January 2016) 10.
372. *Ibid.* 16.
373. UNFCCC, Decision 1/CP.21, Adoption of the Paris Agreement (December 12, 2015), U.N. Doc. FCCC/CP/2015/L.9/Rev/1, Article 6.
374. 1994 Agreement Establishing the World Trade Organization, 1867 UNTS 154; 33 ILM 1144 (1994), Article IX 1.

with climate measures consistent with either the UNFCCC or a climate club agreement[375] without defeating those rules in essence. Next to the development of climate clubs, such WTO climate waiver should promote trade disciplines on fossil fuel subsidies, an innovative approach to green subsidies,[376] and a broadening of GATT Article XX to secure ambitious climate and environmental domestic policies.

Despite bearing undoubted relevance to the intractable conflict between trade and climate concerns, those policy recommendations have remained today fairly impracticable. It would seem that there is a crucial need to develop a pragmatic approach[377] encouraging the flourishing development of club-related plurilateral initiatives before considering further multilateral negotiations. On top of this, the WTO jurisprudence has maintained an extensive interpretation of the environmental exceptions under Article XX(b) and (g), as individual disputes arise.

V CONCLUSION

Unless there is a significant reform in the current laws, policies, and actions regarding climate change, the global levels of GHG emissions will continue to increase. As a response to this problem, many countries have enacted regulatory measures and economic incentives to mitigate climate change at the national level. One of them is the ETS, in which carbon has been given a price and the environmental externality of the industry is internalized with this pricing mechanism. However, these national-level carbon pricing mechanisms create the risks of carbon leakage, reduction of competitiveness of their industry, and free-riding in the countries that have no climate-change mitigation measures.

In order to eliminate these concerns, the structure of the climate club has been proposed. In the club regime, countries sharing the same environmental concerns form a unified regime in the realms of carbon pricing and regulations; and in exchange for this, they provide advantages to each other in the trade sector. Climate clubs can adopt several trade measures and market-based mechanisms to link their members.

One of them is the harmonize ETS regimes among climate-club members. The recognition of the emission units of different club jurisdictions for compliance purposes and exclusion of non-club members units raises the issue of violation of the MFN principle. In order to assess the relevance of the MFN principle for the emission units, the units need to fall under one of the WTO agreements. In this chapter, emission units have been analyzed as a good and a service respectively, in line with the GATT and GATS MFN clauses.

375. Beatriz Leycegui and Imanol Ramirez, "Addressing Climate Change: A WTO Exception to Incorporate Climate Clubs" (E15 Expert Group on Measures to Address Climate Change and the Trade System, May 2015) 15.
376. James Bacchus, "Global Rules for Mutually Supportive and Reinforcing Trade and Climate Regimes" (E15 Expert Group on Measures to Address Climate Change and the Trade System, January 2016) 1.
377. Beatriz Leycegui and Imanol Ramirez, "Addressing Climate Change: A WTO Exception to Incorporate Climate Clubs" (E15 Expert Group on Measures to Address Climate Change and the Trade System, May 2015) 3.

The key point in the MFN analysis is the determination of the likeness of the units of climate-club members and non-club members. Given that the characteristics of the emission units bear notable differences originated from the ETS that they are issued in, the club and non-club emission units are found not to be like products. This conclusion paves the way for further integration of the ETSs and the creation of climate clubs, which reflects the coalition of the willing for carbon abatement efforts.

Moreover, unshared trade benefits within climate clubs would constitute a potential violation of the WTO's core nondiscrimination principles of MFN and national treatment, as provided under Articles I and III of GATT, respectively. However, equilibrium must be sought between further trade liberalization and combating climate change. To avoid such violations, the WTO members would have to explore alternative ways or adopt supportive climate policies. A more promising approach would be for WTO members to establish general permanent exceptions that allow preferential arrangement among climate clubs within the WTO.

Under Article XX of GATT (general exceptions), the existing WTO rules allow members to take action on climate change. Such a possibility exists within Article XXIV of the GATT 1994. Also, exceptions in the WTO regime, such as GATT Article XXIV and the Enabling Clause, could be a framework to design a climate club exception in the WTO. Both constitute an acknowledgement from WTO Members of the necessity to address other legitimate objectives within the organization while departing from certain established principles. Paragraph 2(c) of the Enabling Clause of the GATT provides that differential treatment could be accorded in regional or global arrangements among developing countries for the mutual reduction or elimination of tariffs and non-tariff measures on products imported from each other. The context, in which these exceptions were negotiated and accepted, can shed some light on the feasibility of a WTO exception for climate clubs.

In addition, emissions trading could also be governed by RTAs, which are permitted under WTO law, as the RTAs' climate-change measures and policies intersect with international trade in several ways. Through RTAs, it would be permitted to have exclusive trade benefits among club members without violating the WTO's core principle of nondiscrimination.

The next chapter investigates how free trade agreements and bilateral investment treaties can also play a role in reaching a sustainable future.

CHAPTER 4

Free Trade Agreements and Bilateral Investment Treaties for a Sustainable Future

We argue that international trade law may be the solution to decarbonize the economy and invest in renewable energy, but there must be effective measures agreed by the international community. That would require agreeing on preferential trade agreements (PTAs) with environmental and sustainable development chapters on a unilateral, bilateral, or plurilateral basis, as multilateral agreements have proven to be increasingly difficult to negotiate. This chapter will assess the feasibility of including such chapters through an analysis of existing environmental and climate change content in recent PTAs. Equally, recent developments in newer bilateral investment treaties (BITs) show that there is a positive shift toward explicit inclusion of sustainable development issues into their wording (e.g., states' right to adopt sustainable development-oriented regulation). Reference to sustainable development issues may appear in different parts of a BIT, e.g., as a general objective of the treaty in a preamble, as a separate obligation of the state parties or investors, or as an exception to the substantive protection standards, such as prohibition of unlawful indirect expropriation. Specific position and wording of sustainable development provisions determine to what extent such provisions will be binding and enforceable against state parties. Certain BITs contain only a broad reference to sustainable development without clear practical consequences, while others list sustainable development obligations that may be enforced in practice, e.g., by means of state to state dispute settlement mechanism or, if such obligations are imposed on foreign investors, as a counterclaim brought by a host state against an investor. Finally, we analyze how efforts to modernize the Energy Charter Treaty may contribute to sustainable development.

I INTRODUCTION[1]

Climate change is the most important issue of our time, which, if unmitigated, will present significant dangers to human life and health, as well as social and economic well-being.[2] The United Nations Secretary General, António Guterres stated that "even as we witness devastating climate impact causing havoc across the world, we are still not doing enough, nor moving fast enough, to prevent irreversible and catastrophic climate disruption."[3] His frustration is shared among the legal and academic perspectives, as well as within the most vulnerable groups of countries,[4] as there is a feeling that not much has been achieved through Multilateral Environmental Agreements (MEAs) due to the lack of enforceability of their provisions. The impacts of climate change have no borders and will affect every country in the world.[5]

1. An earlier version of this chapter was published as R. Leal-Arcas et al., "The contribution of free trade agreements and bilateral investment treaties to a sustainable future," *Zeitschrift für Europarechtliche Studien—ZEuS*, Vol. 23, Issue 1, 2020, pp. 3-76.
2. Kati Kulovesi, *Real or Imagined Controversies? A Climate Law Perspective on the growing links between the International Trade and Climate Change Regimes*, 6:1 Trade, L. & Dev. 55 (2014), at 59; Rafael Leal-Arcas, *Solutions for Sustainability: How the International Trade, Energy and Climate Change Regimes Can Help*. Springer Nature, 2019; Walter Leal and Rafael Leal-Arcas (eds), *University Initiatives in Climate Change Mitigation and Adaptation*. Springer, 2018.
3. United Nation Press release, "No Time Left for Limitless Negotiations, Secretary-General Stresses as Climate Talks Open, Warning: 'We Cannot Afford to Fail in Katowice'," December 3, 2018, available at https://www.un.org/press/en/2018/sgsm19386.doc.htm.
4. Among these countries are those near the Equator and small island states, concerned about the impact the sea level rise will have on them. Indeed, it is by now a fact that Greenland is melting. As a result, its ice sheet is rapidly retreating and can hold enough water to see the sea level rise by more than seven meters if it all melts and goes into the oceans. The Economist announced that, according to Polar Portal, "Greenland is currently losing 3bn tonnes of ice a day." That was certainly not the case millennia ago, when Greenland was covered by forests, hence its name. *See* The Economist, "Greenland is melting," June 22, 2019, p. 69. For an analysis of climate change in small island states, *see* Brad Plumer and Lisa Friedman, "Island Nations, With No Time to Lose, Take Climate Response Into Their Own Hands," *The New York Times*, 2017, available at https://www.nytimes.com/2017/11/17/climate/islands-climate-change-un-bonn.html; Timothée Ourbak and Alexandre K. Magnan, "The Paris Agreement and climate change negotiations: Small Islands, big players," Regional Climate Change (November 17, 2017), available at https://link.springer.com/article/10.1007/s10113-017-1247-9; Adaptation Fund in the Small Island Developing States, (November 3, 2017), available at https://www.adaptation-fund.org/wp-content/uploads/2017/11/Adaptation-Fund-in-the-SIDS-1.pdf; Global Environment Facility, "LDCs," available at https://www.thegef.org/topics/least-developed-countries-fund-ldcf; "Inside Kiribati: The Island Being Erased by Climate Change," *Al Jazeera Documentary* (October 10, 2017), available at https://www.youtube.com/watch?v=plzBaNaCIv4; Justin Worland, "The Leaders of These Sinking Countries Are Fighting to Stop Climate Change. Here's What the Rest of the World Can Learn," *TIME* (June 13, 2019), available at https://time.com/longform/sinking-islands-climate-change; Peter Yeung, "Climate Change is Already Battering this West African City," *CityLab* (April 7, 2019), available at https://www.citylab.com/environment/2019/04/climate-change-rising-sea-levels-west-africa-coastal-cities/586651/; Tara John, "While the Rich World Braces for Future Climate Change, the Poor World is Already Being Devastated By It," *CNN* (April 1, 2019), available at https://www.cnn.com/2019/03/31/africa/poorest-hit-the-hardest-climate-change-mozambique-intl/index.html.
5. As a result of human activity, a million species of animals and plants are threatened with extinction. The Economist reported in May 2019 that, according to the Intergovernmental Science-Policy Platform on Biodiversity and Ecosystem Services, "more than 85% of wetlands have been lost. More than 90% of ocean fish stocks are being harvested at or above sustainable levels." *See* The Economist, "Dead end," May 11, 2019, p. 73.

Some dreamers have gone much further and propose that humans should move to the Moon as an insurance policy for the Earth.[6] Therefore, the expectation is that, until that is possible, the international community would make the mitigation of such impacts their number one priority. However, due to countries having different agendas and economic interests, it has been considerably difficult to establish binding commitments in relation to climate mitigation and adaptation.

A good example is that of China's rare earths. In 2010, China decided to restrict its exports of rare earths to protect its environment. The World Trade Organization (WTO) decided against China's restrictions after the US, the EU, and Japan challenged them. China's share of global rare-earth production dropped from over 95% in 2010 to 70% in 2018.[7] China is trying to clean its rivers and, although it is still excavating a lot of rare-earth elements, it is committed to importing most of what it needs in order to protect its own environment.[8]

There is an undeniable connection between climate change and international trade.[9] However, there is no mention in the WTO Agreement about mitigating climate change, despite the fact that the United Nations Framework Convention on Climate Change (UNFCCC) had been concluded two years before the creation of the WTO Agreement. The two regimes need to be intertwined in order to effectively address the major global challenges of the 21st century—climate change, economic and social development, human well-being, sustainable development and global security.[10] Energy concerns are closely linked with trade and climate change, as energy is a commodity that is traded, and it has been generated through fossil fuels in the last 150 years, emitting many greenhouse gases (GHG) into the atmosphere.[11] As population

6. Jeff Bezos has proposed returning to the Moon by 2024. His justification is the mismatch between population growth and the Earth's limited natural resources, given the increasing levels of energy demand. Another dreamer, Elon Musk, would like humans to colonize Mars. *See* The Economist, "Living in Outer Space: Back to the Future," May 18, 2019, p. 74.
7. The Economist, "Magnetic Attraction," June 15, 2019, pp. 51-52, at 51.
8. *Ibid.* at 52.
9. The perception of this connection seems to be often negative. Nicolas Hulot, a former French ecology minister, has consistently argued that free trade deals are the cause of the world's environmental problems. *See* N. Hulot, "Soyons cohérents, rejetons le Ceta," *Le Journal du Dimanche*, June 29, 2019, available at https://www.lejdd.fr/Politique/nicolas-hulot-soyons-coherents-rejetons-le-ceta-3907233. Others have argued along the same lines: J. Watts, "We Must Not Barter the Amazon Rainforest for Burgers and Steaks," *The Guardian*, July 2, 2019 (in relation to the EU-Mercosur Trade Agreement, stating the serious environmental concerns raised by the agreement). Available at https://www.theguardian.com/environment/commentisfree/2019/jul/02/barter-amazon-rainforest-burgers-steaks-brazil. Equally, Dara Calleary, the deputy leader of one of the main opposition parties in Ireland (Fianna Fail), said that the EU-Mercosur Trade Agreement "goes against all of the principles, beliefs and declarations relating to climate change." *See* Pat Leahy, "Hogan Rejects 'Inaccurate' Criticism of EU-Mercosur Deal," *The Irish Times*, July 5, 2019, available at https://www.irishtimes.com/news/politics/hogan-rejects-inaccurate-criticism-of-eu-mercosur-deal-1.3946980?mode=amp.
10. International Institute for Applied Systems Analysis, "Global Energy Assessment—Toward a Sustainable Future," Cambridge University Press, Cambridge UK and New York, NY, USA and the International Institute for Applied Systems Analysis, Laxenburg, Austria.
11. *See* generally Rafael Leal-Arcas and Jan Wouters (eds), *Research Handbook on EU Energy Law and Policy*. Edward Elgar Publishing, 2017; Rafael Leal-Arcas et al., *Energy Security, Trade and the EU: Regional and International Perspectives*. Edward Elgar, 2016; Rafael Leal-Arcas, *The European Energy Union: The Quest for Secure, Affordable and Sustainable Energy*. Claeys &

increases,[12] energy consumption will also increase, creating an urgency in developing clean energy technologies.

International trade law may be the solution to decarbonize the economy and invest in renewable energy, but there must be effective measures agreed by the international community. The theory is that of agreeing on preferential trade agreements (PTAs)[13] with environmental and sustainable development chapters on a unilateral, bilateral or plurilateral basis, as multilateral agreements have proven to be increasingly difficult to negotiate. This chapter will assess the feasibility of including such chapters through an analysis of existing environmental and climate change-content in recent PTAs.

This article is divided into five sections. After the introductory section, section II analyzes the extent to which free trade agreements (FTAs) can become the enforcers of climate-change obligations. Section III provides an analysis of the contribution of bilateral investment treaties to climate action and sustainable energy. Section IV offers an analysis of how current efforts to modernize the Energy Charter Treaty can contribute to the European Union's approach of linking trade and investment policies with climate action. Section V concludes the chapter.

II FREE TRADE AGREEMENTS AS THE ENFORCER OF CLIMATE CHANGE OBLIGATIONS?

1 Introduction

The recent decline in the conclusion of multilateral environmental agreements (MEAs) has prompted scholars to identify the many institutional difficulties that climate negotiations are facing. Considering the continuous stalemate of WTO negotiations and the ineffectiveness of the UNFCCC, the Kyoto Protocol and the Paris Agreement, some scholars argue that multilateral environmental agreements are inadequate to mitigate climate change.[14] Furthermore, the United States' intention to withdraw from the Paris Agreement since 2017 has overturned any prospects of bringing on board all major emitters in order to tackle the environmental challenges faced by today's world. That

Casteels Publishing, 2016; Rafael Leal-Arcas et al., *International Energy Governance: Selected Legal Issues*. Edward Elgar, 2014.

12. Whether population is destiny, where growth is concerned, is the subject of new research. *See* Klaus Desmet, Dávid Krisztián Nagy, and Esteban Rossi-Hansberg, "The Geography of Development," 126:3 *Journal of Political Economy* 903-83 (2018).
13. For the purposes of this chapter, preferential trade agreements (PTAs) and free trade agreements (FTAs) are used interchangeably.
14. *See, inter alia*, Joy. A. Kim, "Harnessing Regional Trade Agreements for the Post-2012 Climate Change Regime," *Climate and Trade Policies in a Post-2012 World* (UNCEP 2009); R. Leal-Arcas, "Using Preferential Trade Agreements to Promote Climate Change Mitigation," in F. Baetens and J. Caiado (eds.), *Frontiers of International Law*. Brill Nijhoff 2014; R. Clémencon, "The Two Sides of the Paris Climate Agreement: Dismal Failure or Historic Breakthrough?," 25:1 *Journal of Environment & Development* 3-24 (2016); G. Bang, J. Hovi and T. Skodvin, "The Paris Agreement: Short-Term and Long-Term Effectiveness," 4:3 *Politics and Governance* (2016).

said, the USA is doing very much at the subnational level toward the mitigation of climate change.[15]

In contrast, the web of bilateral and regional trade agreements is rapidly increasing.[16] The USA has (FTAs) in force with 20 countries,[17] while the European Union (EU) has in place 36 FTAs.[18] Many other FTAs are currently being negotiated or awaiting ratification. Notably, many of the new generation FTAs include a comprehensive chapter on environmental protection with increasingly detailed and far-reaching provisions on environment, climate change, biodiversity and sustainable development.

This section explains why FTAs have the potential to be more effective legal instruments than environmental agreements for environmental protection purposes. It then compares the texts of environmental and climate-related provisions in the new generation of American FTAs to the FTAs concluded by the EU and highlights key differences between them. Particular focus will be given to the legalization of the environmental and climate-related obligations under the selected FTAs. We argue that European FTAs have the potential of becoming the new enforcers of climate change mitigation, but only if the EU starts to link the FTAs' general dispute resolution mechanism to the FTAs' environmental provisions. Finally, this section will look at the EU-Canada Comprehensive Economic and Trade Agreement (CETA), the United States-Mexico-Canada Agreement (USMCA), and the Transatlantic Trade and Investment Partnership (TTIP). More specifically it will argue that PTAs with a combination of CETA-like provisions and USMCA-like enforcement should be considered in future agreements.

2 The Relationship Between Trade and the Environment

Drivers of climate change and its impacts around the world are linked to nearly every industry. For this reason, it is very difficult for the UNFCCC framework to govern climate change in isolation. It is therefore unsurprising that international institutions such as the World Trade Organization (WTO) have been trying to solve the problem.

Trade liberalization has made significant progress since the General Agreement on Tariffs and Trade 1947 (GATT) particularly after the establishment of the WTO in 1994. The UNFCCC framework imposes duties on governments to mitigate and adapt to climate change, including qualified targets for emissions reductions. Both regimes

15. For instance, New York's state legislature passed a bill for the state of New York to have carbon-free electricity by 2040 and to eliminate carbon emissions by 2050. In 2018, California, Colorado, Maine, New Jersey, New Mexico, and Washington are all states with law and policies that aim at decarbonizing electricity generation, bearing in mind that electricity is our comfort. Oregon and New York are heading in the same direction. *See* The Economist, "Climate Change: States' Rights," June 29, 2019, pp. 13-14, at 13.
16. For the purposes of this chapter, Regional Trade Agreements include Custom Unions, Free Trade Agreements, Preferential Trade Agreements, and Association Agreements with trade provisions.
17. https://ustr.gov/trade-agreements/free-trade-agreements.
18. http://ec.europa.eu/trade/policy/countries-and-regions/negotiations-and-agreements/#_in-place.

are incompatible with one another and the agendas must be linked for the fight against climate change to succeed.

The link between trade and environmental law is increasingly relevant, even if not directly expressed in the GATT or WTO agreement.[19] When the GATT was passed, there was no concern over climate change and its adverse effects, however when the WTO agreement was drafted, there was already a lot of evidence regarding the link between climate change and trade.[20] Even though such evidence was already known, the WTO agreement failed to include climate change provisions or mention the UNFCCC in an attempt to link the efforts.

As per the WTO, sustainable development and protection of the environment are included in their fundamental goals, under the Marrakesh Agreement.[21] While there is no specific mention of the environment, members of the WTO can adopt trade-related measures to protect the environment, if such measures are not misused against other WTO members. It has also created a Committee Trade and Environment (CTE), which is a forum dedicated to dialogue between governments on the impact of trade policies on the environment, and of environment policies on trade.[22] The CTE has a working programme since 1994, focusing on issues related to trade rules and environmental agreements, environmental protection, sustainable development and transparency.

The Paris Agreement makes no mention of trade and fails to link trade and climate change.[23] However, there is mention of "market-based mechanisms"[24] such as emission trading and promotion of sustainable development and encouragement of broad participation through public and private sectors. Arguably this calls for the participation of trade mechanisms in the climate change agenda. The problem with the UNFCCC framework is the differentiation between developed and developing countries. The main issue was in the Kyoto Protocol as it only gave obligations to developed countries (or Annex I countries) to reduce their emissions. The Paris Agreement tried to move away from this distinction, however, due to opposition by developing

19. Rafael Leal-Arcas, "Unilateral Trade-Related Climate Change Measures," 13 *J. World Investment & Trade* 875 (2012), at 878.
20. *Ibid.*
21. Marrakesh Agreement Establishing the World Trade Organization (1994); *see also* WTO, "Trade and Environment" (WTO website, published online on: https://www.wto.org/english/tratop_e/envir_e/envir_e.htm.
22. *Ibid.*
23. Yet, there is a growing body of academic literature on the links between international trade and climate change. *See* for instance, R. Leal-Arcas, *Climate Change and International Trade*. Edward Elgar, 2013; Daniel C. Esty. "Toward a Sustainable Global Economy: An Initiative for G20 Leadership," *Journal of Self-Governance and Management Economics* (2017); William Nordhaus. "A New Solution: The Climate Club," *The New York Review of Books*, 2015, available at http://www.nybooks.com/articles/2015/06/04/new-solution-climate-club/; Antoine Dechezleprêtre and Misato Sato, "The Impacts of Environmental Regulations on Competitiveness," *Review of Environmental Economics and Policy* (2017); Dani Rodrik, "Global Trade Needs Rules That Adapt to Economic Diversity," *Financial Times* (August 6, 2018) at p. 9; "Trade and Environment," World Trade Organization, available at https://www.wto.org/english/tratop_e/envir_e/envir_e.htm; Aaron Cosbey, "Climate Policies, Economic Diversification and Trade," 2017, available at http://dx.doi.org/10.2139/ssrn.3135359.
24. Article 6 of Adoption of the Paris Agreement, Decision 1/CP.21, in COP Report No. 21, Addendum, at 2, U.N. Doc. FCCC/CP/2015/10/Add.1 (January 29, 2016).

countries their contribution is still on a voluntary basis. It is therefore not surprising that the United States (U.S.) wants to withdraw from the Paris Agreement, as China, who is the largest emitter of Greenhouse Gases (GHG),[25] is still considered a developing country. Some have argued that regardless of China's economic development, there should still be a distinction between developed and developing countries,[26] as the poorest countries do not have the capacity to mitigate and adapt to climate change without assistance, financial and otherwise.

The U.S. and other developed countries believe that the distinction is unfair as it provides for inappropriate trade advantages for developing countries and because of the growing GHG emissions by developing countries, which will only continue to grow in the next 25 years unless binding commitments are implemented.[27] PTAs can be the solution for including developing countries, as they are proactive in creating them.[28] Climate-related provisions can be added into PTAs agreed between developing countries encourage the reduction of GHG emissions through the facilitation of trade and investment in environmental goods and services.

However, most GHG emissions come from five entities, i.e., China, the U.S., the EU, India,[29] and Russia, as per Figure 4.1.

25. China's large population of almost 1.4 billion people and rapid economic growth have made it the world's major GHG emitter.
26. Maureen F. Irish, "LDCs, Climate Change and Trade," 5 *Law & Development Review* 2 (2012), at 2.
27. Rafael Leal-Arcas, *Climate Change and International Trade*. Edward Elgar, 2013, Chapter 6.
28. Rafael Leal-Arcas, "Climate Change Mitigation from the Bottom Up: Using Preferential Trade Agreements to Promote Climate Change Mitigation," *CCLR* 34-42 (2013).
29. For an analysis of India's climate change policy, *see* Varun Sivaram, "Still Shining? Our Third Annual Review on Solar Scale-up in India," *Council on Foreign Relations*, 2018, available at https://www.cfr.org/blog/still-shining-our-third-annual-review-solar-scale-india; Damien Carrington and Michael Safi, "How India's Battle with Climate Change Could Determine All Our Fates," *The Guardian*, (November 6, 2017), available at https://www.theguardian.com/environment/2017/nov/06/how-indias-battle-with-climate-change-could-determine-all-of-our-fates; Adil Najam, "The View from the South: Developing Countries in Global Environmental Politics," in Regina S. Axelrod and Stacy D. VanDeveer (eds.), *The Global Environment: Institutions, Law, and Policy*, Sage Publishing, 2019; Aaron Atteridge, Manish Kumar Shrivastava, Neha Pahuja and Himani Upadhyay, "Climate Policy in India: What Shapes International, National and State Policy?" AMBIO, 2012, available at http://link.springer.com/article/10.1007%2Fs13280-011-0242-5; Tom Osborn, "Why Developing Countries Are Disproportionately Affected by Climate Change—And What Can They Do About It," *Huffington Post*, 2015, available at https://www.huffingtonpost.com/tom-osborn/why-developing-countries-_b_6511346.html; Barbara Buchner, "Innovative Climate Finance: A Spotlight on India," *Climate Action*, 2016, available at http://www.climateactionprogramme.org/climate-leader-papers/innovative_climate_finance_a_spotlight_on_india; Mendelsohn et al. "The Distributional Impact of Climate Change on Rich and Poor Countries," *Environment and Development Economics* 11(2006): 159-178.

Figure 4.1 Annual Share of Global CO_2 Emission, as of 2016

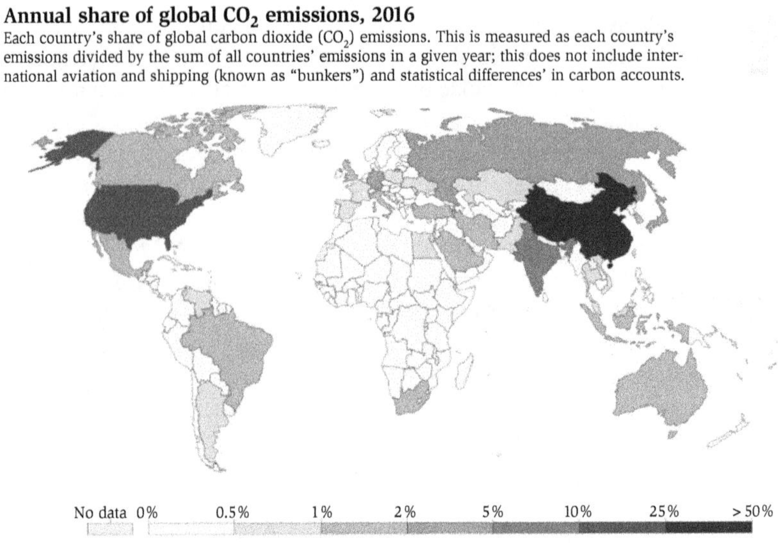

Source: Our World in Data Based on Global Carbon Project (2017).

The best solution to link climate change and trade concerns would be to create PTAs between the largest emitters, regardless of their developed status. MEAs should not be replaced and preferentially both systems would cooperate with each other.

The inclusion of environmental and sustainable chapters in PTAs must not violate WTO rules. WTO members may not discriminate either "like goods" from different countries[30] or between foreign goods and domestic goods;[31] nor may countries impose quantitative restrictions.[32]

Measures banning certain goods that are produced with high pollution technologies or importing higher tariffs on such products can be discriminatory.[33] However, there is a set of admissible justifications within Article XX GATT and Article XIV GATS. The promotion of environmental goods and services can be done through GATT Article XX(g)[34] as, per the WTO Dispute Settlement Body, the atmosphere as well as "clean air" may be seen as exhaustible natural resources insofar as the chemical composition

30. Most favourable Nation, Article I:1 GATT and Article II:1 General Agreement on Tariffs and Services (GATS).
31. National treatment, Articles III:4 GATT and XVII GATS.
32. Articles XI GATT and XVI GATS.
33. Example concerning production or process measures (PPMs) are protectionist and are considered discriminatory: *see* Robert Howse and Donald Regan, "The Product/Process Distinction—An Illusory Basis for Disciplining 'Unilateralism' in Trade Policy," 11-12 *European Journal of International Law* 249 (2000).
34. GATT Article XX(g) reads: "relating to the conservation of exhaustible natural resources if such measures are made effective in conjunction with restrictions on domestic production or consumption."

may be irreversibly modified.[35] For the exception to apply, the resource only has to potentially be exhausted. This is appropriate as it serves the objective of the exception, to conserve natural resource before it has depleted.[36]

Trade can incentivize cleaner energy technologies which can lead to less carbon incentive types of growth. The facilitation of trade on environmental goods and services is needed to advance the fight against climate change.

3 The Potential of Free Trade Agreements for Climate-Change Mitigation

This subsection will examine PTAs that contain provisions to mitigate climate change, such as environmental and sustainable trade chapters. There has been a recent proliferation of PTAs in the international trading system, mostly due to the difficulties in trying to agree on multilateral agreements, as the negotiations at the Doha Round have shown.[37]

This proliferation is mainly due to multilateral negotiations being extremely complex and taking a long time due to the rule that nothing is agreed until everything is agreed by everyone.[38] The case for using a bottom-up approach to mitigate climate change through PTAs is a very common research topic in academia.[39] It tends to promote an alternative structure to the present MEAs, due to their lack of enforcement and the length of negotiations. It is therefore necessary for PTAs to be explored to enforce climate change obligations, as these move much faster due to countries benefitting from extended trading markets. Trade agreements are generally preferred rather than climate change agreements, as acting against climate change may be seen by many as detrimental toward their economic growth while trade opens a new market to countries and offers economic incentives.[40]

35. United States—Standards for Reformulated and Conventional Gasoline (WT/DS2/AB/R).
36. United States—Measures Concerning the Importation, Marketing and Sale of Tuna and Tuna Products WT/DS381/49/Rev.1.
37. Rafael Leal-Arcas, "Proliferation of Regional Trade Agreements: Complementing or Supplanting Multilateralism?," 11:2 *Chicago Journal of International Law* 597-629 (2011). For an in-depth analysis of the EU trading system, *see* Rafael Leal-Arcas, *EU Trade Law*. Edward Elgar, 2019 and Rafael Leal-Arcas, *Theory and Practice of EC External Trade Law and Policy*. Cameron May, 2018.
38. This is the WTO principle called "single undertaking."
39. *See* for example Rafael Leal-Arcas, "Climate Change Mitigation from Bottom Up: Using Preferential Trade Agreements to Promote Climate Change Mitigation," 7:1 *Carbon & Climate Law Review* 34-42 (2013); Rafael Leal-Arcas, "Alternative Architecture for Climate Change: Major Economies," 4 *Eur. J. Legal Stud.* 29 (2011); Rafael Leal-Arcas, "Trade Proposals for Climate Action," 6 *Trade L. & Dev.* 11 (2014); *Addressing Climate Change under Preferential Trade Agreements: Towards Alignment of Carbon Standards under the Transatlantic Trade and Investment Partnership*, 35 Global Environmental Change 514-522 (2015); Jean-Frédéric Morin and Sikina Jinnah, *The Untapped Potential of Preferential Trade Agreements for Climate Governance* (2018).
40. Rafael Leal-Arcas, "Climate Change Mitigation from Bottom Up: Using Preferential Trade Agreements to Promote Climate Change Mitigation," 7:1 *Carbon & Climate Law Review* 34-42 (2013), at p. 35.

The UNFCCC framework is the only platform attempting to mitigate and adapt to climate change internationally and to mitigate the difficulties in establishing ongoing obligations to finance developing countries capacities to combat climate-change phenomena, as well as the slow progress in achieving significant emission reduction, lack of enforcement mechanisms and unfair distinction between certain developed and developing countries. PTAs may be the building blocks to an international climate change agreement, though an "accidental" effort,[41] as they have the potential of including countries with the highest emissions and the major forces in the world's economy. This can influence other countries into agreeing to comply with environmental obligations.

The environmental protection under the new generation of FTAs has become increasingly far-reaching.[42] At the same time, the current fragmented perspectives on how to tackle climate change (in particular among the top emitters of the word, such as China and the USA) makes it highly unlikely that the global community be able to successfully negotiate and conclude an effective global environmental agreement to which all relevant emitters in the world are party. Moreover, one could question the enforceability of environmental obligations imposed on states under the various MEAs currently in force. It is generally assumed that most MEAs, including the Paris Agreement, lack effective dispute settlement mechanisms and do not contain legally binding and enforceable obligations with respect to the parties' mitigation contributions.[43]

In contrast, the very nature of FTAs is to impose binding obligations upon the parties and a breach of such obligations would, in principle, give rise to a legal action under the sanction-backed dispute resolution mechanism of the FTA. Therefore, if FTAs contain legally binding and enforceable provisions on climate change and environmental protection, FTAs—rather than MEAs—could potentially be the enforcer of climate change obligations and thus be at the forefront of climate change mitigation.[44]

This assumption may be counterintuitive for those who presume that trade agreements and environmental protection have conflicting objectives. According to some, trade liberalization leads to increased economic activity and long-distance

41. *Ibid.* p. 36.
42. *See* in more detail: S. Jinnah and J. Kennedy, "Environmental Provisions in US Trade Agreements: A New Era of Trade-Environment Politics," XII:1 *Whitehead Journal of Diplomacy and International Relations* (2011) and S. Jinnah and E. Morgera, "Environmental Provisions in American and EU Free Trade Agreements: A Preliminary Comparison and Research Agenda," 22:3 *RECIEL* (2013).
43. On the Paris Agreement, *see* D. Bodansky, "The Legal Character of the Paris Agreement," 25:2 *RECIEL* (2016).
44. *See* S. Jinnah, "Strategic Linkages: The Evolving Role of Trade Agreements in Global Environmental Governance," 20:3 *Journal of Environment & Development* (2011); R. Leal-Arcas, "Using Preferential Trade Agreements to Promote Climate Change Mitigation," in F. Baetens and J. Caiado (eds.), *Frontiers of International Law*. Brill Nijhoff, 2014; H. van Asselt, "Climate Change and Trade Policy Interaction: Implications of Regionalism," OECD Trade and Environment Working Papers, 2017/03, OECD Publishing (2017), pp. 23-35; R. Leal-Arcas and E. Alvarez Armas, "The Climate-Energy-Trade Nexus in EU External Relations," in S. Minas and V. Ntousas (eds.), *EU Climate Diplomacy: Politics, Technology and Networks*. Routledge, 2018.

transportation, which, in turn, causes additional greenhouse gas emissions and pollution.[45] Think, for instance, of aviation, where it is expected that the number of planes flying will double by 2040 as people, especially in Asia, get richer and the middle class in China and India gets solidified. The result will be an exacerbation of environmental problems.[46] A way to minimize the negative impact of aviation on the environment is by replacing engines with electric motors.[47]

However, academic studies refute this line of argumentation and show that trade agreements *"will not increase the amount of pollution, but in fact may help the environment."*[48] Notably, CO_2 emissions of countries that concluded an FTA with environmental provisions tend to converge and are lower in absolute terms, whereas this is not the case for FTAs without provisions on environmental protection.[49] An in-depth analysis of the environmental impact of environmental provisions in FTAs is beyond the scope of this chapter. Therefore, this chapter continues on the premise supported by scholarly research that FTAs play a significant role in climate change mitigation and environmental protection.

Morin and Jinnah rightly point out that the potential contribution of PTAs to climate governance rests on four distinctive features of trade agreements and their negotiations.[50] First, FTA negotiations involve a limited number of parties addressing a wide array of different issues. Such negotiations encourage bargaining and the conclusion of new agreements (while, in contrast, MEAs are negotiated among a large group of countries on a specific issue area). Second, FTAs are based on direct reciprocity and a breach of the agreement gives rise to a claim under the sanction-based dispute settlement mechanism of the FTA (while MEAs, as discussed, generally lack strong enforcement mechanisms). Third, Morin and Jinnah submit that PTAs offer opportunity for policy experimentation since several new FTAs are negotiated and concluded every year with great diversity in types and membership. As a result, innovative provisions can be designed and tested at a limited scale and among

---

45. K. Conca, "The WTO and the Undermining of Global Environmental Governance," 7:3 *Review of International Political Economy* (2000); D. Rodrik, *The Globalization Paradox: Democracy and the Future of the World Economy*. W.W. Norton & Co, 2011 (where Rodrik argues that full participation in the global economy implies that a country needs to give up a degree of either national sovereignty or democracy. For instance, lowering technical barriers to trade means harmonizing regulatory and trade policies with other countries, which, in turn, means that governments are limited in their ability to dictate national preferences).
46. The Economist, "The Future of Flight," June 1, 2019, pp. 1-12, at 4.
47. *Ibid.* at 7.
48. S. Ghosh and S. Yamarik, "Do Regional Trading Arrangements Harm the Environment?," 6:2 *Applied Econometrics and International Development* (2006). *See also* S. Jinnah and A. Lindsay, "Diffusion Through Issue Linkage: Environmental Norms in US Trade Agreements," 16:3 *Global Environmental Politics* (2016), who conclude that their article *"confirms prior hypotheses in the literature that environmental linkages within PTAs are important mechanisms of norm diffusion."*
49. L. Baghdadi, I. Martinez-Zarzosob, and H. Zitounac, "Are RTA Agreements with Environmental Provisions Reducing Emissions?," 90:2 *Journal of International Economics* (2013). *See*, with a similar conclusion, I. Martínez-Zarzoso W. Oueslati "Do Deep and Comprehensive Regional Trade Agreements Help in Reducing Air Pollution?, 18:6 *International Environmental Agreements: Politics, Law and Economics* (2018).
50. J. Morin and S. Jinnah, "The Untapped Potential of Preferential Trade Agreements for Climate Governance," 27:3 *Environmental Politics* 542-543 (2018).

like-minded countries. Fourth, FTAs are uniquely positioned to address trade-related aspects of climate mitigation (such as the export of low-emission technologies and fossil fuel subsidies).

4 Preferential Trade Agreements with Climate-Related Provisions: Substantive Content and Enforceability

This part will examine the regulatory contribution that PTAs with environmental provisions made in the creation of a global climate governance.[51] These PTAs can be compared and assessed as per the Trade and Environmental database (TREND),[52] and the analysis will focus on the substantive content and the enforceability of such provisions of these PTAs.[53]

For PTAs to be effective in establishing a global climate governance, they must have a wide-variety of climate provisions with specific and strongly worded obligations and enforcement mechanisms. Most climate-related provisions are replicated from previous agreements however innovation in drafting such provisions is needed as the current ones are not achieving the desired objectives.

There are different provisions that relate to climate change mitigation. The most used are the ones which address energy efficiency[54] and renewable energy.[55] To effectively decarbonize the economy and tackle climate concerns such as reduction of

51. One issue with "global" climate governance is that, currently, the benefits of climate change mitigation are global, but the costs are local. If we provide an analogy from Physics, it would be similar to the binary approach between relativity (a theory of the universe) and quantum mechanics (which deals with the behavior of very small things). However, there are great benefits to decarbonizing the economy, many of which are local. For instance, reducing the burden of air pollution on health.
52. Trade and Environment database, "Environmental provisions in Preferential Trade Agreements," available at https://klimalog.die-gdi.de/trend/.
53. Global governance research has shown that trade regulation can indeed contribute to the protection of public goods, e.g., by trading up. The more critical International Political Economy (IPE) literature has also argued that the potential benefits of trade do not only require effective regulation, but have to be weighed against its negative (unintended) consequences. The IPE literature has also explored mega-PTAs as a means to internalize the negative externalities of international trade, showing that their creation and effectiveness hinge on economic, political, and social scope conditions that are rarely met.
54. On energy efficiency, *see* Michael Gerrard, *The Law of Clean Energy: Efficiency and Renewables*. ABA Book Publishing, 2011, pp. 1-22; Sheryl Carter, "Ramping Up Energy Efficiency Key to Address Climate Change," *Natural Resources Defense Council*, September 22, 2016, available at https://www.nrdc.org/experts/sheryl-carter/ramping-energy-efficiency-key-address-climate-change; IEA, "Energy Efficiency 2017," 2017, available at https://www.iea.org/efficiency/; Jim Tankersley, "Can We Nudge People Into Conserving Energy?" *Washington Post*, October 24, 2014, available at https://www.washingtonpost.com/news/storyline/wp/2014/10/24/can-we-shame-people-into-conserving-energy/; Kenneth Gillingham, David Rapson, and Gernot Wagner, "The Rebound Effect and Energy Efficiency Policy," *Review of Environmental Economics and Policy* (2015); Amory Lovins, "Energy Strategy: The Road Not Taken?" *Foreign Affairs*, 1976, available at https://www.foreignaffairs.com/articles/united-states/1976-10-01/energy-strategy-road-not-taken; Suh et al. "Life Cycle Environmental and Natural Resource Implications of Energy Efficiency Technologies," *Journal of Industrial Ecology* (2016), available at https://onlinelibrary.wiley.com/doi/full/10.1111/jiec.12435.
55. J.F. Morin and S. Jinnah, "The Untapped Potential of Preferential Trade Agreements for Climate Governance," (2018), pp. 541-565.

GHG emissions, it is important to have sector-specific provisions. The energy sector produces the most GHG emissions, as per Figure 4.2.

Figure 4.2 Global Manmade Greenhouse Gas Emissions, by Sector (as of 2013)

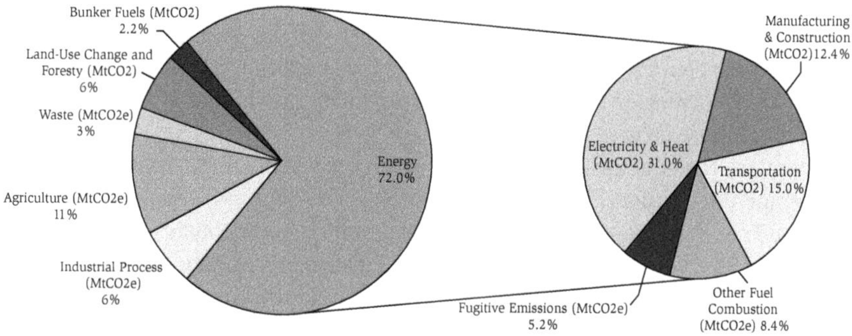

NOTES

Globally, the primary sources of greenhouse gas emissions are electricity and heat (31%), agriculture (11%), transportation (15%), forestry (6%) and manufacturing (12%). Energy production of all types accounts for 72 percent of all emissions.

PTAs can be a mechanism to drive research and development in the energy sector, as cooperation between countries would make it easier to share information and create alternatives to the fossil-fuel electricity generation, promoting innovation in the renewable energy sector. Countries already include provisions for research, cooperation, assistance, project development, and the exchange of information on renewable energy. An example can be found in the Korea-Australia Free Trade Agreement of 2014, where it was agreed to exchange of scientific measures relating to climate change and energy efficiency.[56] This is very general and vague, however a more specific provision can be found on the South Korea-Peru Free Trade Agreement of 2011, which encourages public and private institutions related to small and medium-sized enterprises to cooperate in renewable energy developments.[57]

In relation to general provisions on climate governance, there were provisions before the UNFCCC was created, such as provisions in the EU-Poland and EU-Hungary PTAs, negotiated in 1991.[58] These provisions only "encouraged" cooperation between the parties on climate change matters by establishing a dialogue between them. This could be made more specific, such as including provisions with "trade-related" aspects such as border tax adjustments for countries that do not attempt to reduce their GHG

56. Article 16.14 of the Korea-Australia FTA.
57. South Korea-Peru FTA, Article 20.4.
58. J.F. Morin and S. Jinnah, "The Untapped Potential of Preferential Trade Agreements for Climate Governance," (2018), pp. 541-565.

emissions.[59] Some PTAs have included direct provisions such as in the Indonesia-Japan PTA which refers directly to the Kyoto Protocol's Clean Development Mechanism.[60] Another alternative could be by promoting environmental goods and services and not treating them as "like" highly carbon produced goods and services. The differentiation could be related to the emissions that certain products have in production, and introduction a carbon tax that do not comply with reduction standards. The promotion of environmental goods and services can be seen in the Association Agreement between the EU and Georgia, which asks the parties to facilitate the removal of any obstacles to trade and investment of goods and services which are relevant for climate change mitigation.[61]

There are fewer adaptation than mitigation provisions in PTAs, and even when such are included, there is vagueness for the required cooperation. The South Korea-Peru Free Trade Agreement, for example, only asks for policies and measures to be adopted to evaluate the vulnerability and adaptation to climate change.[62]

PTAs may also include indirect provisions relating to climate change, most frequently, trade-restrictive measure addressing the conservation of natural resources. This can be used to protect the atmosphere and justify measures to reduce air pollution, as clean air is an exhaustible natural resource. Many PTAs include provisions on air pollution and vehicle emissions, which indirectly mitigates climate change, as anthropogenic gases are the main causes for the current climate crisis. Another indirect provision in PTAs that mitigates climate change is that the levels of environmental protection cannot be weakened to benefit trade or attract investment. The EU tends to create PTAs with precise language and specific commitments. For example, in the EU-Montenegro Agreement, there are specific standards for vehicle emissions.[63]

Most PTAs replicate provisions from earlier agreements, keeping the generic, nonbinding language. To have effective climate change-related provisions, the language must be precise, command an obligation and have proper delegation. The extent of the commitment made by parties will be measured by the level of obligations and precision of language, as such this would limit discretion on the interpretation of provisions. Delegation is crucial as provisions must be able to be adjudicated, implemented and enforced, otherwise the obligations will lack significance.[64] Several PTAs appear to have legally binding language, as 64 % of PTAs have at least one provision with a high level of obligation. This is overshadowed by the fact that only 10 % of such PTAs have precise targets and only 30 % provide for third-party settlement disputes.[65]

59. A "trade-related" provision has been included in the Australia-Korea FTA in Article 18.8.
60. J.F. Morin and S. Jinnah, "The Untapped Potential of Preferential Trade Agreements for Climate Governance," (2018), pp. 541-565.
61. Article 231 of the Association Agreement between the EU and Georgia.
62. J.F. Morin and S. Jinnah, "The Untapped Potential of Preferential Trade Agreements for Climate Governance," (2018), pp. 541-565 and Article 19.8 of the South Korea-Peru FTA.
63. J.F. Morin and S. Jinnah, "The Untapped Potential of Preferential Trade Agreements for Climate Governance," (2018), pp. 541-565.
64. *Ibid.*
65. *Ibid.*

Considering the legality and precision issues, a comparison will be drawn between CETA and USMCA to determine the levels of precision, bindingness and enforcement, and the possibility of replication in the TTIP negotiations.

5 The US and EU Approach to Environmental Provisions in Free Trade Agreements

By comparing the text of the environmental and climate-related provisions in the new generation FTAs of the USA and EU to earlier trade agreements, we will ascertain whether the new FTAs have indeed the potential to become the new enforcers of environmental and climate change obligations.

5.1 The Evolving Practice of Environmental and Climate Change-Related Provisions in Free Trade Agreements Concluded by the US and the EU

In the early 1990s, the USA became increasingly concerned that its domestic environmental standards were more stringent than other countries and that this imbalance would affect the county's competitiveness.[66] As a result, the USA started including environmental provisions in trade agreements in order to create a level playing field. The North American Free Trade Agreement (NAFTA) and its side agreement, the North American Agreement on Environmental Cooperation (NAAEC), were the first FTA with extensive provisions on environmental protection and the enforcement of environmental regulations.[67] Noticeably, in case of persistent failure to enforce domestic environmental laws, the NAAEC provides that any signatory state may request the establishment of an arbitral panel to consider the matter, which has the power to impose a monetary fine or, in case of nonpayment, suspend NAFTA benefits.[68]

Subsequent to NAFTA and following the enactment of the 2002 Trade Act,[69] the USA entered into FTAs with Singapore and Chile.[70] These FTAs were the first to include comprehensive chapters entirely devoted to environmental protection, recognizing the importance of MEAs,[71] introducing consultations to resolve disputes under the environmental chapters[72] and creating Environmental Councils to monitor the implementation of the parties' environmental obligations under the FTA.[73] A few years later, these innovative (but one-size-fits-all) provisions on environmental protection were

66. J. Morin and M. Rochette, "Transatlantic Convergence of Preferential Trade Agreements Environmental Clauses," 19:4 *Business and Politics* (2017).
67. North American Free Trade Agreement (December 17, 1992; in force January 1, 1994).
68. Articles 24, 34, and 36 of the NAAEC.
69. H.R. 3009 (107th): Trade Act of 2002.
70. US-Singapore Free Trade Agreement (May 6, 2003; in force January 1, 2004); US-Chile Free Trade Agreement (June 6, 2003; in force January 1, 2004).
71. Article 18.8 USA-Singapore FTA and Article 19.9 USA-Chile FTA.
72. Article 18.7 USA-Singapore FTA and Article 19.6 USA-Chile FTA.
73. Article 18.4 USA-Singapore FTA and Article 19.3 USA-Chile FTA.

replicated in subsequent FTAs with Australia,[74] Morocco,[75] Bahrain,[76] Oman[77] and a multilateral trade agreement with Costa Rica, El Salvador, Guatemala, Honduras, Nicaragua and the Dominican Republic (DR-CAFTA).[78]

The EU's approach to the inclusion of environmental provisions in its FTAs with third countries differs from the USA's approach. Rather than perceiving its trade partners as competitors with whom a level playing field should be established, the EU traditionally aims to achieve greater coherence between its trade, environmental and developmental objectives.[79] Consequently, the EU has always adopted a more cooperative and light-touch approach.[80] The environmental provisions in the early generation EU FTAs had no standardized form and greatly varied in legal force, standards, priority issues and areas for environmental integration. Moreover, the environmental clauses were open-ended, focusing on environmental cooperation and dialogue rather than enforcement.[81] However, it should be noted that the EU has moved in a slightly different direction following the 2006 Renewed EU Sustainable Development Strategy.[82] Post-2006 FTAs include comprehensive and streamlined chapters fully devoted to environmental protection, with explicit references to multilateral environmental agreements (MEAs)[83] and the establishment of committees or councils to monitor the implementation of the FTA's environmental chapter.[84]

Thus, it could be argued that, although the USA and EU initially developed different approaches to environmental provisions in free trade agreements, over time, both regions have borrowed some of the environmental features from each other's FTAs. Over the years, the USA has expanded the scope of environmental-related provisions in its PTAs and started to include a selection of MEAs under the umbrella of the trade agreement as a mechanism to diffuse environmental norm abroad (rather than to create a level playing field).[85] At the same time, the EU PTAs gained depth and

74. USA-Australia Free Trade Agreement (May 18, 2004; in force January 1, 2005).
75. USA-Morocco Free Trade Agreement (June 15, 2004; in force January 1, 2006).
76. USA-Bahrain Free Trade Agreement (September 15, 2004; in force January 11, 2006).
77. USA-Oman Free Trade Agreement (January 19, 2006; in force January 1, 2009).
78. Dominican-Republic-Central America-USA Free Trade Agreement (August 5, 2004; in force March 1, 2006 (El Salvador), April 1, 2006 (Honduras and Nicaragua), July 1, 2006 (Guatemala), March 1, 2007 (Dominican Republic) and January 1, 2008 (Costa Rica).
79. J. Morin and M. Rochette, "Transatlantic Convergence of Preferential Trade Agreements Environmental Clauses," 19:4 *Business and Politics* (2017).
80. S. Jinnah and E. Morgera, "Environmental Provisions in American and EU Free Trade Agreements: A Preliminary Comparison and Research Agenda," 22:3 *RECIEL* 332-335 (2013).
81. On environmental integration in FTAs (in particular on EU association agreements), *see* G. Marin-Duran and E. Morgera, *Environmental Integration in the EU's External Relations: Beyond multilateral dimensions*, Hart Publishing, 2012, in particular Chapter 2.
82. Council of the European Union, 10117/06, Brussels June 9, 2006.
83. *See*, for example, EU-Peru/Colombia Free Trade Agreement (December 21, 2012; provisionally applied since March 1, 2013 (Peru) and August 1, 2013 (Colombia)), Article 270.2; Article 13.5 EU-Singapore FTA. On the reference to MEAs in trade agreements in more detail, *see* J. Morin and C. Bialais, "Strengthening Multilateral Environmental Governance Through Bilateral Trade Deals," *CIGI Policy Brief No. 123* (2018).
84. *See*, for example, Article 13.15 EU-Vietnam FTA.
85. J. Morin and M. Rochette, "Transatlantic Convergence of Preferential Trade Agreements Environmental Clauses," 19:4 *Business and Politics* (2017).

encompassed more enforcement mechanisms such as consultations and panel reports.[86] However, as will be shown in the next paragraph, crucial differences remain.

5.2 *Textual Analysis of Recent US and EU Trade Agreements*

In this subsection, we aim to analyze the legalization of the new generation of American and EU free trade agreements by conceptualizing the issue first and then by providing a textual analysis.

5.2.1 *The Concept of Legalization*

In order to ascertain whether FTAs can be the enforcer of climate-change obligations, one would need to analyze the wording of the FTAs and assess the legal strength of the environmental clauses. For this purpose, the present paragraph will measure the degree of legalization of environmental and climate-related provisions in the new generation of American and EU FTAs.[87] Legalization is defined along three dimensions:

(1) obligation;
(2) precision; and
(3) delegation.

Obligation means that states are legally bound by a rule or set of rules (and thus relates to the strength of the commitment made). Precision refers to rules that unambiguously define the conduct they require, authorize or proscribe (and therefore narrows down the possible interpretations of a rule). Finally, delegation relates to the power of third parties to implement, interpret and apply the rules and to resolve disputes.[88]

Of course, the world is not just black and white. Each of the dimensions of legalization is a matter of degree and gradation, ranging from the weakest form (the absence of any legal obligation, precision or delegation) to the strongest form (a detailed and binding commitment that is fully enforceable in court or arbitration).[89] Moreover, one can plot where a particular arrangement of environmental provisions falls on all three dimensions of legalization.[90] If a provision scores high on all three

86. On this convergence of American and EU FTAs, *see* J. Morin and M. Rochette, "Transatlantic Convergence of Preferential Trade Agreements Environmental Clauses," 19:4 *Business and Politics* (2017).
87. On legalization, *see* K. Abbot et. al., "The Concept of Legalization," 54:3 *International Organization* 401-419 (2000). For a broader analysis of the regulatory contribution that FTAs make to global climate governance (including aspects of legalization), *see* J. Morin and S. Jinnah, "The Untapped Potential of Preferential Trade Agreements for Climate Governance," 27:3 *Environmental Politics* (2018).
88. K. Abbot et. al., "The Concept of Legalization," 54:3 *International Organization* 401-402 (2000).
89. *See* Figure A1 in Annex 1 of this chapter: *The Dimensions of Legalization.*
90. K. Abbot et. al., "The Concept of Legalization," 54:3 *International Organization* 404-408 (2000).

dimensions, the particular clause is classified as highly legalized and has the potential to become instrumental to climate change mitigation.[91]

5.2.2 Textual Analysis

Based on the legalization concept described above, in this section, we identify the level of obligation, precision, and delegation of the main environmental and climate-related provisions in the recently concluded or negotiated American and EU FTAs. All legalization dimensions are measured along a three-degree continuum (low, moderate or high).[92] Levels of obligation range from not binding (a mere recommendation to consider certain environmental issues) to legally binding commitments (generally preceded by the words "shall" or "must"). Precision ranges from general environmental references (such as the mere acknowledgement of a given MEA) to specific and detailed targets (for example, the commitment to act in accordance with certain environmental rules or agreements). Finally, delegation is classified from a political mechanism of enforcement (such as government consultations) to judicial enforcement by an independent court or tribunal that renders a final and binding decision.[93] Based on the level of legalization of their major environmental and climate-related provisions, each of the fifteen analyzed FTAs has received a score between 0 and 6 (with a maximum of 3 points for each of the three legalization dimensions).[94] The findings of the textual analysis are outlined in Annex 3 of this chapter.

From Annex 3, it is immediately apparent that FTAs concluded by the USA receive a higher total score than EU FTAs (and are thus classified as more legalized). The reasons for this higher degree in legalization will be explained in subparagraphs (i) and (ii) below.

(i) American FTAs

The FTAs concluded by the USA generally contain binding obligations (such as the commitment that parties "shall adopt, maintain and implement" laws in order to fulfil their obligations under certain MEAs[95]). However, the same binding obligations have a limited scope since they only relate to MEAs included in the list of covered

91. *See also* J. Morin and S. Jinnah, "The Untapped Potential of Preferential Trade Agreements for Climate Governance," 27:3 *Environmental Politics* 551 (2018).
92. Inspiration is derived from the analysis of J. Morin and S. Jinnah in J. Morin and S. Jinnah, "The Untapped Potential of Preferential Trade Agreements for Climate Governance," 27:3 *Environmental Politics* (2018).
93. *See* Annex 2 of this chapter.
94. It should be noted that the Transatlantic Trade Investment Partnership (TTIP) between the EU and the USA has deliberately been excluded from this part of this research, since negotiations of this Agreement had been suspended and have only been resumed as of 2019. The available (draft) documents most likely do not represent the parties' current position on environmental provisions in TTIP. On the potential enforceability of environmental provisions under TTIP, *see* R. Leal-Arcas and E. Alvarez Armas, "The Climate-Energy-Trade Nexus in EU External Relations," in: S. Minas and V. Ntousas (eds.), *EU Climate Diplomacy: Politics, Technology and Networks*. Routledge, 2018.
95. *See*, for example, Article 18.2 USA-Peru FTA and Article 17.2 USA-Panama FTA.

agreements, annexed to the FTA.[96] These annexes include only seven international agreements that are focused on specific issue areas, such as the International Convention for the Regulation of Whaling 1946. Moreover, when it comes to international environmental agreements not covered by the annex, the commitment is much weaker and merely provides that parties "shall continue to seek means to enhance the mutual supportiveness"[97] of these agreements. As a result of the above, the environmental obligations contained in the American FTAs are classified as "moderate."

American FTAs score high on precision, because the reference to a closed list of covered international agreements (albeit limited in scope) provides great certainty and the agreements clearly define the state's conduct that is required. Furthermore, the FTAs include additional references to environmental rules that the signatory states have to comply with. For example, Annex 18.3.4 (Forest Sector Governance) of the USA-Peru FTA contains an eight-page list of specific and detailed provisions that Peru must implement in its own national laws. In addition, the recent UMSCA agreement is particularly precise, with the incorporation of various specific and detailed targets. For example, art. 24.9 provides that each party shall take measures to control the production and trade in substances that can significantly deplete and otherwise modify the ozone layer. Footnote 6 supplements this provision in more detail by regulating when a party shall be deemed in compliance with this provision.[98]

The FTAs concluded by the USA with Peru,[99] Colombia,[100] Panama[101] and South Korea[102] score high on the delegation dimension. This generation of American FTAs, concluded after the Democrats regained control over the House and the Senate in 2006, links the FTA's environmental provision to the formal dispute settlement mechanism under the FTA, which makes the full environmental chapter directly enforceable. In case of an alleged breach of environmental commitments under the FTA, a Panel established under the dispute resolution chapter can render a binding report.[103] Failure to implement the determinations of the panel gives the complaining party the right to suspend the application to the other party of benefits under the FTA.[104] The utilization of the sanction-based dispute settlement regime in order to address environmental protection led some authors to label the USA approach as "confrontational."[105]

96. *See*, for example, Article 18.3 USA-Peru FTA.
97. *See*, for example, Article 17.13 USA-Panama FTA. Note that Article 28.8 of UMSCA departs from earlier texts of American FTAs, by stating that parties "shall continue to seek means to enhance the mutual supportiveness of multilateral environmental agreements to which they are all party."
98. *See*, in similar vein, the detailed commitments with respect to sustainable fisheries management in Article 24.10 UMSCA.
99. USA-Peru Trade Promotion Agreement (April 12, 2006; in force February 1, 2009). *See* on the USA-Peru FTA and its implications: S. Jinnah, "Strategic Linkages: The Evolving Role of Trade Agreements in Global Environmental Governance," 20:2 *Journal of Environment & Development* (2011).
100. USA-Colombia FTA (November 22, 2006; in force May 14, 2012).
101. USA-Panama TPA (June 28, 2007; in force October 31, 2012).
102. USA-Korea FTA (June 30, 2007; in force March 15, 2012).
103. *See*, for example, Article 30.13 USA-Panama FTA.
104. *See*, for example, Article 22.13 USA-Korea FTA.
105. S. Jinnah and E. Morgera, "Environmental Provisions in American and EU Free Trade Agreements: A Preliminary Comparison and Research Agenda," 22:3 *RECIEL*335 (2013).

(ii) European Union FTAs

Turning to the European Union's FTAs, a distinction should be made between the agreements pre-2018 and those agreements recently concluded or currently being negotiated.

The pre-2018 agreements with South Korea, Colombia, Peru, Moldova, Georgia, Ukraine and Canada predominantly contain vague and ambiguous language on environmental protection (such as the commitment to "reach the ultimate objective" of the UNFCC and the Kyoto Protocol,[106] or the obligation to "address global environmental challenges"[107]). Moreover, the FTAs generally lack precision because they merely refer to general environmental issues without specifying instruments that states have to implement in their national law or targets that have to be reached.[108] The provisions are, therefore, unable to result in measurable environmental objectives. On the other hand, this generation of European FTAs also contains some stronger and more precise obligations. For example, the provisions in which the parties "reaffirm their commitments to the effective implementation"[109] of the MEAs to which they are party might come across as weak at first. However, they are in fact binding obligations to implement environmental instruments because a failure to effectively implement the MEAs might constitute a breach of this environmental provision in the FTA.[110] As a result of the mixed nature of the commitments undertaken by the parties, the level of obligation and precision under this generation of FTAs is classified as "moderate."

The European Union FTAs concluded in or after 2018 contain binding[111] and precise[112] obligations. Notably, and in contrast to previous EU FTAs, art. 12.6 of the EU-Singapore FTA provides that the parties "shall effectively implement" in their laws the MEAs to which they are party.[113] Even more ambitious is Article 5.1 of the

106. *See* Article 13.5(3) EU-South Korea FTA.
107. *See* Article 267(2) EU-Colombia/Peru FTA.
108. With the exception of the EU-Colombia/Peru FTA, which provides in Article 270(2) a detailed list of international agreements that states have to effectively implement in their national laws (including, for example, the Montreal Protocol on Substances that Deplete the Ozone Layer 1987.
109. *See* Article 13.5(2) EU-South Korea FTA; Article 270 EU-Colombia/Peru FTA; Article 24.4(2) CETA.
110. *See* R. Leal-Arcas and E. Alvarez Armas, "The Climate-Energy-Trade Nexus in EU External Relations," in: S. Minas and V. Ntousas (eds.), *EU Climate Diplomacy: Politics, Technology and Networks*. Routledge, 2018.
111. For example, Article 16.6(2) EU-Japan EPA provides that "parties *shall implement* effective measures to combat illegal trade." Article 13.9(2) EU-Vietnam FTA provides that "each party *shall comply* with long-term conservation and management measures."
112. For example, Article 8(3)(1) EU-Mexico (draft) Agreement provides that each party shall implement long-term conservation and management measures "as defined in the main UN and FAO instruments to these issues." Article 18.8(d) EU-Singapore FTA provides that parties will "uphold the principles of the FAO Agreement to Promote Compliance with International Conservation and Management Measures by Fishing Vessels on the High Seas."
113. In previous European FTAs, the parties reaffirmed their commitment to effectively implement the MEAs to which they are party. The commitment under the EU-Singapore FTA that a party "shall effectively implement" MEAs is, of course, much stronger and leaves no room for ambiguity. It should be noted that with respect to major environmental agreements such as UNFCCC and its Kyoto Protocol, the EU-Singapore FTA contains less strong language and parties only "reaffirm" their commitment to the ultimate objectives of these agreements.

EU-Mexico (draft) Agreement, which provides that each party "shall effectively implement" the UNFCC and the Paris Agreement.[114] This (draft) provision contains an unparalleled level of protection since a state's failure to implement the Paris Agreement would constitute a direct breach of art. 5.1 EU-Mexico (draft) Agreement. In addition, the post-2018 European FTAs all contain various precise obligations to act in accordance with specific environmental agreements, rules or protocols. For example, the EU FTAs with Japan, Vietnam and Mexico all include provisions on biological diversity pursuant to with the signatory states shall comply with specific agreements relating to the issue (such as the Convention on International Trade in Endangered Species of Wild Fauna and Flora).[115] As a consequence of the above, the EU FTAs with Japan, Singapore Vietnam and Mexico score high on the obligation and precision dimensions.

However, both the pre-2018 and the recently concluded or negotiated European FTAs score low on delegation because all of them exclude the environmental chapter from the general dispute settlement mechanism of the FTA.[116] The environmental and climate change-related provisions, no matter how strong and precise they are, can therefore not be enforced. Instead, disputes relating to the environmental chapter of the FTA can only be resolved by nonbinding consultations or panel reports.[117]

5.3 *Conclusion and Way Forward*

It would seem that, in climate change, sanctions are in the political sphere not in the legal sphere. That said, the American FTAs analyzed in this chapter contain strong, precise, and enforceable environmental provisions. In theory, the FTAs concluded by the USA could thus be an effective instrument to enforce environmental and climate change obligations. The problem, however, is that most binding commitments under American FTAs only apply to the international agreements included in the list of covered agreements. Unfortunately, this list is rather limited and does not include major climate change agreements such as the UNFCCC, its Kyoto Protocol, and the Paris Agreement on Climate Change.[118] As a result of this restrictive scope, the practical impact of these FTAs on climate change mitigation will be limited.

European FTAs, on the other hand, have gained depth and the most recent agreements with Japan, Singapore, Vietnam, and Mexico contain a wide array of

114. *See also* Article 13.6(1) EU-Vietnam FTA, which provides that the parties reaffirm their commitment to reaching the ultimate objective of the UNFCC, its Kyoto Protocol and the Paris Agreement.
115. *See* Article 16.6 EU-Japan EPA; Article 13.7 EU-Vietnam FTA; Article 6 EU-Mexico (draft) Global Agreement.
116. *See* Article 13.16 EU-South Korea FTA; Article 285(5) EU-Colombia/Peru FTA; Article 12.16(1) EU-Japan EPA; Article 12.16(1) EU-Singapore FTA; Article 12.16(1) EU-Vietnam FTA and Article 15 EU-Mexico (draft) Global Agreement.
117. *See*, for example, Article 12.17 of the EU-Singapore FTA and Article 13.17 of the EU-Vietnam FTA.
118. *See* for instance Chapter Nineteen (Environment) of the US-Australia FTA, where there is no mention of climate change. Available at https://dfat.gov.au/about-us/publications/trade-investment/australia-united-states-free-trade-agreement/Pages/chapter-nineteen-environment.aspx.

binding and precise environment-related issues. Notably, the climate change provision of the EU-Mexico (draft) Global Agreement, pursuant to which the signatory states "shall effectively implement" the UNFCC and the Paris Agreement,[119] might constitute the most far-reaching environmental provision found in any of the analyzed FTAs. However, the other side of the coin is that European FTAs do not include an effective and sanctions-based dispute settlement mechanism that extends to the environmental chapter of their trade agreements. It would seem that the EU does not favor litigation as a way to solve environmental protection-related trade disputes.

The positions of the European Commission and the EU Member States regarding issues covered by trade and sustainable development (TSD) chapters in FTAs, including climate change, have been evolving.[120] France, Spain, and Luxembourg outlined a proposal at the EU Environment Council in March 2019 to include commitments related to the implementation of the Paris Agreement on Climate Change[121] into essential FTA clauses, which means that the EU could suspend an agreement in case these obligations are not met by the partner countries.[122] Such a proposal raises an interesting question: how could the EU (and, more precisely, the European Commission) make an evaluation of other countries' delivery of their obligations without interfering with competences that belong to another international organization (i.e., the UNFCCC and its bodies, in this case)?

In a nutshell, US FTAs offer enforceability of environmental protection provisions via sanctions, whereas EU FTAs make use of international cooperation/dialogue and other soft power mechanisms. Regarding the origins of the EU approach, a decision was made to choose a broad scope and high level of ambition of TSD chapters in FTAs, inclusion of civil society monitoring mechanism, and putting an emphasis on dialogue and cooperation, rather than sanctions or financial penalties (which is the approach of the US and Canada). For example, in the area of labor rights, TSD chapters in EU FTAs commit the parties to ratify and to effectively implement eight fundamental conventions of the International Labor Organization (ILO),[123] while the US FTAs speak only

119. Article 5.1 of the EU-Mexico (draft) Global Agreement.
120. In the context of the ongoing EU negotiations with Chile on the modernization of the trade pillar of their Association Agreement, a report on an impact analysis recommended the inclusion of strong provisions related to climate change and to meeting their obligations under the Paris Agreement on Climate Change in the TSD chapter. *See* http://trade-sia-chile.eu/images/reports/EU-Chile_SIA_draft_final_report_2019-03.pdf, pp. 138 ff. *See also* the EU's proposal for a TSD chapter in the EU-Australia FTA, Articles X.5 (on trade and climate change) and X.13-X.15 (on dispute settlement mechanism), available at http://trade.ec.europa.eu/doclib/docs/2019/april/tradoc_157865.pdf. Equally, in 2018, the European Commission published texts of the trade part of the "agreement in principle" reached between the EU and Mexico for a modernized EU-Mexico Global Agreement. *See* the text of the TSD chapter, Article 5 (on trade and climate change) and Articles 15-18 (on dispute settlement mechanism), available at http://trade.ec.europa.eu/doclib/docs/2018/april/tradoc_156822.pdf.
121. One should add that the credibility of the Paris Agreement would be questionable if the US will withdraw and if countries will not stand up to their pledges.
122. Council of the European Union, "Strengthening coherence between EU free trade agreements and the Paris Agreement on climate change," March 1, 2019, 7016/19, available at http://data.consilium.europa.eu/doc/document/ST-7016-2019-INIT/en/pdf.
123. In the past, the EU has tested an approach similar to that of the US in the application of pre-ratification conditionality regarding a TSD-related area. For instance, in the negotiations

about domestic enforcement. If we take an example from climate change, when the Republic of Korea wanted, and indeed needed, to improve its record on GHG emissions reduction, the EU offered the possibility to discuss lessons learned from its own climate change policy and operation of the Emission Trading Scheme (ETS). That discussion took place during an annual meeting under the EU-Korea FTA TSD chapter in December 2014 and was followed by an assistance project launched in 2016.[124]

It is also pertinent to note that TSD chapters in FTAs have their own dispute settlement mechanism, which can be activated if necessary. It envisages launching government consultations, followed by setting up a panel of experts (arbitrators) if the Parties cannot solve the problem on their own. A case in point is to be found in the EU-Korea FTA, where the Republic of Korea failed to meet its commitments under the TSD chapter[125] in the labor-related part (namely ratification and effective implementation of the ILO fundamental conventions). After several attempts of dialogue and cooperation to encourage progress in this area, the EU's patience came to the end and, in December 2018, in accordance with Article 13.14 of the EU-Korea FTA,[126] the European Commission launched government consultations[127] with Korea as a first step in the dispute settlement mechanism under the TSD chapter of the EU-Korea FTA.[128] Even if the case relates to labor right, not climate change, it is an interesting test case for the enforceability of TSD provisions in FTAs.[129]

with Georgia for an Association Agreement, the European Commission clearly stated that the negotiations would not be concluded, and therefore the Agreement would not enter into force, if Georgia did not change its Labor Code, which, at the time, was not in compliance with Georgia's commitment under the GSP + scheme on effective implementation of ILO fundamental conventions. Georgia was very reluctant to change its Labor Code at the beginning, but, given the incentive of getting into a preferential trade regime (as part of the Association Agreement) with its biggest trading partner (i.e., the EU), eventually it followed the EU's suggestions. The Labour Code was amended and negotiations for an Association Agreement were concluded. Even if this example refers to labor rights, not climate change, it demonstrates that there is space for the enforceability of TSD-related aspects of EU trade relations and trade agreements with third parties.

124. https://eeas.europa.eu/headquarters/headquarters-homepage/14138/eu-korea-emissions-trading-system-cooperation-project-launches-series-activities-benefit_en.
125. *See* Chapter 13, EU-Korea FTA.
126. Article 13.14 of the EU-Korea FTA deals with Government consultations to reach mutually satisfactory resolutions of matters of dispute.
127. http://trade.ec.europa.eu/doclib/press/index.cfm?id = 1961.
128. http://trade.ec.europa.eu/doclib/docs/2018/december/tradoc_157586.pdf. Under the EU approach to dispute settlement, there is no need to prove the link between the failure of a partner country to meet their TSD commitments and the impact on trade between the Parties. This is an important difference between the EU and US approach and, indeed, an EU advantage in enforceability. In the case of labor rights, the only case that has reached arbitration (US v. Guatemala) was not successful for the US because the panel was not able to establish a clear link between Guatemala's poor record on labor rights and trade between the Parties.
129. In February 2018, in a dedicated non-paper, the European Commission outlined its commitment to improve the overall implementation and enforcement of TSD chapters in EU FTAs to better meet expectations. This commitment relates to both labor right and environmental protection, including climate change. *See* http://trade.ec.europa.eu/doclib/docs/2018/february/tradoc_156618.pdf.

6 Lessons Learned from CETA and USMCA: How Can the TTIP Be an Improvement on 21st Century PTAs

When comparing the PTAs agreed between the 1990s and the beginning of the 21st century, it is apparent that the 1990s provisions were vague and nonenforceable. However, this changed in the beginning of the 21st century, where climate provisions within PTAs became more precise; for example, EU-agreements language is more precise and highly binding, with a higher degree of legality than usual.[130] That said, such agreements favor dispute resolution by consultation. In contrast, US agreements tend to include dispute settlement mechanisms with legally binding decisions and sanctions-based enforcement provisions, although with weaker levels of obligation and precision than the EU's.[131] It is argued that a "trade-off" in provisions is either highly enforceable or precise with high levels of obligation.[132]

We also note that CETA offers various dispute settlement mechanisms in different chapters of the Agreement. This shows that there is no universal dispute settlement system in CETA.

6.1 CETA's Climate-Related Provisions and Level of Obligations

CETA is a comprehensive trade and investment agreement between the EU and its member states, on the one hand, and Canada, on the other, which is based on the foundations and values shared between them. Both the EU and Canada have pledged their ambition and commitment to uphold the Paris Agreement and to fight climate change impacts, therefore CETA has included two separate chapters on climate change: one on sustainable development (SD) and trade[133] and one of trade and the environment.[134]

The agreement made a clear link in the SD chapter between the economic growth, social development and environmental protection. It encourages the promotion of trade and economic flow that enhances environmental and labor protection framework.[135] The language used in this chapter is precise; however, it lacks obligation and enforcement, as it uses arguably only voluntary commitments. It starts with obligatory language: "[...] accordingly, each Party shall strive to promote trade and economic flows [...] enhancing [...]environmental protection [...]." However, it continues by saying that Parties should encourage "development and use of voluntary schemes relating to the sustainable production of goods and services[...]."[136] Although provisions lack in obligation, it does specify that Parties can use eco-labelling schemes as a way to promote sustainable practices. The SD chapter has less force than the labor

130. J.F. Morin and S. Jinnah, "The Untapped Potential of Preferential Trade Agreements for Climate Governance," 27:3 *Environmental Politics* 541-565 (2018).
131. *Ibid.*
132. *Ibid.*
133. CETA, Chapter 22.
134. *Ibid.* Chapter 24.
135. *Ibid.* Article 22.2.
136. *Ibid.* Article 22.3.2(a).

chapter, as it provides for a dispute mechanism by consultation, instead of dispute resolution.[137]

The chapter on the environment commits the parties to putting into practice MEAs, while protecting each party's right to regulate on environmental matters, with each side enforcing domestic environmental laws and preventing the relaxation of any of those laws.[138] This chapter is substantial in enforcing climate change obligations as it promotes the trade and investment on environmental goods and services, including the reduction of non-tariff barriers.[139] This is significant as it specifies the promotion of environmental trade and removal of barriers, but it only asks for the reduction and not the removal of said barriers. However, there is an obligation to facilitate the removal of these barriers, and it specifically refers to the importance of climate mitigation and, more specifically, it promotes renewable energy.[140] In addition, the chapter encourages the conservation and sustainable management of forests[141] and fisheries[142] and ensures participation of non-governmental groups. When it comes to dispute mechanisms, it provides for consultation on matters arising from this chapter[143] and dispute resolution (DR).[144] Therefore, CETA has strong binding obligations in relation to trade and the environment, which upholds both parties commitments under their national and international environmental obligations.

CETA set up a Committee on Trade and Sustainable Development,[145] which supervises the matters under Chapters 22, 23, and 24 and their implementation.[146] It has obligations to promote transparency and public participation[147] and must make reports public unless it decides otherwise.[148]

This agreement is a good example of the EU enhancing its environmental commitments in their recent agreements, and such a method can be used in relation to the TTIP in order to further enhance the enforcement of climate-related provisions within international trade law. Since the EU is a global leader in climate change governance through climate change agreements, as it includes climate provisions more

137. *See* Article 22.4 for SD "Civil Society Forum" consultative mechanism meetings; Article 23.11 for dispute resolution. The SD chapter does not mention any forms of dispute resolution.
138. CETA, Articles 24.2 to 24.5.
139. *Ibid.* Article 24.9.1.
140. *Ibid.* Article 14.9.2. This provision carries a high level of obligation as the wording of "shall"; This is arguably dismissed by the fact that it follows with "consistent with their international obligations." However, it could be argued that climate mitigation is an international obligation under the UNFCCC and such promotion of environmental goods and services falls within such obligations.
141. *Ibid.* Article 22.10.
142. *Ibid.* Article 22.11.
143. *Ibid.* Article 24.14.
144. *Ibid.* Article 24.16, in accordance with Chapter 29 and rule 42 of the Rules of Procedures for Arbitration set out in Annex 29-A.
145. *Ibid.* Article 26.2.1(g).
146. *Ibid.* Article 22.4.1.
147. *Ibid.* Article 22.4.4.
148. *Ibid.* Article 22.4.4 (a). This may hinder transparency, as they can decide not to make such reports of implementation public.

frequently with a high variety of provisions,[149] other countries may follow their lead and replicate such provisions in their future PTA negotiations.

6.2 USMCA's Environment Provisions

This part will briefly examine provisions related to environmental protection in USMCA and the correspondent dispute mechanisms.

Chapter 24 of the USMCA concerns environmental protection and establishes an obligation on the Parties to improve their levels of environmental protection but allows parties to establish and regulate their national legislation.[150] When analyzing the scope and objectives of the agreement, the language is aspirational and has a passive voice, as it uses terms such as "recognizes," "taking into account," "promote," etc. However, the Agreement establishes binding obligations in the enforcement of environmental laws[151] and prohibits the relaxation of environmental laws for the encouragement of trade and investment.[152] The agreement also obliges parties to promote public awareness and participation in environmental matters.[153]

The USMCA goes beyond the climate-change protection in CETA, as it includes indirect environmental protection under the Environmental chapter such as the protection of the marine environment from ship pollution,[154] the protection of the Ozone layer,[155] promotes the improvement of air quality,[156] a chapter on trade and diversity[157] and protection against marine litter,[158] among others. However, most of the provisions still have weak content in comparison when establishing recognition and cooperation, with no high degree of obligation as the language is unspecific.[159]

It is important to note that the dispute mechanism under USMCA is binding and provides for dispute alternatives. Parties should try to agree on the interpretation and application of the environmental chapter through dialogue, consultation, exchange of information and cooperation.[160] If the initial consultation fails, there are provisions for a Senior Representative consultation[161] and, if still unresolved, a Ministerial Consultation.[162] Only if the consulting parties fail to resolve the matter under any of the three consultations, can they request a meeting with the designated Commission to try to

149. *See* generally J. Morin and S. Jinnah, "The Untapped Potential of Preferential Trade Agreements for Climate Governance," 27:3 *Environmental Politics* (2018).
150. USMCA Article 24.3.
151. *Ibid.* Article 24.4.1.
152. *Ibid.* Article 24.4.3.
153. *Ibid.* Article 24.5.
154. *Ibid.* Article 24.10.
155. *Ibid.* Article 24.9.
156. *Ibid.* Article 24.11.
157. *Ibid.* Article 24.15.
158. *Ibid.* Article 24.12.
159. *Ibid.* For example, Article 24.12.3 only recognizes that Parties should act to address marine litter through cooperation.
160. *Ibid.* Article 24.29.
161. *Ibid.* Article 24.30.
162. *Ibid.* Article 24.31.

mediate or conciliate,[163] and thereafter request the establishment of a Panel of Experts.[164] It is also interesting to note that, before the ratification of USMCA by US Congress, Democrats in the House of Representatives were asking for tougher protections for the environment.

6.3 *TTIP: What Should the Agreement Include to Better Enforce Climate-Related Provisions*

As Polanco *et al* (2017) argue, PTAs would be more effective than MEAs, as they only involve the negotiation between a limited amount of countries, who will deliberate in a multitude of issues and agree on new commitments.[165] PTA disputes can be resolved within the WTO's dispute settlement procedure through sanction-based mechanism with the potential of enhancing compliance between parties. The U.S. is an example of dispute settlement mechanisms that offer strong incentives of compliance through sanction-based mechanisms.

Countries should use this method when creating new PTAs with climate change related provisions. Most recent PTAs have included climate-related provisions; however, most lack substantive content and enforcement, the TTIP could improve on this.

Currently, the U.S. and the EU are negotiating a new PTA, the TTIP, which is a significant trade deal due to them being the second and third largest GHG polluters, and such trade deal amounts for 50% of global trade.[166] The initial proposal paper on the TTIP on Trade and SD[167] includes the proposition of a chapter covering aspects of climate change and SD. These chapters can be included to prevent a "race to the bottom" on environmental provisions. This would promote the cooperation on trade-related issues of SD.

It is important to note that climate-related provisions do not necessarily have to be drafted in their own separate chapter and can be included in all chapters. For example, there could be a provision for tariff differentiation based on the carbon footprint of products, which would be made as part of the tariff schedules included in a chapter for trade in goods. The TTIP can liberalize trade in environmental goods and services of any kind, as opposed to "like" services that would not be produced in an environmental manner. Polanco argues this would be a breakthrough and would play a positive role in the environment, as the impact of such products is positive to climate change. This measure is cheaper to implement and therefore likely to happen, and such would lead to almost immediate environmental benefits. An argument could be made that preferential treatment of service suppliers that meet environmentally relevant

163. *Ibid.* Article 31.5.
164. *Ibid.* Article 31.6.
165. R. Polanco et al., "TTIP and Climate Change: How Real Are Race to the Bottom Concerns," 3 *CCLR* 206-222 (2017).
166. Rafael Leal-Arcas, "Climate Change Mitigation from the Bottom Up: Using Preferential Trade Agreements to Promote Climate Change Mitigation," *CCLR* 34-42 (2013), at 41.
167. European Commission, "EU-US Transatlantic Trade and Investment Partnership: Trade and Sustainable Development. Initial EU Position Paper" (2013).

criteria such as emission standards would improve the overall environmental performance.

The TTIP should include the ratification of MEAs and other environmental instruments, to ensure global and transboundary environmental challenges are being addressed, especially in relation to climate change combatting. The agreement should also include chapters on trade and SD and trade and environment, such as in CETA. This is a common practice not only in the EU, but also in the U.S., as the country is directly mandated by Congress under the Trade Act 2002 to include such provisions with legally binding and enforceable commitments. Therefore, to enhance the already ongoing efforts by the EU and the US to globalize enforcement of climate-change obligations though trade, the TTIP would further enhance the precision and ambition on its provisions, introduce specific commitments to include the promotion of renewable energy and environmentally friendly goods and services, such as a possible carbon taxes on goods which exceed the limit on emission levels during production. Finally, it should include a strong enforcement mechanism, comparable to the one found in the USMCA. The TTIP can replicate good practices of environmental and climate-change provisions.

Research shows that countries that include more climate provisions in their FTAs tend to be low GHG emitters.[168] For trade to be an enforcer of climate change obligations successfully, future PTAs should be negotiated with oil-exporting countries. For example, Norway includes fewer provisions in its PTAs; however, this attitude may change soon, as Norway has recently diverted oil funds to subsidize renewables, which marks a shift from their oil exploration to the expansion of clean-technology markets.[169] This move of transferring even a small percentage (i.e., 10%) of fossil-fuel subsidies to the renewables industry could be the catalyzer to pay for a global transition to clean energy.[170] In fact, ending the fossil-fuel subsidies "would cut global emissions by about a quarter, the IMF estimates, and halve the number of early deaths from fossil fuel air pollution."[171]

Let us now turn to the contribution of bilateral investment treaties to sustainability.

168. *See* generally J. Morin and S. Jinnah, "The Untapped Potential of Preferential Trade Agreements for Climate Governance," 27:3 *Environmental Politics* (2018).
169. D. Carrington, "'Historic Breakthrough': Norway's Giant Oil Fund Dives into Renewables," *The Guardian* online, April 5, 2019.
170. *See* D. Carrington, "Just 10% of Fossil Fuel Subsidy Cash 'Could Pay for Green Transition'," *The Guardian*, August 1, 2019, available at https://www.theguardian.com/environment/2019/aug/01/fossil-fuel-subsidy-cash-pay-green-energy-transition.
171. *Ibid.*

III HOW CAN BILATERAL INVESTMENT TREATIES CONTRIBUTE TO CLIMATE ACTION AND SUSTAINABLE ENERGY?

1 Introduction

Bilateral investment treaties (BITs) are treaties entered into by two contracting states with the aim to protect and promote investment. These treaties typically include substantive protections for the investors and investments from one contracting party in the territory of another contracting party, such as protection against unlawful expropriation or unfair and inequitable treatment, as well as a dispute resolution mechanism which allows investors to bring claims regarding their investment against the host state in international arbitration outside national courts, subject to specific conditions and potential limitations.[172] Arbitration forums which may be referred to in BITs differ, from International Centre for Settlement of Investment Disputes (ICSID) through arbitral institutions, such as the Stockholm Chamber of Commerce or International Chamber of Commerce, to ad hoc arbitration under UNCITRAL Rules.[173] Statistics show that, as of 2017, a total of 2,946 BITs had been signed.[174]

Compared to the older BITs with emphasis on the protection of foreign investment by providing substantive protection standards and allowing investors to bring a claim against host states through investor-state dispute settlement mechanism, the recent developments in newer BITs show that there is a positive shift toward explicit inclusion of sustainable development issues into their wording (e.g. states' right to adopt sustainable development-oriented regulation).[175] It is clear from the wording of new BITs that reference to sustainable development issues may appear in different parts of a BIT, e.g., as a general objective of the treaty in a preamble, as a separate obligation of the state parties or investors, or as an exception to the substantive protection standards, such as prohibition of unlawful indirect expropriation. Specific position and wording of sustainable development provisions determine to what extent such provisions will be binding and enforceable against state parties. Certain BITs contain only a broad reference to sustainable development without clear practical consequences, while others list sustainable development obligations that may be enforced in practice, e.g., by means of state-to-state dispute settlement mechanism or, if such obligations are imposed on foreign investors, as a counterclaim brought by a host state against an investor. Since this chapter emphasizes climate action and sustainable energy, we will focus on the protection of the environment as one of the key aspects of sustainable development.

This section starts with the question of compatibility of traditional objectives of BITs and environmental protection goals, before addressing three potential solutions

172. Gary Born, *International Arbitration: Law and Practice*, 2nd edn. Wolters Kluwer, 2016, 421-422; Rafael Leal-Arcas, *International Trade and Investment Law: Multilateral, Regional and Bilateral Governance*, Edward Elgar, 2010.
173. Gary Born, *International Arbitration: Law and Practice*, 2nd edn. Wolters Kluwer, 2016, 422.
174. UNCTAD, "Recent developments in the international investment regime" (UNCTAD, May 1, 2018) https://unctad.org/en/PublicationsLibrary/diaepcbinf2018d1_en.pdf, p. 2.
175. *Ibid.* 5.

that might increase BITs' contribution to climate action and environmental protection, namely (1) imposition of environmental obligation on state parties or investors,[176] (2) freedom of state parties to adopt environmental regulation and exemption of environmental measures from BITs' substantive protection standards, and (3) counterclaims brought by host states for violation of environmental protection standards.

2 Are BITs Reconcilable with Environmental Protection Goals?

The main objectives of BITs may be traditionally found in their preamble. Most BITs focus on three main objectives, namely desire to increase economic cooperation between state parties, stimulation of economic growth and creation of favorable environment for investment.[177] Looking at these objectives, one may argue that they are not necessarily compatible with the environmental protection. However, despite existing challenges, recent state practice and pertaining case law of investment tribunals show that the existing system of BITs may be adapted in order to accommodate environmental protection goals.[178]

Although far from being common, some recent BITs provide for additional objective to the above-mentioned traditional ones in their preamble, namely concept of sustainable development.[179] This language may be found, e.g., in 2018 Argentina-Japan BIT or in 2018 Netherlands Model BIT. The preamble of an international treaty does not necessarily impose enforceable obligations on states, but it has an important role in interpretation of the treaty, in accordance with Article 31(2) of the Vienna Convention on the Law of Treaties.[180] This rather recent development shows that states view their BITs as a chance to promote more environmentally friendly investment framework that expands the traditional perception of the investment protection as being strictly focused on economic development and creation of more favorable conditions for foreign investment. Furthermore, this development also contributes to the balancing of the inequality of rights and obligations between investors and host states under BITs, with host states typically being the sole bearers of obligations and

176. It is interesting to see the difference in expectations between shareholders of big oil companies on both sides of the Atlantic. In 2018, shareholders of Royal Dutch Shell persuaded Royal Dutch Shell to pledge the reduction of GHG emissions from its operations and products. What is more, in 2019, BP's shareholders voted to require BP to disclose how its plans match with the goals of the Paris Agreement on Climate Change. By contrast, shareholders of ExxonMobil and Chevron (both big US oil companies) have voted against any sort of climate change resolutions. *See* The Economist, "Big oil and climate change: Back to the well," June 1, 2019, p. 59. In fact, it has been reported that ExxonMobil intends to increase oil and gas production by 25% between 2017 and 2025 (*see* The Economist, "The Seven Ages of Climate Man," May 25, 2019, p. 70) and that the 24 biggest listed oil companies only invested 1.3% of capital expenditure in 2018 in renewable technologies (*see ibid.*).
177. Marie-Claire Cordonier Segger and Andrew Newcombe, "An Integrated Agenda for Sustainable Development in International Investment Law," in Marie-Claire Cordonier Segger et al. (eds.), *Sustainable Development in World Investment Law*, Global Trade Law Series, Volume 30. Kluwer Law International, 2011, p. 126.
178. *Ibid.* 124.
179. *Ibid.* 125.
180. Vienna Convention on the Law of Treaties (adopted May 23, 1969, entered into force January 27, 1980) 1155 UNTS 331 (VCLT) Article 31(2).

investors the sole bearers of rights, by imposing certain obligations also on investors,[181] some of them directly enforceable under the BITs, as discussed below in subsection 3.

One of the ways in which BITs reconcile their object and purpose with the environmental-protection goals is by including specific provisions on conflict with other international treaties in their wording.[182] Such provision may be found, e.g., in 2016 India Model BIT or in 2018 Netherlands Model BIT, and it solves potential conflicts that may arise between the BIT and other international treaties to which a state may be a party. Such other international treaties may include *inter alia* environmental protection agreements or human rights conventions. Article 6(5) of the 2018 Netherlands Model BIT may serve as an example of such conflict provision. This article states that "*[w]ithin the scope and application of this Agreement, the Contracting Parties reaffirm their obligations under the multilateral agreements in the field of environmental protection [...] to which they are party, such as the Paris Agreement[...].*" Some other international treaties that include investment chapters, such as NAFTA, go even further and provide a list of environmental agreements that prevail over the wording of NAFTA, with the option to add other environmental agreements on the list upon parties' agreement.[183] The language of these provisions narrows down the otherwise broad object and purpose of BITs, by imposing certain limitations on states with regard to their environmental obligations when encouraging and protecting foreign investment in their territories.

Embedding environmental language into BITs is not entirely unknown in state practice, although it has only been gaining its momentum in recent years. The first states to add environmental provisions into their BIT were China and Singapore in 1985.[184] The development since then has been monitored in 2011 OECD study of 1,623 international investment agreements (1, 593 BITs and 30 other non-BIT investment agreements, mainly free trade agreements), which showed that only 8.2% or 133 out of 1,623 treaties include some sort of reference to environmental issues.[185]

Interestingly, whereas only 103 of 1,593 reviewed BITs include such reference, all 30 reviewed non-BIT investment agreements include such reference.[186] The same survey found that relatively high number of BITs containing environmental language is entered into by the same small number of states, which at some point began systematically include this language in all their treaties (e.g. Canada, Mexico, USA, or

181. Herbert Smith Freehills, "Is the Recently Signed Morocco-Nigeria BIT a Step Towards a More Balanced Form of Intra-African Investor Protection?" (Herbert Smith Freehills, May 23, 2017) https://hsfnotes.com/arbitration/2017/05/23/is-the-recently-signed-morocco-nigeria-bit-a-step-towards-a-more-balanced-form-of-intra-african-investor-protection/.
182. Marie-Claire Cordonier Segger and Andrew Newcombe, "An Integrated Agenda for Sustainable Development in International Investment Law," in Marie-Claire Cordonier Segger, Markus W. Gehring, et al., *Sustainable Development in World Investment Law*, Global Trade Law Series, Volume 30. Kluwer Law International, 2011, 127.
183. 1994 North American Free Trade Agreement, Article 104(1).
184. Kathryn Gordon, Joachim Pohl, "Environmental Concerns in International Investment Agreements: A Survey" (OECD, June 1, 2011) http://www.oecd.org/daf/inv/internationalinvestmentagreements/WP-2011_1.pdf, p. 8.
185. *Ibid.* 7.
186. *Ibid.* 5.

Luxembourg).[187] However, a more recent survey shows that the number of recently concluded or renegotiated BITs which adopt pro-environmental language is increasing, leading to greater environmental protection through BITs.[188] This approach shows a clear intent of some states to subject future foreign investments to certain environmental standards, which in turn reflects the effort of these states to tackle climate change and other environmental issues by other means than just by international environmental agreements.

Therefore, it may be concluded that objectives pursued by both BITs and environmental protection regulations are in fact reconcilable, proof of which may be found in more recent treaty practice. Considering latest developments, it may be assumed that even more BITs will include environmental protection provisions in the future.

3 How Can BITs Contribute to Environmental Protection?

Given the number of existing BITs and a dramatic rise of investor-state disputes in recent years,[189] there are several ways how BITs can contribute to protection of environment. This section addresses three different proposals which the authors consider most effective.

3.1 Imposition of Environmental Obligations in BITs

The first and probably most effective option would be for state parties to undertake environmental protection obligations in their BITs, such as an obligation to enforce BIT standards on investors by requiring high levels of environmental protection.[190] Since the primary aim of BITs is to promote and protect investments, it is clear that the nature of environmental obligations imposed by BITs will always be linked to foreign investments in some way.

Certain BITs, mainly the more recent ones, already impose concrete environmental obligations on state parties, but at the same time go even further by providing additional remedy to investors in case of a host state's breach of its environmental obligations. The example of such treaty is 2012 US Model BIT. In Article 12.3 it states that "*each Party shall ensure that it does not [...] fail to effectively enforce [domestic environmental] laws through a sustained or recurring course of action or inaction as an*

187. *Ibid.*
188. UNCTAD, "Recent developments in the international investment regime" (UNCTAD, May 1, 2018) https://unctad.org/en/PublicationsLibrary/diaepcbinf2018d1_en.pdf, 5, 7.
189. According to official statistics of the International Centre for Settlement of Investment Disputes available at https://icsid.worldbank.org/en/Documents/resources/ICSID%20Web%20Stats%202019-1(English).pdf, ICSID had registered 401 new cases in the last nine years (i.e. between 2010 and 2018), compared to 305 new cases registered in the previous 38 years (i.e. between 1972, when the first ICSID case was registered, and 2009).
190. Graham Mayeda, "Sustainable International Investment Agreements: Challenges and Solutions for Developing Countries," in Marie-Claire Cordonier Segger et al. (eds.), *Sustainable Development in World Investment Law*, Global Trade Law Series, Volume 30. Kluwer Law International, 2011, p. 545.

encouragement for the establishment, acquisition, expansion, or retention of an investment in its territory."[191] This provision constitutes the specific international treaty obligation of state parties to enforce national environmental laws on foreign investments and investors.

At the same time, this provision clearly connects the environmental protection with different stages of the investment, namely its establishment, acquisition, expansion, or retention. Furthermore, Article 12.6 of the same BIT provides for a state party's right to initiate "*consultations with the other [p]arty regarding any matter arising under [article 12 on Investment and Environment].*"[192] Therefore, an investor under the 2012 US Model BIT has the opportunity to turn to his home state government and ask them to convince the host state's government to take particular environmental action, in case the investor seeks restitution of his interest through an "*adequate environmental remediation*" rather than through damages, should the latter not be an appropriate remedy for the investor in a particular case.[193] This constitutes a potential additional remedy for investors besides the standard BIT remedy of monetary damages.[194] Moreover, it is clear from the wording of Article 12.6 that this remedy is primarily designed for use by BIT's state parties, so it could also be the other state party who enforces these environmental obligations against the host state, not just an investor.

BITs do not have to impose environmental obligations solely on state parties. There are few BITs that impose direct obligations with regard to protection of environment also on investors. An example of such a treaty is the 2016 Morocco-Nigeria BIT, which takes particularly pro-environmental and pro-sustainability approach throughout its whole wording.[195] In its Article 14(1) Morocco-Nigeria BIT explicitly requires investors to conduct a pre-investment environmental impact assessment, by saying that "*[i]nvestors [...] shall comply with environmental assessment screening and assessment processes applicable to their proposed investments prior to their establishment, as required by the laws of the host state [...] or the laws of the home state [...], whichever is more rigorous [...].*"[196] This is a significant step toward sustainable development.

However, the crucial question with regard to investors' environmental obligations under BITs is their enforceability vis-à-vis investors, as investors are not generally bound by the BITs. In this regard, state parties in Article 20 undertake to subject investors to civil liability in their home jurisdiction for any breach resulting in "*significant damage, personal injuries or loss of life in the host state.*" This concept of investors' liability may, however, in practice prove to be problematic, as it depends on

191. 2012 U.S. Model Bilateral Investment Treaty, para. 12.3.
192. *Ibid.* 12.6.
193. Greg Hicks, "Environmental Protection in International Investment Agreements" (Center for Strategic and International Studies, April 7, 2015) https://www.csis.org/analysis/environmental-protection-international-investment-agreements.
194. *Ibid.*
195. Herbert Smith Freehills, "Is the Recently Signed Morocco-Nigeria BIT a Step Towards a More Balanced form of Intra-African Investor Protection?" (Herbert Smith Freehills, May 23, 2017) https://hsfnotes.com/arbitration/2017/05/23/is-the-recently-signed-morocco-nigeria-bit-a-step-towards-a-more-balanced-form-of-intra-african-investor-protection/.
196. 2016 Morocco-Nigeria BIT, Article 14(1).

investor's home state to bring an independent action against its own national (either natural or juridical person).[197]

There is another potential way how states may try to enforce these obligations against investors, namely counterclaims. As will be addressed in further detail in subsection 3.3 below, counterclaims are available to a host state only if there is a claim brought against the host state by an investor and thus, they are of limited use. Therefore, it may be assumed that the enforceability issues will most likely hinder the practical impact of these otherwise ambitious BIT provisions imposing environmental protection obligations on investors.

Although states may include specific pro-environmental obligations in their future BITs, it might be problematic to include such obligations in the already existing BITs, unless parties agree to renegotiate them. Due to the large number of existing BITs, this would require to reopen negotiations between a large number of states, placing great burden especially on states with high number of BITs, such as UK, Switzerland, Netherlands, France or Luxembourg.[198] Moreover, unless both parties agree on renegotiation of a particular BIT, there is no possibility to include such provisions in the BITs unilaterally by a state wishing to take a more pro-environmental approach. This may be a problem as many BITs are concluded between developed and developing countries, whereas developed countries may try to shield their investors from burdensome environmental obligations under the BITs or from broad regulatory discretion of the host states with regard to environment, as their investors are most likely the beneficiaries of the BIT protections, since the biggest flow of investments goes from developed to developing countries.[199] Despite hurdles state parties may have to overcome when faced with renegotiation of BITs, statistics show that renegotiation of existing treaties is in fact gaining in popularity, with at least 27 outdated BITs being replaced by more modern treaties since 2012.[200]

Even if state parties do not agree on specific environmental obligations in their BITs, some investors have tried to bring an action against host states for the failure to comply with the environmental protection standards under very generally worded provisions of the BITs. An example was the case of *Allard v. Barbados*, in which a Canadian investor brought claims against Barbados for breach of Article II(2)(a) on the fair and equitable treatment (FET) and full protection and security (FPS) and Article VIII on expropriation of the 1996 Canada-Barbados BIT, claiming that Barbados has *"failed to take reasonable and necessary environmental protection measures and [...] has directly contributed to the contamination of the Claimant's eco-tourism site,*

197. *Ibid.*
198. Statistics on concluded BITs by economy at: https://investmentpolicyhub.unctad.org/IIA/IiasByCountry#iiaInnerMenu.
199. Historically, the main reason for entering into a BIT between a developed and a developing country was to protect investment of investors from a developed country in a developing country, although some authors argue that this trend is shifting toward different reason—liberalization of the investment flows. *See* Kenneth J. Vandevelde, "A Brief History of International Investment Agreements" 12 *U.C. Davis Journal of International Law & Policy* 157, p. 183 (2005).
200. UNCTAD, "Recent Developments in the International Investment Regime" (UNCTAD, May 1, 2018) https://unctad.org/en/PublicationsLibrary/diaepcbinf2018d1_en.pdf, 7.

thereby destroying the value of his investment."[201] Both Article II(2) and Article VIII of the Canada-Barbados BIT represent generally formulated BIT standards, and neither these articles nor other provisions of the BIT imposed any specific environmental obligations on the state parties. Eventually, Allard's claims were dismissed due to his failure to bear the burden of proof that there was: (i) sufficient changes in the environmental conditions of the eco-tourism site and (ii) causal link between such changes and actions or inactions of Barbados.

However, despite the fact that Allard's claims were dismissed, the tribunal's reasoning shows that in principle, the tribunal did not refuse to entertain such environmental claim under the generally worded substantive protection standards of the BIT, but it concluded that such claims must still meet specific requirements of particular substantive protection standards, such as existence of investor's reasonable legitimate expectations and reliance on them in case of FET standard.[202] Had Allard proven the fulfilment of these requirements in establishing the breach of the substantive protection standards under the Canada-Barbados BIT with regard to his claim, it is possible that the tribunal would have found in his favor even under such generally worded substantive protection standards without specific environmental obligations imposed on the state parties. However, the investor's burden of proof with regard to the breach of general substantive standard (such as FET) by means of the host state's environmentally harmful conduct is higher than in case of violation of specific environmental obligation of a host state explicitly formulated in the BIT. Moreover, one could argue that the success of such environmental claim based on a generally worded substantive protection standard depends on a tribunal's assessment if such standard in fact encompasses protection of environment. If the tribunal decides that it does not, the investor's environmental claims would be dismissed as falling outside the BIT's scope.

3.2 Freedom to Adopt Environmental Regulations and Exemptions from Substantive Protection Standards under BITs

Even if state parties fail to include concrete environmental obligations in their BITs, there is another way how BITs can contribute to climate action and sustainable development. As recent practice shows, some BITs provide for the provisions that allow state parties to adopt any nondiscriminatory proportionate and reasonable pro-environmental legislation without violating provisions of the BITs aimed at the protection of foreign investors, in particular provisions on prohibition of unlawful (indirect) expropriation.[203] Such provisions could prevent investors from bringing claims against host states as a result of passing of legislation for the protection of environment, such as shutting down of nuclear power plants or imposition of quotas on production of carbon emissions. Recently, there have been some cases brought

201. *Peter A. Allard v. The Government of Barbados*, PCA Case No. 2012-06, Award (June 27, 2016), para. 3.
202. *Ibid.* 194.
203. Stephan W. Schill, "Do Investment Treaties Chill Unilateral State Regulation to Mitigate Climate Change?," 24 *Journal of International Arbitration* 469, 470 (2007).

under different investment treaties (although not precisely under BITs) where investors challenged the host states' legislation adopted with the intention of tackling greenhouse gas pollution and protection of environment, such as EUR 4.7 billion Energy Charter Treaty claim for unlawful expropriation brought by Swedish investor Vattenfall against Germany over the compulsory phase out of its nuclear power plants in Germany imposed by the 2011 amendment to the German Atomic Energy Act in the aftermath of the Fukushima disaster in March 2011.[204]

Similarly, in another case brought by the US investor Bilcon of Delaware against Canada under NAFTA, there have been concerns as to whether investor-state dispute settlement mechanisms included in investment and trade agreements undermine the environmental policies of states, by allowing investors file claims against environmental measures implemented by governments, if such measures interfere with investors' profits.[205] Therefore, it seems to be appropriate for states to include in their BITs provisions that would allow states to adopt *bona fide* nondiscriminatory and proportionate environmental protection measures without violating BIT's substantive protection standards, or even completely exempt these measures from the scope of the substantive protection standards, thus avoiding their liability vis-à-vis foreign investors or, if the regulatory measure is adopted in nondiscriminatory, fair and equitable manner, avoiding payment of compensation to the foreign investors.[206]

Examples of BITs already adopting this approach are 2012 US Model BIT or 2018 Netherlands Draft Model BIT. Both the US and the Netherlands Model BITs include specific provisions on sustainable development and protection of environment, the former in Article 12 (Investment and Environment) and the latter in section 3 (Sustainable development). Moreover, the US Model BIT in par. 4(b) of its Annex B (Expropriation) states that "*[e]xcept in rare circumstances, non-discriminatory regulatory actions by a Party that are designed and applied to protect legitimate public welfare objectives, such as public health, safety, and the environment, do not constitute indirect expropriations.*"[207] Article 13(8) of the Netherlands Investment Agreement contains almost identical provision, holding that "*[e]xcept in the rare circumstance when the impact of a measure or series of measures is so severe in light of its purpose that it appears manifestly excessive, non-discriminatory measures [...] designed and applied in good faith to protect legitimate public interests, such as [...] environment or public*

204. *Vattenfall AB and others v. Federal Republic of Germany*, ICSID Case No. ARB/12/12, Decision on the Achmea Issue (August 31, 2018). For an in-depth analysis of the Energy Charter Treaty, *see* Rafael Leal-Arcas (ed.), *Commentary on the Energy Charter Treaty*, Edward Elgar, 2018.
205. Scott McAnsh and Amir Attaran, "Investor-State Dispute Settlements: Great for Business, Risky for the Environment" (ecojustice blog, February 14, 2017) https://www.ecojustice.ca/isds-settlements-great-for-business-risky-for-the-environment/ accessed March 13, 2019; Joseph A. Strazzeri, "A Lucas Analysis of Regulatory Expropriations under NAFTA Chapter Eleven," 14 *Georgetown International Environmental Law Review* 837, p. 859 (2002); David A. Gantz, "Potential Conflicts Between Investor Rights and Environmental Regulation Under NAFTA's Chapter 11," 33 *George Washington International Law Review* 651, p. 656 (2001).
206. David A. Gantz, "Potential Conflicts Between Investor Rights and Environmental Regulation Under NAFTA's Chapter 11," 33 *George Washington International Law Review* 651, p. 656 (2001).
207. 2012 U.S. Model Bilateral Investment Treaty, Annex B, Article 4(b).

morals [...] do not constitute indirect expropriations."[208] Similar provisions may be found in 2017 Israel-Japan BIT.

The above-mentioned provisions of the US and the Netherlands Model BITs both allow state parties to adopt in good faith nondiscriminatory and reasonable measures designed to protect the environment without breaching the BITs and at the same time exempt these measures from the scope of certain substantive protections under BITs, particularly protection against indirect expropriation. These provisions may serve as a template for the drafters of future BITs, potentially also extending the exemption of regulatory measures designed to promote sustainable development also to other substantive protection standards, especially FET.

3.3 *Counterclaims for Breach of Environmental Regulations by Investors*

The final proposal discussed in this section is host states' counterclaims brought under BITs against investors who do not observe the environmental standards formulated in BITs, resulting in harm to the host states.[209] The counterclaims would allow a host state to defend itself against the claims for violation of an investor's economic interests under a BIT, by either mitigating the amount of compensation to be paid to the investor as a result of such violation or by not being obliged to pay the damages to the investor at all, should the host state prove the investor's involvement in the conduct harmful to the environment in breach of the BIT.[210] However, this proposal presupposes the imposition of specific requirements on investors to comply with environmental norms directly in the BITs, otherwise it might not be possible for a host state to bring counterclaims against an investor who causes environmental harm, unless domestic laws of the host state include such norms and the BIT requires investments to be made in compliance with domestic law of the host state.[211] In recent practice, there have been few cases where host states have brought environmental counterclaims against investors in investment arbitration by claiming breach of their domestic environmental laws, e.g., *Burlington v. Ecuador*[212] or *Perenco v. Ecuador*.[213]

While the latter case is still pending, the former represents one of the very few instances when a host state successfully brought environmental counterclaims against an investor. In 2017, an ICSID tribunal awarded almost 42 million USD to Ecuador after

208. 2018 Netherlands Draft Model Bilateral Investment Treaty, Article 13(8).
209. Graham Mayeda, "Sustainable International Investment Agreements: Challenges and Solutions for Developing Countries," in Marie-Claire Cordonier Segger et al. (eds.), *Sustainable Development in World Investment Law*, Global Trade Law Series, Volume 30. Kluwer Law International, 2011, p. 545.
210. *Ibid.* 556-557.
211. Kate Parlett and Sara Ewad, "Protection of the Environment in Investment Arbitration—A Double Edged Sword" (Kluwer Arbitration Blog, August 22, 2017) http://arbitrationblog.kluwerarbitration.com/2017/08/22/protection-environment-investment-arbitration-double-edged-sword/.
212. *Burlington Resources Inc. v. Republic of Ecuador*, ICSID Case No. ARB/08/5, Decision on Ecuador's Counterclaims (February 7, 2017).
213. *Perenco Ecuador Ltd. v. The Republic of Ecuador and Empresa Estatal Petróleos del Ecuador (Petroecuador)*, ICSID Case No. ARB/08/6, Interim Decision on the Environmental Counterclaim (August 11, 2015).

finding Burlington liable toward Ecuador for the costs of restoring the environment in certain areas for exploration and exploitation of oil in Ecuador, caused by the investor's breach of certain Ecuadorian domestic environmental regulations.[214] This example demonstrates that in practice states may actually be successful in holding investors responsible for the harmful conduct to the environment thought investor-state dispute settlement provisions of the BITs. However, by their very nature, counterclaims may be brought only if an investor initiates investment arbitration by bringing claims against a host state, which makes them not particularly an effective tool in the hands of host states who cannot resort to them every time they deem bringing an environmental action against an investor appropriate. Nevertheless, counterclaims are an example of how BITs can contribute to climate action.

4 Conclusion

Latest development in treaty practice shows that BITs may play a vital role in environmental protection and sustainable development. As the global community becomes increasingly concerned about the environment some states are starting to re-evaluate the traditional objectives of BITs and include sustainable development as condition of establishment and operation of foreign investments in their territory as part of newly negotiated BITs. This trend is starting to gain more weight especially in recent years, with several recently concluded BITs referring to the sustainable development at least in their preambles.[215] Moreover, some BITs also include specific provisions aimed at resolution of conflicts between obligations under the BITs and under domestic or international environmental protection regulations in favor of the latter, sending clear message that environmental protection has precedence over the protection of investments. All this indicates a shift toward promotion of more sustainable investments and greater scrutiny of environmental and social impact of investments by the host states in the future.

We propose three potential solutions how the BITs can in fact contribute to climate action and sustainable energy. Our first proposal would be to impose specific investment-related environmental obligation on states or investors in the BITs that could be enforced either by the other state party to the BIT or by investors. Second, BITs could include specific provisions allowing states to adopt *bona fide* nondiscriminatory environmental regulation without violating substantive protection standards under BITs, such as FET or prohibition of expropriation, or even completely exempting adoption of environmental measures from the scope of the substantive protection standards, thus exempting host states from any liability to pay damages for a breach of investors' economic interests caused by adoption of reasonable nondiscriminatory environmental measures. Finally, the third proposal is to embed specific provisions allowing host states to bring counterclaims against investors who cause harm to the

214. *Burlington Resources Inc. v. Republic of Ecuador*, ICSID Case No. ARB/08/5, Decision on Ecuador's Counterclaims (February 7, 2017) 468.
215. *See*, for example, 2018 Argentina-Japan BIT, 2018 Brazil-Guyana BIT, or 2018 Netherlands Draft Model BIT.

host-states environment, by, e.g., producing excessive amounts of greenhouse gas or by polluting waters. However, due of the limited nature of counterclaims that can be brought only when foreign investors bring their claims against host states, this third solution seems to be the least effective.

All three proposals are being tested in practice to different extent, but in order to have a real impact, more states would have to adopt this pro-environmental language in their BITs. One way or another, recent practice shows that there is a shift in treaty practice toward higher degree of environmental protection, which is demonstrated by a number of newly concluded BITs as well as by states' attempts to renegotiate older BITs.

Whereas the unilateral and bilateral efforts could contribute to the sustainable development and mitigation of greenhouse gas emissions, one must agree with some of the publicists who claim that such actions would *"only provide second-best solution"* and the only effective way in order to tackle climate change would be a truly global mechanism that imposes strict enforceable obligations on states, as opposed to mere best efforts obligations imposed by some existing international conventions, such as the Paris Agreement.[216] However, due to different political and economic interests of the major global players, such as USA, China or EU, an agreement on such global mechanism seems unlikely. Therefore, until a global solution is found, BITs, together with regional FTAs, could bridge this gap and contribute to climate action and broader environmental protection.

IV HOW CAN CURRENT EFFORTS TO MODERNIZE THE ENERGY CHARTER TREATY CONTRIBUTE TO THE EUROPEAN UNION'S APPROACH OF LINKING TRADE AND INVESTMENT POLICIES WITH CLIMATE ACTION?

1 The European Approach to Linking Trade and Investment to Climate Action

1.1 Introduction

The European Union (EU) proclaims to be a frontrunner with regard to climate protection[217] and recently reinforced its commitment to the Paris Agreement on Climate Change. With this renewed commitment to the Paris Agreement, the EU at the same time has to further align its policy efforts in trade and investment to the Paris goals. As action is urgent and trade and investment have significant effects on the environment and on the level of greenhouse gas emissions, those two policy areas are

216. Stephan W. Schill, "Do Investment Treaties Chill Unilateral State Regulation to Mitigate Climate Change?," 24 *Journal of International Arbitration* 469, p. 477 (2007).

217. *See* European Commission (2012). Address by Herman Van Rompuy, President of the European Council & José Manuel Durão Barroso, President of the European Commission. "From war to peace: a European tale." Available at: http://europa.eu/rapid/press- release_SPEECH-12-930 _en.htm.

crucial leverages in contributing to the Paris Agreement goals and, at the same time, working toward the Sustainable Development Goals (SDGs), which were set by the United Nations General Assembly in 2015. Particularly SDG number 7, namely affordable and clean energy, as well as SDG number 13, namely climate action, are relevant in this regard.

To reach a further alignment to the Paris Agreement goals, the EU is intensifying its efforts to not only invest in renewable energy and climate protection on a stand-alone basis, but also to specifically link its investment and trade policy to development and climate protection purposes. This materializes in both the EU's external trade as well as its external investment policy and is underlined by two concrete proposed strategies, namely (1) Commissioner Malmström's 15-point plan and (2) the EU's reformed approach on investment protection. Both will be introduced in the first part of this section.

After that, the second part of this section turns attention to the Energy Charter Treaty (ECT) which is undergoing a modernization process, as suggested by the European Commission and authorized by the Council as of July 2019. Since the Energy Charter Treaty is an agreement with both a trade and an investment component and is targeting the energy sector and hence one of the most important sectors in terms of climate action, its reform proposal could contribute to a closer link of the EU's trade and investment policy to the Sustainable Development Goals of climate action and clean energy and therefore requires close examination. Therefore, the second part of this section seeks to evaluate what potential the ECT reform holds.

1.2 Trade and Climate: The 15-Point Plan

As the World Trade Organization (WTO) and with that the multilateral pathway has not made significant progress since its Doha Round stalemate in terms of incorporating a stronger link between trade and sustainable development goals, bilateral action is becoming an important leverage in connecting climate and trade. When concluding Free Trade Agreements (FTAs), the engagement of the agreement partners on a bilateral basis in terms of including climate protection provisions therefore is an important factor which decides how climate-compatible a trade agreement becomes. The EU as an important weight both in global trade as well as in international negotiations around climate governance has a great role to play in this regard which is why the recently proposed strategy with the aim of linking climate and trade can have significant trickle-down effects if it can serve as model strategy for other countries.

The so-called 15-point plan recently published by the European Union, and more concretely by Commissioner Malmström,[218] has the goal of aligning the European trade policy with the Sustainable Development Goals by particularly ensuring the successful

218. European Commission (2018). Feedback and way forward on improving the implementation and enforcement of Trade and Sustainable Development chapters in EU Free Trade Agreements. Available at: http://trade.ec.europa.eu/doclib/docs/2018/february/tradoc_156618.pdf.

enforcement and implementation of Trade and Sustainable Development (TSD) chapters in European Free Trade Agreements. Although this rather broad goal also includes issues relating to labor standards, the EU explicitly mentions "climate action" and the Paris Agreement fulfilment as core points. The 15-point plan can, therefore, contribute to advancing an integrated thinking of climate and trade, thereby aiming to boost climate-positive effects of trade like the technique effect, i.e., the improvement of the diffusion of climate-friendly technologies to trade partner countries, and reducing potential detrimental developments like shifting of climate-damaging business activities to countries with less stringent environmental regulations. In the 15-point plan, the EU commits to the integrated approach and argues that trade alone cannot solve the current challenges, therefore it strives to "put values and principles such as high social and environmental standards at the core of EU trade policy."[219] This underlines how the EU is seeking to mainstream climate considerations into its trade and investment policy in order to leverage all resources to reach the Paris goals.

The 15 "concrete and practicable actions"[220] put forward by the EU (see Annex 4 of this chapter) target the four areas of: (1) working together, (2) enabling civil society including the Social Partners to play their role in implementation, (3) delivery, and (4) transparency and communication. The more concrete subchapters tackle, for example, making sure that countries comply with their commitments via more assertive enforcement, improving the monitoring role of civil society and the mobilization of EU resources to support the implementation of sustainable development chapters in trade agreements.[221] Furthermore, the Commission puts forward in the plan that it will "encourage early ratification of core international agreements,"[222] explicitly mentioning the Paris Agreement as an important multilateral environmental agreement, and "step up efforts to ensure early ratification on the course of trade negotiations, using all available tools."[223] By using its weight as trade power[224] to incentivize the ratification of the Paris Agreement, the EU is also requiring other countries to pursue a similar integrated approach. From its statement in the 15-point plan, the EU has now even gone a step further by requiring the ratification of the Paris Agreement as a prerequisite for market liberalization through a trade agreement. After the French Minister for Foreign Affairs stated that if the US will go through with the withdrawal from the Paris Agreement, no trade and investment agreement with the US will ever be signed, Commissioner Malmström confirmed the demand and made clear that a reference to

219. *Ibid.* p. 1.
220. *Ibid.* p. 2.
221. European Commission (2018). Commissioner Malmström unveils 15-point plan to make EU trade and sustainable development chapters more effective. Available at: http://trade.ec.europa.eu/doclib/press/index.cfm?id = 1803.
222. European Commission (2018). Feedback and way forward on improving the implementation and enforcement of Trade and Sustainable Development chapters in EU Free Trade Agreements. Available at: http://trade.ec.europa.eu/doclib/docs/2018/february/tradoc_156618.pdf, p. 8.
223. *Ibid.*
224. The European Union is the largest trader of manufactured goods and services around the world. Compare European Commission (2019). EU position in world trade. Retrieved from EU position in world trade. Available at: http://ec.europa.eu/trade/policy/eu- position-in-world-trade/.

the Paris Agreement is needed in all trade agreements.[225] This shows that the values at the core of the 15-point plan are taken seriously in international bargaining.

To conclude, the 2018 15-point plan offers a concrete way to tackle problems related to sustainable development and climate provisions in current European FTAs and highlights how the EU is seeking a structured approach to integrating its trade and climate policies more closely going forward and therefore contribute to SDG 13. Also, as already discussed before, the integrated chapters particularly related to advancing technology transfer and renewable energies, very prominent and comprehensive for example in the Agreement between the EU and Vietnam, can enable developing states to reach the goals of clean and affordable energy as put forward by SDG 7.

1.3 Investment and Climate: The Approach to Investment Protection

The European Commission states that one of the principal objectives of its investment policy is to "promote investment that supports sustainable development, respect for human rights and high labour and environmental standards. This includes encouraging corporate social responsibility and responsible business practices."[226] The development toward mainstreaming environmental concerts into the EU's investment strategy was also reflected by the headline of its trade strategy "Trade for All: Towards a more responsible trade and investment policy," which was presented by the European Commission in October 2015. The 2012 EU regulation for BITs sets the condition that concluded investment agreements are "consistent with the EU's principles and objectives for external action."[227]

However, unlike the frequent inclusion of sustainable development provisions in Free Trade Agreements, the included chapters on investment usually in the best case refers to sustainable development or environment in the preamble without making any direct links between investment and climate as an analysis based on the UNCTAD policy hub database shows.[228] Also, in the 2010 presented paper on the reformed investment approach by the EU labelled "Towards a comprehensive European international investment policy," it is only vaguely mentioned that "finally, it should be recalled that the Union's trade and investment policy has to fit with the way the EU and its Member States regulate economic activity within the Union and across our borders. Investment agreements should be consistent with the other policies of the Union and its Member States, including policies on the protection of the environment, decent work, health and safety at work, consumer protection, cultural diversity, development policy and competition policy. Investment policy will continue to allow the Union, and the Member States to adopt and enforce measures necessary to pursue public policy

225. Climate Action (2018). "No Paris Agreement, no EU trade deal," says France to US. Retrieved from http://www.climateaction.org/news/no-paris-agreement-no-eu- trade-deal-says-france-to-us.
226. European Commission (2018). Investment. Available at: https://ec.europa.eu/trade/policy/accessing-markets/investment/.
227. *Ibid.*
228. UNCTAD (2019). Investment Policy Hub. Treaties with Investment Provisions. Retrieved from https://investmentpolicyhub.unctad.org/IIA/CountryOtherIias/78#iiaInnerMen u.

objectives."[229] Current Bilateral Investment Treaties (BITs) of the European Union, therefore, lag behind in terms of putting environmental values in general, and climate considerations in particular, at the core of its external strategy.

2 The Energy Charter Treaty—Then and Now

2.1 Current Set-Up and Goal of the Energy Charter Treaty

Apart from the clearly circumcised areas of trade, reflected by the World Trade Organization framework and by concluded trade agreements, and investment, reflected by the BITs, the Energy Charter Treaty is an important multilateral framework targeting both trade and investment which is important to mention in the context of climate action and clean energy promotion. The Energy Charter Treaty was signed in 1994, came into effect in 1998 and counts 54 signatories as of today, including the EU and the European Atomic Energy Community (EURATOM). Its initial objective was the provision of legal protection for those large oil and gas companies which were seeking to make an investment in the states constituting the former communist bloc.[230] The multilateral agreement is legally binding and covers the four areas of: (1) trade and transit of energy materials and products, (2) investment protection in the energy sector, (3) dispute resolution with regard to matters covered by the ECT, and (4) the promotion of energy efficiency and attempts to minimize the environmental impact of energy production and use. Since it includes not only the principles of the General Agreement on Tariffs and Trade (GATT), but also non-WTO members, it furthermore importantly extends the reach of the WTO framework to non-WTO members.[231]

When evaluating the ECT in terms of climate considerations, particularly point four becomes interesting since the component of energy efficiency has the potential to directly advance climate protection goals if technology transfer to fossil-fuel dependent states comes with a reduction of their energy demand. As Figure 4.3 illustrates, energy efficiency provisions were, together with renewable energy promotion, among the first important climate-related provisions to be included in free trade agreements around the world.[232] This comes despite the fact that energy efficiency provisions could include a potential source of conflict with the WTO if energy efficiency standards bases on processes and methods of production are regarded as disguised barriers to trade

229. European Commission (2010). Communication from the Commission to the Council, the European Parliament, the European Economic and Social Committee and the Committee of the Regions. Towards a Comprehensive European International Investment Policy. Available at: https://eur-lex.europa.eu/LexUriServ/LexUriServ.do?uri=COM:2010:0343:FIN:EN:PDF, p. 9.
230. Frédéric Simon, EU Asserts "Right to Regulate" as Part of Energy Charter Treaty Reform (2019). Available at: https://www.euractiv.com/section/energy/news/eu-asserts-right-to-regulate-as-part-of-energy-charter-treaty-reform/.
231. For further information *see* Yodogawa, "Energy Charter Treaty: Standing Out Beside the WTO." In M. Matsushita and T.J. Schoenbaum (eds.), *Emerging Issues in Sustainable Development* [online] Springer Japan, 2016, 279-288. available from http://link.springer.com/10.1007/978-4-431-56426-3_13 [April 7, 2019].
232. *See also* J.F. Morin and S. Jinnah, "The Untapped Potential of Preferential Trade Agreements for Climate Governance," 27:3 *Environmental Politics* 541-565 (2018).

leading to discrimination between "like" products.[233] However, a sharp increase in energy efficiency provisions only took place around the turn of the millennium. Therefore, the ECT was a pioneer as a legally binding treaty with an energy-efficiency component, which, importantly, extended the reach to countries beyond the WTO. Since energy efficiency provisions feed into the EU's proclaimed goal of staying at the forefront and highly competitive with its green economy,[234] energy efficiency is an important provision for the EU to push internationally.

Figure 4.3 Growth of Climate Provisions in RTAs Over Time

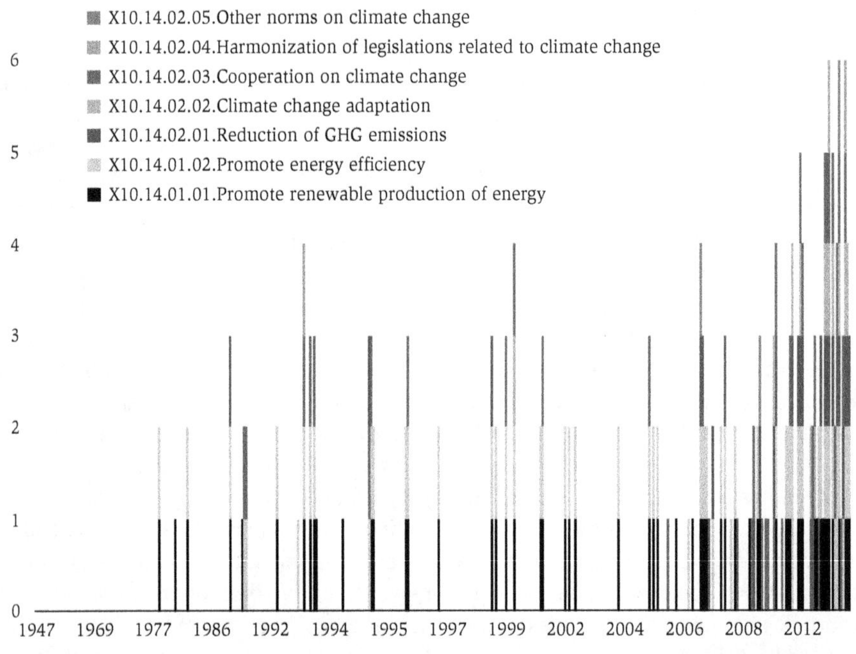

Source: Author's own illustration based on Berger, A., Brandi, C., Bruhn, D., & Morin, J. (2019). TREND analytics—Environmental Provisions in Preferential Trade Agreements. (German Development Institute/Université Laval). DOI: 10.23661/trendanalytics_2017_1.0.

As the energy sector is responsible for a high amount of greenhouse gas emissions, the legal framework governing investments in that area has a very important role to play when it comes to reaching the Paris goals. Since the ECT comprises a high number of signatories and is legally binding, it becomes a crucial leverage in slowing down climate change.

233. C. Carraro and C. Egenhofer (eds.), *Climate and Trade Policy: Bottom-Up Approaches Towards Global Agreement*. Edward Elgar Pub, 2007.

234. S. Bogojević, "Climate Change Law and Policy in the European Union," in K.R. Gray, R. Tarasofsky and C. Carlarne (eds.), *The Oxford Handbook of International Climate Change Law*, Oxford University Press, 2016, 674-691.

2.2 *ECT-Reform Suggestions Related to Clean Energy and Climate Action*

Although many cases were raised on the basis on the Energy Charter Treaty and despite the importance of the sector in reaching the Paris goals, the ECT has hardly been revised in its roughly 30 years of existence. Therefore, the European Commission recently went ahead to label the Energy Charter Treaty as outdated. In its proposal of May 14, 2019, the Commission stated explicitly: "Since the 1990s (most of) the ECT provisions have not been revised. This became particularly problematic in the context of the ECT provisions on the protection of investment, which do not correspond to modern standards as reflected in the EU's reformed approach on investment protection. Those outdated provisions are no longer sustainable or adequate for the current challenges; yet it is today the most litigated investment agreement in the world,"[235] with a total of at least 121 investment notified disputes.
The Recommendation for a Council decision authorizing the entering into negotiations on the modernization of the Energy Charter Treaty includes a list of topics covering investment protection provisions, pre-investment commitments, transit, the economic integration agreements clause and dispute resolution provisions. Although all of those issues need further analysis, this section focuses on the points which are linked to advancing the SDGs 7 and 13, whereas the other topics will be neglected. The proposed EU negotiating directives aim to ensure greater legal certainty and clarify the rules applying to the protection of foreign investments. The European Commission, therefore, calls "the ECT modernisation [...] a logical step in pursuing the EU's reformed approach on investment protection."[236] At the core of the discussion is also the Investment-State Dispute Settlement (ISDS) clause which the EU seeks to fundamentally reform, including a more narrow definition of investors to "prevent that mailbox companies bring disputes under the ECT,"[237] although they do not operate in their country of legal residence with substantive business activities.

Two points mentioned in the proposal are particularly important:

(1) First, the European Commission proposes that "the Modernised ECT should include appropriate standards of protection for investors and investments, in line with the EU law, in particular (and in a non-exhaustive manner): Right to regulate; Most favored nation treatment provision, which under the ECT covers also national treatment; Clarification of 'most constant protection and security'; Fair and equitable treatment and full protection and security, which are appropriately circumscribed for interpretation purposes; Expropriation, covering direct and indirect expropriation, and appropriately defined to clarify the nature of indirect expropriation; Umbrella clause; Transfers:

235. *See* European Commission (2019). Recommendation for a Council Decision authorising the entering into negotiations on the modernisation of the Energy Charter Treaty. Available at: https://trade.ec.europa.eu/doclib/docs/2019/may/tradoc_157884.pdf, p. 1.
236. *Ibid.*
237. Frédéric Simon, EU Asserts "Right to Regulate" as Part of Energy Charter Treaty Reform (2019). Available at: https://www.euractiv.com/section/energy/news/eu-asserts-right-to-regulate-as-part-of-energy-charter-treaty-reform/.

allowing free transfers relating to an investment, together with appropriate exceptions and safeguards for financial difficulties or crises; and Denial of benefits."[238]

(2) Second, the recommendation includes the suggestion that "the Modernised ECT should include provisions on sustainable development, including on climate change and clean energy transition, in line with recently concluded agreements and EU positions in ongoing negotiations."[239]

On July 15, 2019, the EU member states approved the European Commission's proposal to start the negotiation to modernize the investment protection standards contained in the ECT.[240] The Council particularly supported the modernization of the ECT in the area of the "right to regulate" regarding health, safety and the environment and insisted on the reflection of climate change and clean energy transition as important factors in the modernized ECT.[241]

The negotiating directives approved by the Council for the first time connect the ECT explicitly to clean energy and climate change, thereby showing that the European Commission is trying to actively inscribe the Sustainable Development Goals into not only its Free Trade Agreements, but also other multilateral agreements. It therefore has an important signaling effect in showing to trade and investment partners that current agreements are updated in terms of climate and environmental considerations. If the treaty gets expanded to more signatory states which are still starting their renewable energy transition, technology transfer in terms of energy efficiency technologies could result in decreasing the costs for renewable energies in those countries and therefore contribute to the goals of (a) clean and (b) affordable energy simultaneously.

2.3 Criticism

Although the reform suggestions regarding the ECT show that the urgency of linking the ECT to the energy transition and climate protection goals has been recognized, the modernization process is being accompanied by harsh critique.

Particularly prominent is the extremely critical paper "One treaty to rule them all: The ever-expanding Energy Charter Treaty and the power it gives corporations to halt the energy transition" which was published by Corporate Europe Observatory and the Transnational Institute in June 2018. The authors label the ECT "an obscure international agreement"[242] which is "granting corporations enormous powers over our

238. European Commission (2019). ANNEX to the Recommendation for a Council Decision authorising the entering into negotiations on the modernisation of the Energy Charter Treaty. Available at: http://trade.ec.europa.eu/doclib/docs/2019/may/tradoc_157885.pdf, p. 2.
239. *Ibid.*
240. European Commission (2019). Energy Charter Treaty modernisation: Commission welcomes Council's mandate. Available at: http://trade.ec.europa.eu/doclib/press/index.cfm?id=2049.
241. *Ibid.*
242. P. Eberhardt,C. Olivet and L. Steinfort, "One Treaty to Rule Them All. The Ever-Expanding Energy Charter Treaty and the Power It Gives Corporations to Halt the Energy Transition." Brussels/Amsterdam, June 2018, p. 6.

energy systems including the ability to sue governments, which could obstruct the transition from climate-wrecking fossil fuels towards renewable energy. And the ECT is in the process of expansion, threatening to bind yet more countries to corporate-friendly energy policies."[243] They particularly criticize the ISDS clause, claiming that, although similar policy mechanisms have achieved great public criticism, thinking for example of the protests surrounding the Transatlantic Trade and Investment Partnership (TTIP), the ECT has escaped such a public outrage. This comes despite very prominent cases which have been raised under the ECT. Here, the authors, for example, refer to Vattenfall which "sued Germany over environmental restrictions on a coal-red power plant and for phasing out nuclear power"[244] and to "Oil and gas company Rockhopper [which] is suing Italy over a ban on offshore oil drilling."[245] Also the lacking transparency and public awareness regarding the agreement is under sceptical scrutiny by the authors.

Furthermore, the ECT Secretariat itself is under heavy criticism. Masami Nakata, the former Energy Charter's Assistant Secretary General, prepared a report in which she blames failed management and structuring methods for the currently difficult situation. Those concerns are reflected in the suggestions regarding transparency improvements by the European Commission. The Assistant Secretary General also questions that the modernization will actually succeed in aligning the ECT with the Paris Agreement on Climate Change. Here, Energy Charter's Assistant Secretary General Masami Nakata stated that "it is unlikely that Contracting Parties would reach an agreement to align the Treaty with the Paris Climate Agreement"[246] since the member states are in very different phases regarding their internal clean energy transition and have different takes on fossil fuels, with the EU,[247] Japan and Switzerland trying to implement decarbonization strategies while other contracting parties either heavily rely on fossil fuel export or present transit countries.

2.4 *Reform Potential for Clean Energy and Climate Action*

The harsh criticism first of all shows that the ECT is an agreement of great importance and that a modernization of the agreement comes at a time which is overdue. Although many activists call for a withdrawal of the treaty instead of seeking to modernize it, it is questionable whether this would be the solution since the sunset clause means that

243. *Ibid.*
244. *Ibid.*
245. *Ibid.*
246. *See* the views of Masami Nakata in Frédéric Simon, Leaked Report Reveals "Misfunctioning" of Energy Charter Treaty Amid EU Reform Calls (2019). Available at: https://www.euractiv.com/section/energy/news/leaked-report-reveals-misfunctioning-of-energy-charter-treaty-amid-eu-reform-calls/.
247. The Commission recently presented the EU's 2050 long-term strategy. The vision aims at decarbonizing the EU's energy supply through electrification on a large scale and to become climate-neutral by 2050. The paper outlines the steps necessary to achieve this vision. Although it has not been poured into legal actions yet, energy outlooks forecast that by 2050 renewable energies will make up 87% of the EU energy mix (BNEF 2018). New Energy Outlook (BNEF, July 2018). Available at: https://www.bnef.com/core/insights/18819/view.

companies are able to sue countries even after they have pulled out from the treaty, as the example of Rockhopper shows which is suing Italy although the country withdrew from the agreement in 2016. As existing investments will stay protected until 2036 after Italy's withdrawal, it means that leaving the treaty would only have a very lagged effect which is detrimental as the next two decades are very important years in terms of climate action and an acceleration of the energy transition. In this sense, a reform of the treaty seems more reasonable than a withdrawal which will not have an effect until it might already be too late to accelerate climate action.

In terms of reform potential, a substantive provision clarifying the host states' right to regulate would be able to prevent investors suing host states for advancing their clean energy transition by restricting fossil fuel investments on their ground. This would be a crucial alteration particularly in the years ahead, in which the EU member states will have to revise their own internal regulations in order to keep up with the greenhouse gas emission goals defined by the European Union. Many countries will in this sense have to restrict fossil fuel investments and could potentially face a challenge under the umbrella of the ECT.

At the same time, adding the proposed provisions on climate change and clean energy in line with recently concluded agreements and EU positions in ongoing negotiations would be a game-changer as the EU's proclaimed position in the climate regime, as already outlined, is quite ambitious and since it declared to make an alignment to the Paris Agreement a prerequisite for investment or trade treaties. As the scope of the ECT is planning to be extended to more signatory countries, the ECT modernization could add another layer of enforcement to the Paris Agreement.

Those potentially positive reform effects, however, have to be evaluated in terms of their likelihood. Since, as outlined, the positions of the signatory states are very far apart, it seems not very probable that a renegotiation of the agreement will result in substantive provisions that would then restrict the opportunity for power of coal,[248] oil and gas companies to sue host states in order to slow down, or even bring to a stop, clean energy transition ambitions of host states. Hence, if the Treaty in its current state with only vague provisions on sustainable development gets expanded to other developing countries around the globe, it might result in increasing investments in fossil fuels and could in this sense potentially impede clean energy ambitions in fast-growing industrializing states.

Therefore, unless substantive provisions are reached in the modernization process, the EU has to consider if pulling out from the treaty could not have an important signaling effect and will prevent an outdated treaty to be expanded to host

248. A note on coal deserves to be made. Asia accounts for 75% of the world's coal demand, where China consumes half of it. There are risks to using coal: 1) environmental, in that emissions from coal plants, both already built and planed ones, will mean that there will be more CO_2 emissions and therefore temperatures will go up; 2) economic, in that banks have announced that they will stop funding new coal plants; and 3) politics, in that voters in liberal democracies are increasingly concerned about climate change. *See* The Economist, "Coal and Climate Change: Betting on Black," August 24, 2019, p. 8. The good news is that the number of new coal-fired plants receiving investment approval in Asia has been declining sharply since 2016. *See* The Economist, "Power Generation: Down and Dirty," August 24, 2019, pp. 39-40, at 40.

states in which the judicial system is not capable to take up big investment disputes with large multinational corporations. As the member states of European Union represent around 50% of the Energy Charter Conference membership as well as of the Signatory Parties to the ECT, leaving the treaty would make it less attractive for potential new members to sign up and could prevent an insufficient treaty being expanded to more countries.

Apart from the provisions in the Treaty itself, it is fundamental that the suggested improvements regarding transparency are implemented. Whereas the EU has made the involvement of the Civil Society an integral part of all its newer generation Free Trade Agreements, the ECT is yet excluded from this framework. However, including civil society actors such as nongovernmental organizations or experts into the process could lead to both an improvement of the ECT itself and a higher legitimacy of the Treaty. Since civil society actors constitute an important monitoring "watchdog" role, their involvement in the ECT would serve as an additional layer of detecting breaches of the new sustainable development clauses and would, therefore, prevent a further criticism from NGOs.

2.5 Evaluation and Outlook

The modernization debate around the ECT, although overdue, shows that, in European trade and investment governance, climate considerations have now reached even agreements that have long escaped public attention. The reform process offers the chance to improve the long-neglected Treaty and might serve to trigger a public debate that would improve concerns with regard to transparency. If the EU manages to include substantive provisions regarding the host states' right to regulate as well as a connection to multilateral agreements like the Paris Agreement, the ECT could become a treaty advancing the clean energy agenda. This gets important particularly as more states still in the beginning of their clean energy journey are expected to become signatory states for the ECT.

However, it remains to be seen what the European Commission finally manages to negotiate as signatory states have widely differing views on the fossil fuel sector. If the EU does not reach the incorporation of substantive climate protection provisions, pulling out from the Treaty might, despite negative effects of the sunset clause, have an important signaling effect as the ECT is expanded to developing states. In any case, the modernization process itself has the potential to transform a sector that is a crucial component for the EU to reach its 2050 clean energy goals and to start a debate around a treaty which could become an important leverage in contributing to reaching SDGs 7 and 13. The further modernization process, therefore, requires close monitoring and an update as soon as negotiations have started.

V FINAL CONCLUSIONS

In 1748, the French philosopher Montesquieu wrote in *The Spirit of the Laws* that "two nations that trade with each other become reciprocally dependent ... so the natural effect of commerce is to lead to peace." In as much as we are in support of free trade, striking a balance between economic prosperity, thanks to trade, and environmental protection is not easy. We believe this dualism does not have to be perceived as a trade-off. In fact, one can grow economically and cleanly at the same time. How? The answer lies in technology and clean trade, namely trading environmental goods and services. And the evidence is in California, whose GHG emissions between 2000 and 2016 fell by 9%, despite the fact that its economy and population grew.[249]

Based on the findings in this chapter, we conclude that EU FTAs have the potential to be the true enforcer of environmental and climate change obligations. The EU leads the world as signatory to the largest number of FTAs. The recently concluded or negotiated European FTAs contain detailed, precise and binding commitments, with references to crucial MEAs such as the Kyoto Protocol and the Paris Agreement. However, for these binding and precise obligations to have any impact on environmental protection and climate change mitigation, the environmental provisions would need to be directly enforceable by means of a binding and final decision of a third party. Unfortunately, European FTAs continue to favor dispute settlement by means of soft mechanisms, such as consultations and (nonbinding) panel reports. Consequently, the European FTAs remain toothless and favor soft power. Without strong enforcement mechanisms, the environmental provisions in the EU's FTAs are nothing more than artifices to "greenwash" trade agreements. The only way for the EU to be at the forefront of climate change mitigation is to follow the American approach of including the FTA's environmental chapter within the scope of the general dispute mechanism of the FTA.

There is a need to globalize climate change obligations to successfully decarbonize the economy and fight climate change. The MEA and trade mechanisms have been negotiated in parallel; however, the two should interlink for the fight against climate change to have a discernable effect. The MEAs were a step in the right direction, as they have established adaptation and mitigation climate change obligations. The multilateral system in the WTO and MEAs shares the problems relating to lengthy and complex negotiations due to its extensive membership.

Therefore, the recent proliferation of PTAs can be used to successfully enforce climate change obligations through trade, as countries are more willing to enter into trade agreements due to the benefits of extending their markets. PTAs could lead to the multilateralism of the climate agenda as there could be a replication of agreements with existing climate-related provisions within major economies. These are easier and faster to negotiate and could be a breakthrough in tacking climate change. It would be an opportunity to bring together major GHG emitters and promote the cooperation in innovating new cleaner technologies and improving the environment.

Recent EU FTAs offer very solid substantive wording to protect the environment. However, it is most surprising that EU FTAs do not cover the environmental protection chapter in their dispute settlement chapter, arguably because international agreements are the outcome of a political compromise. For an actor that claims to lead on

249. The Economist, "Environmental Policy: The Great Divide," June 29, 2019, pp. 34-35, at 34.

environmental protection (and more specifically, on climate change mitigation), the EU should make it possible to enforce the environmental protection chapters in all its FTAs. Doing so will certainly make the EU a champion in environmental protection.

The next chapter will provide three case studies of climate clubs linked to energy to reach sustainability more effectively.

VI ANNEXES

Annex 1

Figure A1 The Dimensions of Legalization

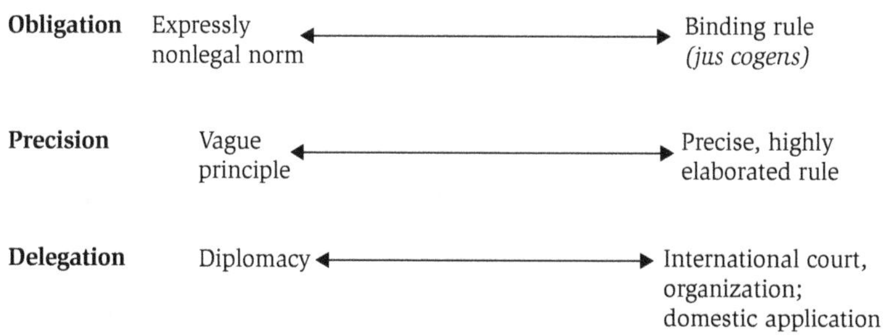

Source: K. Abbot et. al., "The Concept of Legalization," *International Organization*, 54/3 (2000), pp. 401-404.

Annex 2

	Obligation	*Precision*	*Delegation*
0. Low	Not binding	General references	Political mechanism
1. Moderate	Optional commitment	Broad areas of discretion	Limited judicial mechanism
2. High	Binding commitment	Specific targets	Full (independent and binding) judicial mechanism

Annex 3

FTA	*Obligation*	*Precision*	*Delegation*	*Total*
USA-Peru TPA (2006)	1	2	2	5
USA-Colombia TPA (2006)	1	2	2	5
USA-Panama TPA (2007)	1	2	2	5
USA-Korea FTA (2007)	1	2	2	5

FTA	*Obligation*	*Precision*	*Delegation*	*Total*
UMSCA (2018)	1	2	2	5
EU-South Korea FTA (2010)	1	1	0	2
EU-Peru/Colombia FTA (2012)	1	2	0	3
EU-Moldova Agreement (2014)	1	0	0	1
EU-Georgia Agreement (2014)	1	0	0	1
EU-Ukraine Agreement (2014)	1	1	0	2
CETA (2016)	1	1	0	2
EU-Japan EPA (2018)	2	2	0	4
EU-Singapore FTA (2018)	2	2	0	4
EU-Vietnam FTA	2	2	0	4
EU-Mexico (draft) Agreement	2	2	0	4

FTA	*Score*	
USA-Peru TPA	5	Moderate obligation (1), because strong use of mandatory language and commitments. For example: "shall adopt, maintain and implement" (art. 18.2); "shall not fail to effectively enforce" (art. 18.3); "shall ensure" (art. 18.4 and 18.13). However, no maximum score on obligation because of the limited number and scope of the overed agreements. Also, when it comes to non-covered environmental agreements to which both signatory states are party, the commitment is weak (art. 18.13 merely provides that parties "shall continue to seek means to enhance the mutual supportiveness of multilateral environmental agreements which they are all party"). High precision (2), with references to a closed list of precisely identified covered agreements (art. 18.2) and confirmation that each party shall not fail to effectively enforce *its own* environmental laws (art. 18.3). High delegation (2), because pursuant to article 18.12(6) parties have full recourse to the Dispute Settlement provisions of the Agreement.

FTA	*Score*	
USA-Colombia TPA (2012)	5	Moderate obligation (1), because of strong use of mandatory language. For example: "shall adopt, maintain and implement" (art. 18.2); "shall not fail to effectively enforce" (art. 18.3); "shall ensure" (art. 18.4 and 18.13). However, no maximum score on obligation because of the limited number and scope of the covered agreements. Also, when it comes to non-covered environmental agreements to which both signatory states are party, the commitment is weak (Article 18.13 merely provides that parties "shall continue to seek means to enhance the mutual supportiveness of multilateral environmental agreements which they are all party"). High precision (2), with references to a closed list of precisely identified covered agreements (art. 18.2) and confirmation that each party shall not fail to effectively enforce *its own* environmental laws (art. 18.3). High delegation (2), because pursuant to article 18.12(6) parties have full recourse to the Dispute Settlement provisions of the Agreement.
USA-Panama TPA (2012)	5	Moderate obligation (1), because of strong commitments. For example: "shall adopt, maintain and implement" (art. 17.2); "shall not fail to effectively enforce" (art. 17.3); "shall ensure" (art. 17.4). However, no maximum score on obligation because of the limited number and scope of the covered agreements. Also, when it comes to non-covered environmental agreements to which both signatory states are party, the commitment is weak (by stating in Article 17.13 that parties "shall continue to seek means to enhance the mutual supportiveness of multilateral environmental agreements which they are all party"). High precision (2), with references to a closed list of precisely identified covered agreements (art. 18.2) and confirmation that each party shall not fail to effectively enforce *its own* environmental laws (art. 18.3). High delegation (2), because pursuant to article 18.12(6) parties have full recourse to the Dispute Settlement provisions of the Agreement.

FTA	*Score*	
USA-Korea FTA (2012)	5	Moderate obligation (1), because of consistent use of "shall" (for example. art. 20.2 and art. 20.4. *See also* art. 20.3, which provides that "neither party shall fail to effectively enforce its environmental laws"). However, no maximum score on obligation because of the limited number and scope of the covered agreements. Also, when it comes to non-covered environmental agreements to which both signatory states are party, the commitment is weak (by stating in art. 20.10 that parties "shall continue to seek means to enhance the mutual supportiveness of multilateral environmental agreements which they are all party"). High precision (2) as a result of multiple references to the covered agreements in Annex 17.2 of the FTA (see art. 20.2 and 20.3). High delegation (2), because, pursuant to article 20.9(4), parties have recourse to all provisions of the Dispute Resolution chapter of the FTA.
UMSCA (2018)	5	Moderate obligation (1), because, although some of the provisions contain strong commitments (for example, art. 24.4(1) which provides that "no Party shall fail to effectively enforce its environmental laws" or art. 24.9(1) pursuant to which each party "shall take measures"), other provisions lack strength and merely acknowledge existing commitments (for example, art. 24.3(2) provides that each party "shall strive to ensure" that its environmental laws provide for high levels of protection). Also, with respect to MEAs, parties only agree to "affirm" their commitments to implement the MEA to which they are party (see art. 24.8). Art. 24.8 is a departure from earlier texts of USA-FTAs on MEAs, which historically provided that parties "shall continue to seek means to enhance the mutual supportiveness of multilateral environmental agreements which they are all party." Notably, compared to earlier USA FTAs, UMSCA includes more detailed provisions/protection on specific environmental issues (e.g., Article 24.9; 24.10; 24.17).

FTA	*Score*	
		High precision (2). UMSCA contains various specific targets, in particular in relation to important issue areas such as the ozone layer (see art. 24.9, which provides that each party shall take measures to control the production and trade in substances that can significantly deplete and otherwise modify the ozone layer. Footnote 6 supplements this provision in more detail by regulating when a party shall be deemed in compliance with this provision—and when not), the Protection of the Marine environment (see art. 24.10) and sustainable fisheries management (see art. 24.18, with footnote 13 including a specific reference to instruments). High delegation (2), because pursuant to article 24.32, parties have recourse to all provisions of the Dispute Resolution chapter of the FTA (including the establishment of a Panel that has renders a binding report).
EU-South Korea FTA (2011)	2	Moderate obligation (1), because the parties merely reaffirm their existing commitments. Although art. 13.5(2) refers to reaffirming the commitments to "effective implementation" of the MEAs to which they are party, art. 13.5(3), which relates to particular climate change MEAs such as the UNFCC, only refers to the parties' commitment to reach "the ultimate objective" of the ENCCC and the Kyoto Protocol. No higher score, because (contrary to for example EU-Japan) the FTA does not include further, stronger binding commitments. Moderate precision (1): art. 13.5(3) merely refers to the parties' commitment to reach "the ultimate objective" of the UNFCC and the Kyoto Protocol. Low delegation (0), because art. 13.16 explicitly confirms that with respect to any matter arising under the "Trade and Sustainable Development" chapter, the parties only have recourse to Government Consultations and the Panel of Experts. The report of the Panel is, however, not binding (see art. 13.15(2)).

FTA	*Score*	
EU-Colombia/Peru FTA (2013)	3	Moderate obligation (1) Although art. 270 refers to reaffirming the commitments of the parties to "effective implementation" of certain identified MEAs, other provisions on climate change and environmental issues contain weak commitments (for example, art. 267(2), which provides that Parties "reiterate their commitment to address global environmental challenges," and art. 275, which bears in mind the UNFCC and the Kyoto Protocol, but only to subsequently state that Parties "recognise that climate change is an issue of common and global concern that calls for the widest possible cooperation." However, the FTA unfortunately fails to impose strong and binding commitments on the parties to address these global climate concerns. High precision (2), because art. 270 includes a detailed list of applicable MEAs that parties should effectively implement in their laws. It should be noted, however, that none of the agreements referred to (except for the Kyoto Protocol) relates to climate change. Furthermore, art. 272 contains a specific reference to the parties' commitment to conserve, and to the sustainable use of, biological diversity "in accordance with the CBD and other relevant international agreements to which the Parties are party." Low delegation (0), because art. 285(5) explicitly provides that the chapter on Trade and Sustainable Development is not subject to the formal dispute settlement chapter of the FTA.
EU-Moldova Agreement (2014)	1	Moderate obligation (1), because the (albeit few) environmental provisions included in this Agreement use strong wording referring to binding commitments (for example, art. 86 and 92 provide that the parties "shall" develop and strengthen their cooperation on environmental issues and to combat climate change). However, many other provisions dealing with environmental issues focus on cooperation and dialogue, rather than binding commitments of the parties (see, for example, art. 87, which confirms that "cooperation shall aim at preserving, protecting, improving and rehabilitating the quality of the environment").

FTA	*Score*	
		Low precision (0); although relatively strong on the obligation dimension, the Agreement lacks precision and merely acknowledges the environmental problems and climate change. Low delegation (0), since no enforcement or dispute settlement mechanism is in place, but parties merely agree that "a regular dialogue will take place" on the issues covered by the Environmental Protection chapter (art. 96).
EU-Georgia Agreement (2014)	1	Moderate obligation (1), because the (albeit few) environmental provisions included in this Agreement use strong wording referring to binding commitments (for example, art. 301 and 307 provide that the parties "shall" develop and strengthen their cooperation on environmental issues and to combat climate change). However, many other provisions dealing with environmental issues focus on cooperation and dialogue, rather than binding commitments of the parties (see, for example, art. 302, which confirms that "cooperation shall aim at preserving, protecting, improving and rehabilitating the quality of the environment"). Low precision (0), because the Agreement lacks precision and specific targets. It merely acknowledges the environmental problems and climate change. Low delegation (0), since no enforcement or dispute settlement mechanism is in place, but parties merely agree that "a regular dialogue will take place" on the issues covered by the Environmental Protection chapter (art. 305).
EU-Ukraine Agreement (2014)	2	Moderate obligation (1). Art. 269(1) contains a binding commitment by stating that the Parties "shall not fail to effectively enforce its environmental laws." Moreover, art. 269(2) provides that a party "shall not weaken or reduce" the environmental protection afforded by its laws to encourage trade or investment. Also, art. 292 refers to reaffirming the commitments of the Parties to "effective implementation" of the MEAs to which they are parties. However, other obligations are more ambiguous (for example, art. 295 only provides that parties "undertake to work together" in order to take effective measures to monitor and control fish and other aquatic resources).

FTA	*Score*	
		Moderate precision (1), because art. 292(2) refers to the implementation of obligations under the MEAs to which the signatory states are party. Moreover, this provision specifies that the Parties should implement these obligations "in their laws and practices." Also, art. 269(1) provides that Parties shall not fail to effectively enforce *its* environmental laws. However, other provisions lack precision and only refer to general issues such as the obligation of parties to "taking effective measures to monitor and control fish and other aquatic resources" (art. 295). Low delegation (0), since no enforcement or dispute settlement mechanism is available other than the (nonbinding) Consultation and reports from the Group of Experts (see art. 300 and 301).
CETA (2017)	2	Moderate obligation (1). CETA contains a mix of weak and stronger commitments. For example, art. 24.2 merely states that parties "recognise" that the environment is a fundamental pillar of sustainable development. However, some other articles are stronger and contain binding commitments (for example, art. 24.5(2) provides that a Party "shall not" waive or otherwise derogate from its environmental law to encourage trade). Also, art. 24.4(2) refers to reaffirming the commitments of the Parties to "effective implementation" of the MEAs to which they are parties). Moderate precision (1), because the Agreement, to a large extent, only refers to general environmental issues and obligations, without specifying an elaborated rule or setting detailed targets. However, art. 24.4(2) refers to all MEAs to which the signatory states are party and is thus precise in the action the state is expected to take. Low delegation (0), because, under the environmental protection chapter of CETA, the parties have only recourse to Consultations (24.14) and a Panel (art. 24.15). The final report of the Panel is nonbinding and parties only commit to "engage in discussions" on the findings in such a report (see art. 24.16(1)).

FTA	*Score*	
EU-Japan EPA (2018)	4	High on obligation (2). Art. 16.4(2) "reaffirms" the parties' commitment to effectively implement the MEAs to which they are parties (including the Paris Agreement, see art. 16.4(4)). The FTA contains various other, stronger commitments. For example, a number of detailed provisions on biological diversity contain binding commitments (for example, art. 16.6(2) provides that each party "shall" implement effective measures to combat illegal trade in endangered species). Moreover, pursuant to art. 16.8(2) each party "shall adopt and implement" their respective effective tools for combating illegal fishing. High precision (2), because the FTA contains many detailed rules that unambiguously make clear what is expected from the state. For example, art. 16.8(2) refers to compliance with, *inter alia*, the UN Convention on the Law of the Sea relating to the Conservation and Management of Straddling Fish Stocks and Highly Migratory Fish Stocks. Also, this provision provides that parties take measures to achieve the objectives and principles of the Code of Conduct for Responsible Fisheries adopted by the Conference of the Food and Agriculture Organisation on October 31, 1995. Notably, the FTA also specifically refers to the effective implementation of the Paris Agreement (see art. 16.4(4)). Low delegation, since art. 16.17 excludes the Environmental chapter from the scope of the formal dispute settlement mechanism under the FTA.
EU-Singapore FTA (2018)	4	High obligation (2). In contrast to previous EU FTAs, art. 12.6 of the EU-Singapore FTA contains a strong and binding reference to the implementation of MEAs to which the signatory states are party ("the parties shall effectively implement in their respective laws (...)"). However, it should be noted that with respect to the environmental UNFCCC and Kyoto Protocol, parties again merely "reaffirm" their commitment to reaching the ultimate objectives of these agreements. Moreover, the FTA contains several other strong commitments with a binding nature (for example, art. 18.8(a) provides that parties "undertake to comply" with long-term conservation measures and sustainable exploitation of fish stocks).

FTA	Score	
		High precision (2), because the FTA contains various references to specific agreements, rules or protocols that regulate certain issues in a detailed manner. For example, art. 18.8(a) provides that parties "undertake to comply with long-term conservation measures and sustainable exploitation of fish stocks as defined in the international instruments ratified by the respective Parties." Moreover, art. 12.8(d) provides that parties will "uphold the principles of the FAO Agreement to Promote Compliance with International Conservation and Management Measures by Fishing Vessels on the High Seas and respect the relevant provisions of the FAO Agreement on Port State Measures to Prevent, Deter and Eliminate IUU Fishing." Furthermore, art. 12.6(3) explicitly mentions the UNFCCC and its Kyoto Protocol. Low delegation (0), because, pursuant to art. 12.16(1), the environmental protection chapter is excluded from the general Dispute Settlement chapter of the FTA.
EU-Vietnam FTA	4	High obligation (2). Although art. 13.5(2) only "reaffirms" the parties' commitment to effectively implement the MEAs to which they are parties and respect to the Kyoto Protocol and Paris agreement, the parties merely reaffirm their commitment to reaching "the ultimate objective" of the respective instruments, the FTA contains other more binding commitments. For example, art. 13.7 on biological diversity provides that parties "shall adopt and implement appropriate effective measures" which are consistent with its commitments under international treaties to which it is a party, leading to a reduction of illegal trade in wildlife. Moreover, art. 13.9 provides that each party "shall comply with long-term conservation and management measures." High on precision (2) because the FTA contains various references to specific agreements, rules or protocols that regulate certain issues in a detailed manner (including the Paris Agreement). For example, art. 13.9(2) provides that parties shall comply with long-term conservation and management measures and sustainable exploitation of marine living resources "as defined in the United Nations Convention on the Law of the Sea of 1982."

FTA	*Score*	
		Low delegation (0), since art. 12.16(1) excludes the FTAs environmental chapter from the scope of the general dispute settlement mechanism.
EU-Mexico (draft) Agreement	4	High obligation (2). In contrast to all other EU FTAs, art. 5.1 contains a strong and binding reference to the implementation of the UNFCC and the Paris Agreement ("each party shall effectively implement the UNFCC and the Paris Agreement (...)"). Moreover, with respect to other MEAs to which the signatory states are party, art. 4(2) equally provide that the parties "shall effectively implement" these agreements. High precision (2), because the FTA refers to various specific agreements, rules or protocols that regulate certain issues in a detailed manner (including the Paris Agreement). For example. art. 8(3)(a) provides that each party shall "implement long-term conservation and management measures and sustainable exploitation of marine living recourses as defined in the main UN and FAO instruments relating to these issues" (notably, in footnote 2 of the FTA a detailed list is provided of agreements that are classified as "main UN and FAO instruments"). *See also* art. 8(3)(b), which provides that each party shall "act consistently with the principles of the UN Convention on the Law of the Sea of 1982" (and many other agreements relating to aquaculture). Low delegation (0), since art. 15 provides that "In case of a disagreement between the Parties regarding the interpretation or application of this Chapter, the Parties shall have recourse exclusively to the dispute resolution procedures established under Article 16 [consultation] and Article 17 [experts]."

Annex 4: 15-point action plan for improvement in the implementation and enforcement of Trade and Sustainable Development chapters in EU FTAs

1. Working Together
   - Partnering with Member States and the European Parliament
   - Working with international organizations
2. Enabling civil society including the Social Partners to play their role in implementation
   - Facilitate the monitoring role of civil society including the Social Partners
   - Extend the scope for civil society, including the Social Partners, to the whole FTA
   - Take action regarding responsible business conduct

3. Delivering
   - Country priorities
   - Assertive enforcement
   - Encourage early ratification of core international agreements
   - Reviewing the TSD implementation effectiveness
   - Handbook for implementation
   - Step up resources
   - Climate action
   - Trade and labor
4. Transparency and Communication
   - More transparency and better communication
   - Time-bound response to TSD submissions

CHAPTER 5

Three Case Studies: Generalized System of Preferences, the Arab World, and EU-Russia Energy Trade Relations

This chapter analyzes three case studies of climate clubs related to energy to reach sustainability more effectively. The first explains two of the main Generalized Systems of Preferences (GSP)—the US program and the EU's GSP. It indicates the reasons behind conditionality and possible future impacts. It discusses the European Union's Generalized System of Preferences, the objectives, the types of GSP offered and the various conditions to benefit from the scheme, focusing on environmental provisions. It then examines the US's GSP and the reasons for the lack of environmental conditionality. It then offers a discussion about the similarities and differences between the two systems' environmental conditions and an examination of what could be done better concerning the environmental conditionality in the GSP. This chapter then provides some insights on how the COVID-19 pandemic may affect the GSP and globalization as we know it.

The second case study explores the trading energy in the Arab world. It provides an analysis of the ongoing efforts at promoting Pan-Arab energy trade by highlighting the role and benefits of promoting intra-regional trade in electricity and gas, with the view to improving energy security in the MENA region, which is fundamental for the development of any country. In fact, modern energy access is a human right. The case study highlights the importance of regional cooperation in energy in MENA and makes the case of a bottom-up approach: enhanced cooperation in the short term before eventual regional integration in the longer term. Such regional cooperation should be beneficial for every party involved in MENA, without paternalizing. Special importance is given to the link between suppliers, transit countries and consumers of electricity and gas.

The third case study examines the EU—Russia energy trade relations. It aims to provide an analysis of the EU external relations in the energy sector, focusing on the EU—Russia energy trade relations and the Energy Community. The hypothesis we have undertaken is that securing reliable flows of diverse sources of renewable energy would foster the EU's energy security and decrease its vulnerability toward major fossil fuel exporters. There are currently efforts toward much greater investment in renewables in the EU, and major oil companies will need to learn from renewables companies if they wish to enter this new field. In fact, the pandemic created by COVID-19 has offered the world what a renewable's future might look like.

I THE GENERALIZED SYSTEM OF PREFERENCES

This section was concluded in June 2020 and therefore written prior to the US elections of 2020. The political changes resulting from the 2020 US elections are therefore not reflected here.

1 Introduction[1]

In 1971, the General Agreement on Tariffs and Trade[2] (GATT) enacted an enabling clause to the most-favored-nation principle as enshrined in Article I:1 of the GATT. The Enabling Clause permitted tariff preferences to be granted to goods originating in developing countries. In 1979, the GATT enabled this preferential treatment permanently, with the condition that the system had to be generalized, nondiscriminatory, and non-reciprocal. This waiver enabled the so-called Generalized System of Preferences (GSP).

The GSP is a program in which tariffs are reduced or not imposed on imports coming from certain developing economies. The main objective of such a programme is to foster economic growth in developing countries by increasing revenue from exports. Every country understands the need to make their economies grow. It is particularly important in developing countries; it creates jobs and alleviates poverty.

However, not everyone takes into account the negative effects that unbalanced economic growth may have on the environment and sustainability at large. Climate change is one of the biggest challenges humanity faces today. Its mitigation has become increasingly important, which is why, in 2015, the Paris Agreement (PA) on Climate Change was concluded by the parties to the UN Framework Convention to Climate Change. In November 2019, the United States notified formally its intention to withdraw from the PA, to come into effect a year later.

The European Union (EU), on the other hand, the third largest emitter of greenhouse gases (GHG) after China and the US, was eager to sign the PA. The PA requires its participants to establish a mechanism to contribute to the mitigation of greenhouse gases and support sustainable development. Academics have argued that legally binding trade agreements are better at improving wanted values such as human rights and environmental protection than softer environmental or human rights agreements.[3] The EU has realized the importance of this too and implemented

1. Sections of this case study have been co-written with Saana Saveljeff.
2. General Agreement on Tariffs and Trade (entered into force 1948) https://www.wto.org/english/res_e/booksp_e/agrmntseries2_gatt_e.pdf.
3. *See* for instance Rafael Leal-Arcas, *Solutions for Sustainability: How the International Trade, Energy and Climate Regimes Can Help*. Springer International Publishing, 2019; Rafael Leal-Arcas et al., "The Contribution of Free Trade Agreements and Bilateral Investment Treaties to aSsustainable Future," 23:1 *Zeitschrift für Europarechtliche Studien—ZEuS* 3-77 (2020).

environmental provisions in its trade agreements. Trade is a useful tool to foster and incentivize sustainable development, especially in developing countries. This is why the EU has also included heavy environmental conditionality in its GSP, among other trade agreements.

The EU's and USA's GSP programs are considered to be the two strongest and most conditional tariff-preference systems in the world. They both offer tariff reductions to developing countries complying with the rules imposed on them. The EU currently offers three different programs according to the needs of the developing country in question; the US, on the other hand, provides one GSP.

According to the United Nations Sustainable Development Goals, there is a strong link between climate change and sustainable development. The goals also state that developing countries are among the most adversely affected and least able to cope with the shocks to their economic, social and natural systems brought by climate change.[4] The EU Commission has indicated that promotion of environmental sustainability is a fundamental component of EU development policy.[5] Thus, the environmental conditionality was added to the GSP+ ensure any development is sustainable and can continue for extended period of time and to ease the shocks of climate change. The US, on the other hand, does not have any environmental conditionality in its GSP at the moment, which is not surprising considering the current US administration's view on climate change and its policy to put the American economy first. It remains to be seen whether this approach will change in future American FTAs and under the Biden administration.

This case study describes the two GSP systems—the US program and the EU's GSP—indicating the reasons behind conditionality and possible future impacts on the environment. In section 2, we will discuss the European Union's GSP, its objectives, the types of GSP offered and the conditions to benefit from the scheme, focusing on environmental provisions. Section 3 of this case study examines the US's GSP and the reasons for the lack of environmental conditionality. In section 4, there is a discussion about the similarities and differences between the two systems' environmental conditions and an examination of what could be done better concerning the environmental conditionality in the GSP. Section 5 provides some insights on how the COVID-19 pandemic may affect the GSP and globalization as we know it. Finally, the case study concludes by summarizing the main findings.

2 The European Union's Generalized System of Preferences

The European Union's GSP consists of three arrangements: standard GSP, GSP+ and Everything But Arms (EBA). The EU imposes heavy environmental criteria on countries seeking to be eligible for the benefits offered by its GSP+ programs. This section will focus on GSP+ due to the environmental conditionality, but will briefly mention the standard GSP and EBA.

4. https://sustainabledevelopment.un.org/topics/climatechange.
5. https://www.oecd.org/dac/EC-Briefing-Note.pdf.

Regulation 978/2012,[6] which applies a scheme of generalized tariff preferences (the GSP Regulation), lays down the criteria and legal basis for all three preferential tariff programs under the GSP. Countries regarded as developing countries automatically achieve the standard GSP as long as they have not achieved high- or upper-middle-income status. EBA, on the other hand, is granted to every country considered as a least developed country by the UN Committee for Development Policy.

Instead of tariff reduction offered in the standard GSP, the GSP+ program eliminates tariffs for products. Tariffs are fully removed for products regarded as non-sensitive (NS) in Annex V, according to Article 7(1) of the Regulation. These products include some agriculture and food products, as well as handicrafts.[7] The tariff added to all products listed as sensitive in Annex V will be reduced by 3.5 percentage points as stated in Article 7(2). However, Article 12 of the Regulation suspends all ad valorem duties on products from GSP+ beneficiary countries that are listed in Annex IX of the GSP regulation. These include a long list of food products and livestock.

Article 9 of the Regulation lays down the necessary requirements to benefit from GSP+. The GSP+ is an extension of the GSP system. To benefit from the GSP+, a country must meet the standard GSP requirements as well as vulnerability and sustainable development-criteria. Thus, the GSP+ includes developing countries that have proven their commitment to sustainable development and good governance by complying with the requirements of Article 9. As part of the sustainable development criteria, a country must ratify 27 international conventions on human and labor rights, environmental protection, and good governance. All the relevant conventions are stated in Parts A and B of Annex VIII, which also shows that eight of these conventions relate to environmental protection. These conventions include the United Nations Framework Convention on Climate Change (1992) and the Convention on International Trade in Endangered Species of Wild Fauna and Flora (1973).[8] Currently, there are eight GSP+ beneficiaries.

The EU has also set a monitoring body, ensuring the ratification of the above-mentioned conventions, effective implementation, and compliance with the reporting requirements. The monitoring body organizes monitoring missions to the beneficiary countries to assess whether they are following the conventions. The beneficiaries are required exchange information about their shortcomings on each of the 27 conventions. In addition, the beneficiaries are expected to provide information on their progress and details of future plans to overcome the shortcomings.[9] Another tool used by the monitoring body is called the "GSP+ dialogue." This tool describes the interaction between the EU and the beneficiaries to assist them to reach sustainable development. A GSP+ monitoring cycle lasts two years and the findings are published.

6. Regulation (EU) No 978/2012 of the European Parliament and of the Council of 25 October 2012 applying a scheme of generalised tariff preferences and repealing Council Regulation (EC) No 732/2008.
7. Annex V of the GSP Regulation.
8. Annex VIII of the Regulation.
9. Second report on the effects of GSP and the special incentive arrangement for sustainable development and good governance (GSP+) covering 2016-2017, available at https://ec.europa.eu/commission/presscorner/detail/en/MEMO_18_350.

The latest one is the Report on the Generalised Scheme of Preferences covering the period 2018-2019.[10] If the conventions are not effectively implemented, or if severe and systematic violations occur, the benefits can be withdrawn according to Article 16 of the Regulation. A 2018 midterm evaluation of the EU's GSP found out that the GSP+ beneficiaries saw an economic incentive for making progress in ratifying the conventions. This finding could indicate that developing countries are actively seeking to develop their economies and willing to comply with rules to access benefits.

The possible removal of benefits as a result of a failure to implement can also create an incentive to comply with the conditions thoroughly. The removal of benefits of a country has happened before. Recently, Cambodia's benefits were temporarily withdrawn due to violations of human-rights principles. Even though the withdrawal was not environment-related, it is an indication that it could happen in the future.

Many Europeans share the view that climate-change mitigation is the number one long-term priority and, without it, there will not be future trade agreements between the EU and its partners.[11] The EU drives the agenda that it is essential to mitigate climate change as early as possible to ensure the continuation of trade between countries and the welfare of individuals across the globe. The EU has had this long-term vision in mind when drafting the GSP+, as well as its other trade agreements, of which many also include environmental conditions.

Another goal of the PA is to promote inclusive and sustainable economic growth. As a result of sustainable economic growth, societies will need to create conditions to allow people to have quality jobs that stimulate the economy without compromising the environment. President Emmanuel Macron of France stated that France refuses to sign commercial agreements with countries that do not respect the Paris Agreement. This statement shows just how serious the leaders of EU Member States are about its policy of "climate-change mitigation first, trade second." The GSP+ reflects this view by including environmental conditions in it. The environmental conditions indicate that the EU is willing to sacrifice short-term economic benefits for the greater good as well as sustainable and environmentally friendly trade.

Besides, the EU also provides financing for beneficiaries to implement conventions. Among other projects, the EU supported the "SMART Myanmar" project, which promotes sustainable consumption and production of garments in Myanmar. Even though the project is not part of the GSP program, it shows that the EU is involved, and willing to help, in environmental protection and climate-change mitigation in non-EU countries. Therefore, it is no surprise that the EU has implemented such heavy environmental conditionality in its GSP and has the intention to toughen environmental conditionality in its other trade agreements. Yet, even if pollution and consumption

10. https://trade.ec.europa.eu/doclib/docs/2018/january/tradoc_156536.pdf.
11. "Climate crisis 'most important issue' for public, poll reveals," available at https://www.sharecast.com/news/international-economic/climate-new-bidscrisis-most-important-issue-public-poll--7027195.html.
https://www.theguardian.com/environment/2019/jul/26/climate-more-pressing-than-brexit-say-71-of-britons-poll.

were to be severely reduced tomorrow, it would take millennia to unwind, as pointed out by David Farrier in his new book.[12]

3 The Generalized System of Preferences of the United States

Established by the Trade Act of 1974, the US GSP promotes economic development by eliminating tariffs on products imported from one of the 119 designated beneficiary countries.[13] According to the US Trade Representative Office (USTR), the GSP promotes sustainable development in developing countries by aiding these countries to increase and diversify their exports.[14] The qualifying conditions have changed over time to reflect the US's economic and global needs, but some criteria have been in place since the start.

Currently, the US's GSP includes mandatory and discretionary criteria. The mandatory criteria are in line with the EU's criteria and include some core conditions, such as that the beneficiary cannot be a communist country and must have implemented commitments to eliminate child labor. The US's GSP also includes discretionary criteria, which the US President may take into account when deciding whether a country should have preferential tariff treatment. The criteria include the obligation that the beneficiary country should provide "equitable and reasonable" access to its market. Currently, unlike the EU, the US's GSP does not include environmental conditions in its mandatory or discretionary criteria.

The US GSP is considered to be the most conditional GSP in the world, together with the EU GSP + . Environmental conditions are lacking from the US GSP, but they are replaced with economic conditions, such as access to markets and respect for US intellectual property rights. The US GSP could be said to be more economy- than environmental-led. The US view on the importance of growing the economy has affected the lack of environmental conditionality in the GSP. Traditionally, the thinking of climate change, environmental protection, and economic growth through trade has been that they contradict each other. President Trump and his administration share this view; thus, it is unlikely that the conditions will change toward climate-change mitigation, especially under his tenure. However, even if most countries may share the view that climate change and economic development are at odds with each other, they are not opposing climate change mitigation as drastically as the US is.

In November 2019, the Trump administration formally announced its decision to withdraw from the Paris Agreement on Climate Change, although the Agreement had already been signed by the previous US President, Mr Obama. The decision to withdraw has made the US the only country formally opposing the Agreement. According to President Trump, the reason for the desire to withdraw is that the Paris Agreement would undermine the US economy.

12. David Farrier, *Footprints: In Search of Future Fossils*. Farrar, Straus and Giroux, 2020.
13. Office of the United States Trade Representative, "Generalized System of Preferences (GSP)" https://ustr.gov/issue-areas/trade-development/preference-programs/generalized-system-preference-gsp, visited December 18, 2020.
14. *Ibid*.

The Obama and Trump administrations have had opposite views on environmental policy. While President Obama's aimed at conserving the environment for future generations, the Trump administration has scrapped many environmental regulations that he has referred to as impediments to US economic and energy output. President Trump has deregulated auto-emission standards in the US that were issued under the Obama administration, questions whether climate change is anthropogenic, and has stated that his political focus is elsewhere, rather than climate-change mitigation. Moreover, the White House under President Trump has indicated that any climate change policies that may hinder the US economic growth will not be pursued. This clearly shows how drastically different the views of the current US administration and those of the EU are on the topic, and why the EU's and US's GSPs are so different when it comes to the environment.

That said, and even if the current Trump administration seems to have no interest in climate change mitigation, at the subnational level in the US, there is a lot going on regarding the implementation of the Paris Agreement. For instance, Texas now meets 20% of its electricity demand with wind power.[15] Were Texas a country, it would be the fifth biggest in the world in wind energy production, where 55% of its electricity comes from wind and solar energy.[16] Paradoxically, the US's renewable-energy boom has been very prominent in Republican-controlled states like Texas, presumably because decarbonization might be perceived as a great business opportunity. Nevertheless, there are many more jobs still in the fossil-fuel industry than in renewable energy in Texas.

Despite the view of the Trump administration on climate change and environmental policies, the US GSP has the promotion of sustainable development as one of the main objectives of its GSP. The United Nations has defined sustainable development as development that meets the present needs without compromising the ability of future generations to meet their needs.[17] Therefore, a sustainable-development goal under the GSP should also include environmental protection and climate-change mitigation. However, since President Trump questions that climate change is anthropogenic and the existence of global warming, it is understandable why current US trade policies do not include conditions furthering these causes.

The Trump administration has followed the "America First" policy. The implementation of this policy can be seen in the recent withdrawal of India's preferential tariff treatment under the US GSP. India had imposed restrictions on dairy products imported from the US to India. The argument behind the restriction was that the feed given to the cows was not in line with Indian culture. This argument was seen as a restriction on access to India's market by US decision-makers. Equitable and reasonable access to the GSP beneficiary's market is one of the discretionary conditions under US GSP. Although the original purpose of the GSP, when adopted at UNCTAD II, was to include unique, non-reciprocal, measures in favor of developing countries, the US

15. "Green Texas" *The Economist* (March 14, 2020), pp. 40-41, at p. 40.
16. *Ibid.* at p. 41.
17. United Nations Department of Economic and Social Affairs, *The United Nations Development Agenda: Development for All* (United Nations, 2007) https://www.un.org/esa/devagenda/UNDA_BW5_Final.pdf accessed April 28, 2020.

GSP has moved toward a reciprocal trade agreement, where immediate economic benefits are expected in return.

Although the US economy would benefit in the short run from the lack of environmental conditionality in its GSP system, it would suffer in the long run. Without the environmental conditions, more developing countries can access the benefits which allows more and cheaper products to be sold in the US. However, without the environmental conditions, the development in the beneficiary countries in unlikely to be sustainable and thus hurting the US economy in the long run. There seems to be tension between the short-term economic gain of individual nations and the long-term interest of sustainable development and potential economic gains of the international community. Instead of combining its trade policies with climate-change mitigation, which would help not only developing countries but also the US in the long term, US trade policies seem to be more focused on short-term economic gain.

4 Comparing the American and European Approach

This section of the case study concentrates on comparing the two GSP systems, describing the possible short-term and long-term impacts of GSP conditionality as well as what could be done better. As noted earlier, the European and American programs have their differences when it comes to conditionality. However, both programs expect something in return for the preferential tariff treatment.

The difference in conditionality between the GSP of the EU and that of the US derives from the divergent views on the importance of climate-change mitigation. Greenhouse gas concentrations are reportedly at the highest levels in three million years. The EU Commission published plans in March 2020 to halve its carbon emissions within a decade. This new climate change law will oblige EU member states to stop their CO2 emissions on a net basis by 2050.[18] Even EU countries with center-right parties in power favor shifting to a greener economy. Moreover, the EU believes that trade should be conducted only on such terms so that sustainable development is achieved. The impacts of the lack of sustainable-development policies on the GSP can be radical, especially in developing countries.

Environmental degradation tends to hit developing countries the hardest. The reason for that is the extensive production of products dependent on natural resources, such as textiles, as well as the lack of environmental-protection laws and programs in those countries. Many developing countries do not see environmental protection and climate-change mitigation as urgent a problem to solve as poverty or adequate labor rights. However, it is essential to ensure that any development in developing countries does not come at the expense of the environment. In the long term, the lack of environmental protection can have a massive impact on the economy of developing countries. Assuming that any development comes at the expense of the environment, the resources will finish faster than the country develops.

18. One irony here is that countries that have benefited the most from turning carbon into cash are now are asking others not to do so. *See* "Minority Report: Brussels Edition" *The Economist* (March 7, 2020), p. 33.

The lack of environmental-protection clauses, especially in a GSP program that benefits the majority of the developing world, might result in developing countries focusing too much on economic growth, simultaneously exploiting the environment. This would mean that, in the long term, the developing world would be unable to use its resources to trade in the future and the developed world would have to find other ways to import the products or find alternative products. This process can be expensive and time-consuming. If the resources become scarce, according to the law of supply and demand, they will also become more expensive. This would mean that imports to developing countries would become more expensive. The lack of sustainable development is a lose-lose situation for everyone.

Making developing countries comply with environmental clauses in exchange for more beneficial tariff treatment makes it economically attractive both in the short term and in the long term for the protection of wild species and their ecosystems. Incentivizing GSP beneficiaries to implement environmental provisions that mitigate climate change and allow for sustainable development would also benefit the GSP offeror in the future. Sustainable development would ensure that there will be natural resources to produce from in the future. This again would allow the developing world to keep trading with developing countries, thereby growing their economies sustainably.

On the other hand, the EU GSP+ environmental conditionality only affects the current eight beneficiaries. Under the standard GSP and EBA, there is no environmental conditionality. The EU should try to incentivize more developing countries to ratify the specific 27 conventions because it is beneficial for everyone in the long run. However, the current impact of it is minimal, as only eight countries enjoy GSP+ treatment in exchange for complying with all the environmental standards. Nevertheless, it is safe to say that the EU is on the right path to sustainable development. As mentioned before, the US has a different view of trade and the GSP supporting sustainable development.

In EU competition law, a dominant firm on the market has a special responsibility to consider how its actions might affect competitors and, ultimately, consumers. A similar policy could and should be applied in global trade, especially when the country with the largest economy in the world is acting irresponsibly toward the future of the globe.

As mentioned before, the US short-sighted trade policy can harm world trade in the long term, especially in the developing world. The lack of environmental conditionality in the US GSP and the intention to withdraw from the Paris Agreement have led the leaders of the world to criticize US trade policy. For example, EU leaders have threatened President Trump by imposing tariffs on all US exports to the EU. It could be a way to condemn the behavior of strong developed countries such as the US. In terms of GSP conditionality, climate-change mitigation is unlikely to change with the current Trump administration. This has consequences for the developing world because the US is one of the main trading partners of many developing countries, and it could impact the future of trade and sustainable development.

5 Pandemics, Anti-globalism, Parochialism, and the GSP

The irony of today's world is that, at a time when international cooperation is needed most due to the COVID-19 pandemic that has created a world health and economic crisis, there seems to be, at worst, a regression in globalization that is heading toward parochialism in dealing with transnational issues. At best, it feels like "the dancing procession of Echternach" where pilgrims would take three steps forward and two steps backwards, thereby making five steps to only advance one step. There are analyses that state that it will take USD 10 trillion to prevent another Great Depression of the 1930s.[19]

The trading system was already suffering from protectionism prior to COVID-19. In 2018, President Trump started (rightly or wrongly) a trade war against China on the grounds that there is a serious US trade deficit in the US-China trade relationship. His administration put an end to the World Trade Organization's Appellate Body in 2019. Overall, as a result of COVID-19, we are seeing higher air-cargo prices, more time spent in transit for goods because they need to be screened more stringently at the border, and travel restrictions in trade in services.

A recent book by Martin Caparros questionably implies that trade is a major cause of high food prices and Mr Caparros is arguably against programs such as the GSP because they tend to perpetuate poverty in that it keeps the poor from rebelling.[20] Trade and investment are stagnated, creating thereby slowbalization, a term coined by The Economist. International trade in goods may shrink by 10-30% in 2020, according to The Economist.[21] This depression and precipitous fall in global trade has nothing to do with new technical barriers to trade; rather, with lack of demand. Foreign direct investment will be reduced by 30-40%.[22] Historians may identify the COVID-19 pandemic as one of the consequences of globalization that precipitated a new era in global trade. In fact, Prime Minister Modi of India told his nation in May 2020 that we are entering an era of economic self-reliance. As a result, life will become more expensive with less freedom.

How will all this affect the EU's and US's GSP? Who will be most hurt by this new trend across the globe? One lesson we have learned from the COVID-19 pandemic is that countries need each other to solve global, transnational problems. A good example comes from the scientific community, which has been working very hard since the beginning of 2020 to find a vaccine against the coronavirus. Equally, supply chains become more resilient through diversification and globalization, not by nationalizing them. It is through a globally interconnected world that the global economy will recover more quickly, as opposed to the current fragmented approach due to the rise of economic nationalism. The current trend will accelerate the rise of regionalism and regionalization, to the detriment of globalization. In the supply chain of the car

---

19. Derek Thompson, "We Can Prevent a Great Depression. It'll Take $10 trillion" *The Atlantic* (May 18, 2020) https://www.theatlantic.com/ideas/archive/2020/05/we-can-prevent-a-great-depres sion-itll-take-10-trillion/611749/.
20. Martin Caparros, *Hunger: The Oldest Problem*. Melville House, 2020.
21. "Goodbye Globalisation" *The Economist* (May 16, 2020), p. 7.
22. "Torn Apart" *The Economist* (May 16, 2020), pp. 57-59, at 57.

industry, 59% of trade is already intra-regional, replacing suppliers further afield with the ones within a given region.[23]

With the rise of economic nationalism/regionalism as well as xenophobia, localization, and protectionism, there is a danger of weakening or even eliminating the GSP. Many people around the world have been left behind for a long time. According to the World Bank, COVID-19 will push around 60 million people into extreme poverty, getting rid of all the recent efforts done to alleviate poverty.[24] Equally, the medical and pharmaceutical sectors may experience pressure to produce at local level. In many developing countries, the healthcare system is already under strain. The COVID-19 outbreak should be embraced as an opportunity for governments and businesses to create a fairer economy for all.

For example, the scale and cost of investment in solar, wind, and energy storage projects via the GSP in the developing world is a great business opportunity. Ion Yadigaroglu estimates that around USD 20 trillion needs to be invested in the developing world during the next 10 years.[25] To make these infrastructure investments more affordable via the GSP, innovation in financing these projects is necessary.

The COVID-19 pandemic may mean the rise of regionalism and deglobalization. Countries are closing their borders because of COVID-19. The US is proposing reshoring supply chains, thereby bringing factories back to the US and showing an anti-globalist perspective. Many American politicians may believe it is the right thing to do by showing patriotism, but US companies do not see it that way. COVID-19 has shown the deficiencies and importance of governance, where there is much to be improved. That said, although it is on the frontline, COVID-19 and its consequences is not as big a problem as environmental protection, and more specifically climate change.

Issues of social justice, green capitalism, dematerialization, and the Anthropocene versus the Capitalocene remain on the table. It remains to be seen whether COVID-19 will have implications for GSP programs, whether it is possible to have prosperity without economic growth, how we can best address inequality, and the dichotomy of wellbeing versus growth.

6 Conclusion

Transnational cooperation is today more relevant than ever. Despite positive impacts of COVID-19 on the environment (such as a drastic drop in GHG emissions from transportation), there are potential negative implications of COVID-19 for the GSP programs.

23. *ibid.* at 58.
24. Jonathan Wheatley, "Virus Will Push up to 60m into Extreme Poverty, World Bank Warns," *Financial Times* (London, May 19, 2020) https://www.ft.com/content/85882871-1b61-49e2-b170-cc3f159b8f88?segmentId=3f81fe28-ba5d-8a93-616e-4859191fabd8, visited December 18, 2020.
25. The Economist, March 14, 2020, p. 14.

Arguably the two most crucial GSP programs in the world have very contradicting views when it comes to environmental provisions. The EU's has heavy environmentally led conditionality when it comes to the GSP+ program, as can be seen from the eight environmental conventions that need to be ratified to access its preferential treatment. If a country does not ratify these conventions, among others human and labor rights conventions, it cannot benefit from the EU's GSP+ program. Besides, the EU has created a body that monitors the effective and adequate implementation of these conventions. These benefits can be withdrawn if severe and systematic failures to implement occur.

Although the GSP+ is significantly more conditional than the standard GSP, it is economically more beneficial. Although it is hard to quantify the impact that environmental conditionality has had on the environment in the beneficiary countries, incorporating conditionality in the GSP is the right way forward in order to support sustainable development across the globe. It would act as a standard, ensuring that economic growth does not come at the expense of environmental protection. The EU has included environmental provisions in its GSP program, as it firmly believes that economic development should not come at the expense of the environment. Further uneven economic growth could be catastrophic not only in developing countries, but also in the developed world.

The United States, on the other hand, does not have any environmental conditionality in its GSP. Its GSP program, including the African Growth and Opportunity Act, is highly conditional in every way, apart from the environment. Although the EU and the US programs are otherwise in line with each other, they greatly differ in this respect.

The current US trade policies, including its GSP, are extremely limited in terms of climate-change mitigation and, thus, sustainable development. This is the result of the Trump administration's view on climate change. The Trump administration has scrapped many previous US environmental policies and has focused on growing the US economy, despite coming at the expense of sustainable development. This can be seen not only from the conditions in the US GSP, but also from the position of the Trump administration in relation to the Paris Agreement on Climate Change.

Finally, the environment-led conditionality in the EU GSP is there to ensure that there will be coexistence between trade and environmental protection in the future, while the current US GSP seems to indicate that short-term economic gains are more important than sustainability.

II TRADING ENERGY IN THE ARAB WORLD

1 Introduction[26]

This section analyzes energy trade in the Arab world in the broader geographical context of the Middle East and North Africa. In an era where regional blocs are gaining

26. Sections of this case study have been co-written with M Akonda.

greater prominence in the international political climate, the Middle East and North Africa (MENA) region—which is where the Arab world is located—is undoubtedly one of the remarkable regions in the world today. While many view the region with a degree of cynicism, given its wide-spread religious tensions and not-so-stellar human rights reputation, the fact remains that sections of MENA are among the world's wealthiest regions in per capita terms and, despite the recent political instability and decline in global crude prices, the region is poised for further socioeconomic growth. A natural corollary of such socioeconomic growth is a greater demand for energy to meet both domestic and industrial needs. Needless to say, governments within the region have recognized the need to avert a potential energy supply crisis which might threaten the positive economic outlook of the region. To this end, countries within the region are looking to harness the potential of electricity and gas trade within the MENA region. It is in this vein that the countries within the region have teamed up, largely through the efforts of the League of Arab States (LAS) and under the auspices of the World Bank, to attempt to create a Pan-Arab Electricity Market (PAEM).

This case study provides an analysis of the ongoing efforts at promoting Pan-Arab energy trade by highlighting the role and benefits of promoting intra-regional trade in electricity and gas, with the view to improving energy security in the MENA region, which is fundamental for the development of any country. In fact, modern energy access is a human right. The case study highlights the importance of regional cooperation in energy in MENA and makes the case of a bottom-up approach: enhanced cooperation in the short term before eventual regional integration in the longer term. Such regional cooperation should be beneficial for every party involved in MENA, without paternalizing. Special importance is given to the link between suppliers, transit countries and consumers of electricity and gas.

The PAEM is envisioned as a shared vision to increase power trade in the region and potentially export power to the European electricity market. Despite having signed a Memorandum of Understanding (MoU) committing to the establishment of the PAEM,[27] several national and regional actions and steps are required to ensure a successful implementation and optimization of the integration program. It is therefore important to scale up the momentum for building political support to further deepen and widen trust in regional cooperation. The case study acknowledges the challenge of speeding up political buy-in for gas trade and highlights the complexity of its nature due to the lack of recognized regional champions.

Further, the proposed integration plan under the PAEM only seeks to achieve progressive regional integration of electricity production. It, however, fails to provide for trade in gas, which, arguably, has the potential to address the region's energy supply needs. A trade-oriented approach to the regional-integration efforts would go a long way to facilitate the achievements of the PAEM, while ensuring the sustainability of energy provision within the region. This case study makes the point that market

27. Saudi Arabia Signs MOU to Establish Arab Market for Electricity (*Asharq Al-Awsat English*—April 6, 2017) http://english.aawsat.com/asharq-al-awsat-english/business/saudi-arabia-signs-mou-establish-arab-market-electricity, accessed on April 13, 2017.

liberalization is a prerequisite for market integration and that, with the exception of movement of labor, trade and other economic linkages have remained weak in MENA.

Despite the lack of consensus on the exact extent of the geopolitical area referred to as MENA, it is generally accepted that the region covers the area between Morocco in northwest Africa, Iran in southwest Asia, and Sudan in northeast Africa. The World Bank recognizes Algeria, Bahrain, Djibouti, Egypt, Iran, Israel, Iraq, Jordan, Kuwait, Lebanon, Libya, Morocco, Oman, Qatar, Saudi Arabia, Syria, Tunisia, United Arab Emirates (UAE), Palestine, and Yemen as the constituent countries of the region.[28] While other commentators sometimes include Ethiopia, Mauritania, Sudan and Turkey in this list.[29] Although the countries of MENA or "the region" refer to Algeria, Bahrain, Djibouti, Egypt, Iran, Israel, Iraq, Jordan, Kuwait, Lebanon, Libya, Morocco, Oman, Qatar, Saudi Arabia, Sudan, Syria, Tunisia, UAE, Palestine, and Yemen, for the purposes of this case study, the two non-Arabic-speaking countries, namely Iran and Israel, are not included in the analysis.

In recent years, the region has been devastated by civil wars in Syria, Iraq, Libya and Yemen. This has resulted in loss of human lives, damage to physical infrastructure and a refugee crisis that have threatened the economic potential of the region. The United Nations High Commissioner for Refugees (UNHCR) reports that approximately five million Syrian refugees have fled to various parts of the MENA region, including Egypt, Iraq, Jordan, Lebanon and Turkey.[30] The oil-rich countries within the region have also had to grapple with low oil prices, youth unemployment and largely undiversified economies. Consequently, these challenges have become an immutable consideration for policy in, and in respect of the region.

Despite these challenges, the region holds enormous potential both for economic growth and political cooperation. Indeed, cooperation within the region could, arguably, be the most promising avenue for addressing the challenges in the region. It is indeed surprising that the MENA region has not been able to capitalize on its wealth and power position stemming from oil and gas to become a strong regional trading bloc. This potential, in turn, may be the key to overcoming the hurdles to cooperation, especially if one takes into account that the Arab countries of the MENA regions share the same language, culture, religion and history.

After this introduction, section 2 of this case study contributes to the debate on energy security within the MENA region by highlighting various aspects of the energy sector, specifically electricity and gas, which are relevant to the discussion on regional integration. It outlines current and future efforts at developing these markets within the region. The case study makes the case that, to achieve security of energy supply, two

28. The World Bank, Regional Overview MENA available at http://www.worldbank.org/en/region/mena, accessed on April 7, 2017.
29. Middle East and North Africa—MENA (*Investopedia*) http://www.investopedia.com/terms/l/lion-economies.asp accessed on April 7, 2017; OECD-MENA Ministerial Conference 2016; United Nations, Office of the High Commissioner for Human Rights, Mena Region Index, available at http://www.ohchr.org/EN/Countries/MENARegion/Pages/MenaRegionIndex.aspx accessed on April 7, 2017.
30. Syria Regional Refugee Response—United Nations High Commissioner for Refugees (UNHCR) available at http://data.unhcr.org/syrianrefugees/regional.php, accessed on May 17, 2017.

main factors are important: diversification to minimize risk and regional cooperation. It is suggested that the MENA region should reduce or eliminate the technical barriers to energy trade. Section 3 concludes the case study with some suggestions on how to move forward.

2 Energy Trade Situation in the Arab World

2.1 Regional Electricity Trade

Engaging in electricity trade is beneficial for countries, but very little is actually traded (see Figure A1 of the Annex in this chapter). This section provides an overview of electricity trade and an analysis of future trends.

2.1.1 Initiatives Overview

Parts of the MENA region have experienced a spurt in economic growth in recent years. Attendant with this level of economic activity has been a growing demand for energy, which has outstripped current supply levels. It is a well-known fact that when there is less demand, there is more supply. Figure 5.1, however, shows that demand in the region was expected to double by 2020, requiring the provision of additional generation capacity. Growing energy consumption, both by industry and households, necessitates a plethora of measures to ensure sustainable provision of energy to meet every day needs. Attempts to address this problem have included increased conservation policies and tariff adjustments. However, it is axiomatic to say that the solution to the problem lies in expanding generation capacity and supply of energy in an environmentally sound and sustainable manner.

Figure 5.1 Supply and Demand Projections

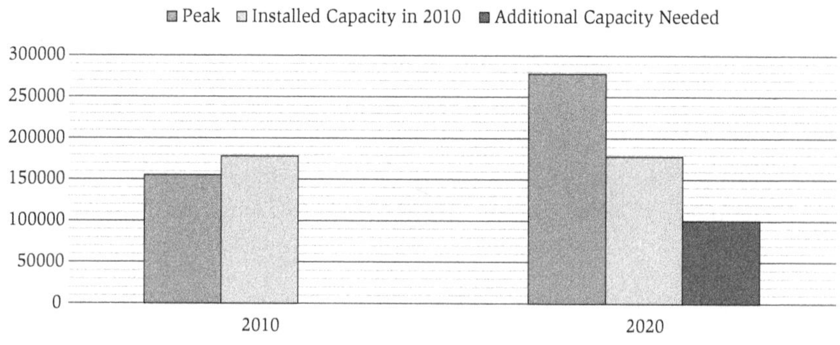

Source: Data reflects values provided by World Bank (2013).

This situation has proven to be a challenge for most countries in the region. Despite oil and gas reserves, domestic production of electricity in Jordan, Saudi Arabia, Kuwait, the UAE, and others has in recent years proven a challenge in meeting the needs of their respective power sectors.[31] This reality has resulted in the need to import both gas and electricity to augment domestic supply. This has, in turn, required the need to construct cross-border infrastructure to facilitate international trade. However, these facilities have proven insufficient to promote the required level of trade to address the energy challenge within the region.

In addition, in recognition of the benefits of regional cooperation, several bilateral and subregional initiatives have been adopted to enable countries within the region to exchange and trade in electricity. These include:

a. The Maghreb regional interconnection, which includes Morocco, Algeria, and Tunisia, and which was initiated in the 1950s and evolved into multiple high-voltage transmission interconnections between the three countries. Morocco was connected to Spain in the late 1990s, and Morocco, Algeria, and Tunisia are now all synchronized with the pan-European high-voltage transmission network.
b. The eight-country regional interconnection (Egypt, Iraq, Jordan, Libya, Lebanon, Palestine, Syria, and Turkey [EIJLLPST]), which was initiated in 1988 by Egypt, Iraq, Jordan, Syria, and Turkey as part of an effort to upgrade their electricity systems to a regional standard.[32] Lebanon, Libya, and the Palestinian Authority later extended the agreement to eight countries.[33] Turkey has been fully synchronized to the European grid since 2012, which will further efforts to standardize the EIJLLPST electricity network with the grids in Turkey and Europe.
c. The regional power interconnection of the Gulf Cooperation Council (GCC) allows electricity exchange among its six Member States—Kuwait, Saudi Arabia, Bahrain, Qatar, the UAE, and Oman—under an agreement signed in 2009. The interconnection is targeted at sharing capacity reserve and improving supply reliability to reduce the need for investment in new generation investment costs in capacity. The GCC Member States established an Interconnection Authority (GCCIA) in 2001 and signed a power exchange and trading agreement in 2009. The GCCIA is the most advanced in institutionalizing regional cooperation among all the MENA subregions. It has a clear legal and institutional framework with a 12-member Board from each GCC member country and a chief executive officer who oversees four operational divisions:

31. The World Bank, *Middle East and North Africa Integration of Electricity Networks in the Arab World Regional Market Structure and Design* (Report No: ACS7124, December 2013) 12.
32. Although the EIJLLPST Interconnection Agreement was signed in the 1980s, physical interconnections between Syria and Jordan and between Syria and Lebanon were constructed in the 1970s.
33. According to the Minutes of the Eighteenth Annual Meeting of Medelec, the EIJLLPST interconnection will be renamed the Electrical Interconnection of Three Continents (EITC) at the next ministerial meeting.

system operations and maintenance, engineering and planning, finance and accounting, and support services. Recently, more focus has been given to accelerating electricity exchange and trade.

d. Saudi Arabia-Egypt new asynchronous interconnection (HVDC, 3000MW). Project implementation is well underway, with commercial operation scheduled in 2017/2018. There are also plans to construct interconnections between Saudi Arabia and Jordan, and between Saudi Arabia and Yemen.

2.1.2 *Future Trends*

Following various research studies undertaken both by the LAS and the World Bank, it became clear that energy supply in the MENA region would benefit immensely from a fully integrated electricity grid. The PAEM is a result of efforts at improving the interdependencies of electricity systems in the region. It is hoped that it will enable countries within the Region exploit their electricity trade potential and eventually expand to the European and Asian markets.

The PAEM envisages a four-stage transitional plan toward the ultimate goal of regional market integration. It is hoped that the phased approach to regional integration would give countries enough time to address any obstacles, and mitigate any risks that might impinge the integration effort. The first stage would focus on existing interconnections with the view to improving their infrastructure and increasing trade opportunities in the existing subregional interconnections.[34] The second stage would see the opening of access to the market to wholesalers and large consumers who are connected to the transmission system, both of whom are important players to make the market work. This would improve competitiveness in the electricity market and improve on sustainability and supply security.[35] The third stage would be the introduction of full wholesale competition in the three subregional markets. This will enable full participation of distributors, suppliers and large customers in the markets and thereby improve the efficiency, economy and transparency of the market.[36] The final stage will see a fully functional, interconnected and synchronized regional market, complete with governing institutions and rules.[37]

2.2 *Regional Gas Trade*

2.2.1 *Initiatives Overview: Liquefied Natural Gas in the Arab World*

The MENA region has currently no shared vision for regional gas trade. This is fairly surprising given the promise of the region in terms of gas resources. Indeed, the MENA

34. The World Bank, *Middle East and North Africa Integration of Electricity Networks in the Arab World Regional Market Structure and Design* (Report No: ACS7124, December 2013).
35. *Ibid.*
36. *Ibid.*
37. *Ibid.*

region holds 41% of world gas reserves and is responsible for 20% of world gas production. However, only 10% of the MENA exported gas is traded within the region.[38] The main reason why MENA nations export much-needed gas to the outside world, rather than within the region itself, lies on the energy-price subsidies in several of these countries. Fuel-price subsidies bring about price distortions, inflate domestic demand and hinder investment in trading opportunities.[39] In view of the rapidly rising demand for gas, together with the abundant export surpluses (*e.g.* Qatar and Algeria) and increasing demand displayed by several countries, there is a case to be made for the promotion of cross-border gas trade in the MENA region. Other reasons that explain why MENA nations export gas to the outside world include: geopolitics (e.g., the Dolphin Gas pipeline project not reaching Bahrain and Kuwait); rivalry behavior (e.g., between Saudi and Qatar, or Egypt and Qatar), and lack of trust in the security of supply from neighbors (e.g., the Arab Gas Pipeline project and an increasing use of liquefied natural gas (LNG)).

Even though the MENA region is endowed with tremendous gas reserves, production has failed to quench demand growth. The "thirst" for natural gas motivates the exploration of alternative ways to access supplies other than through existing pipelines in the region. In this context, LNG emerges as a suitable solution. Indeed, LNG presents the invaluable advantage of transport flexibility which allows its effective geographic diversification.[40] This point is of vital significance in the light of the political convulsion that is sweeping the region. Indeed, geographically confined pipelines are more easily exposed to terrorist outbursts. Investing in floating storage and regasification units (FRSU), as opposed to permanent LNG terminals, is also a further avenue worth exploring for countries that are bearing the brunt of the current crisis.

The MENA region counts with the world's largest LNG exporter since 2006, namely Qatar.[41] Qatar's ascendancy on this energy source is striking to say the least. This is even more the case in the MENA region. For instance, imports by consumer countries in the region amounted to 10.5 billion cubic meters (bcm) in 2015, 40% of which were single-handedly supplied by Qatar.[42] Indeed, the region is experiencing a pressing shortage of gas which has prompted several countries to seek LNG cargoes. Egypt and Jordan received their first LNG shipments in 2015; Kuwait and Bahrain are

38. World Bank, "Pan-Arab Regional Energy Trade Platform," available at https://www.ief.org/_resources/files/events/ief15-ministerial/world-bank-ief15-parallel-roundtable-2-presentation.pdf.
39. World Bank, "Opportunities for Gas Trade in the MENA Region," p. 4, available at https://www.ief.org/_resources/files/events/5th-ief-igu-ministerial-gas-forum/alexander-huurdeman--session-2.pdf.
40. G. Brunekreeft and F. Guliyev, "Gas Supply Security and the Competitiveness on the European Gas Market," in C.J. Jepma (ed.), *Gas Market Trading*. Energy Delta Institute, 2009, p. 103.
41. US Energy Information Centre, "Qatar," p. 12, available at http://www.europarl.europa.eu/meetdocs/2009_2014/documents/darp/dv/darp20140213_04_/darp20140213_04_en.pdf.
42. Arab Investment Petroleum Corporation, "MENA: LNG's top growth target," p. 1, available at http://apicorp-arabia.com/Research/EnergyReseach/2016/APICORPEnergyResearch_V01_N09_2016.pdf.

studying the construction of permanent import terminals; while the UAE is chartering FRSUs as a temporary and lower-cost solution, for instance.[43]

LNG exports in countries such as Egypt, Oman, the UAE and Yemen have been decreasing or have stagnated altogether, while the region's LNG imports have risen to meet domestic demand.[44] One of the reasons for this increase lies in the limited level of regional pipeline gas trade. For example, the Dolphin Gas Pipeline project alone, through which Qatari gas is transported to the UAE and Oman, represents the majority of intra-Arab gas trade.[45] While Qatar is the undisputed champion in the region, Iran could arguably also become a major LNG gas exporter in the medium or long term.

The case of Iran is truly singular, as it holds the second-largest proved natural gas reserves in the world, behind Russia.[46] Yet, the country has been sorely lashed by war, political unrest and international sanctions to such an extent that it has been unable to harness its momentous energy wealth.[47] Following the lifting of the sanctions in January 2016,[48] Iran has made a promising resurgence, at least in terms of crude oil production.[49] The question now is whether Iran can do the same with its natural gas output. Before the sanctions, Iran had lined up six LNG production facilities that were expected to export 95 bcm. However, these projects were cast aside as the sanctions regime proved to be an excessive burden for Iran to attract foreign investment.[50]

This long-lasting political stand-off has also taken its toll on Iran's energy infrastructure. Indeed, Iran is more likely to grant less priority to LNG ventures due to the high costs, technological gaps and cheaper pipeline alternatives.[51] However, the fact remains that Iran has the potential to become a key gas supplier in the region once the adequate infrastructure is in place. Its swift economic recovery suggests that it could well decide to broaden its horizons by investing in LNG as a means to access alternative markets. For instance, Asia-Pacific markets such as China, India or Japan are particularly attractive as demand for LNG is unwaveringly high.[52] These countries are the driving force for natural gas, even if LNG is still a minor form of natural gas. Such a development could arguably entail positive prospects for the MENA region as

43. *Ibid.*, p. 2.
44. Arab Investment Petroleum Corporation, "Competition stiffens for MENA LNG exporters," p. 1, available at http://apicorp-arabia.com/Research/EnergyReseach/2016/APICORPEnergyResearch_V02_N01_2016.pdf.
45. *Ibid.*
46. US Energy Information Administration, "Iran," p. 1, available at https://www.eia.gov/beta/international/analysis_includes/countries_long/Iran/iran.pdf.
47. R. Leal-Arcas, J. Alemany Ríos, and Grasso, C. "The European Union and Its Energy Security Challenges," 8:4 *Journal of World Energy Law and Business* 307 (2015).
48. BBC, "Iran Nuclear Deal: International Sanctions Lifted," available at http://www.bbc.co.uk/news/world-middle-east-35335078.
49. *See* Arab Investment Petroleum Corporation, "Iran's Impressive Return: More to Follow?," p. 1, available at http://www.apicorp-arabia.com/Research/EnergyReseach/2016/APICORPEnergyResearch_V02_N03_2016.pdf.
50. Arab Investment Petroleum Corporation, "Competition stiffens for MENA LNG exporters," p. 1, available at http://apicorp-arabia.com/Research/EnergyReseach/2016/APICORPEnergyResearch_V02_N01_2016.pdf.
51. *Ibid.*, p. 4.
52. Frost and Sullivan, "LNG Industry Is Set for a Transformation in the Middle East," *LNG Industry Magazine* 5-8 (2015).

well. Indeed, as depicted in Figure 5.2, Iran could emerge as a major regional supplier in addition to the well-established Qatar, given the increasing role of LNG in energy security, bearing in mind that energy security for energy exporters means diversification of the energy market, whereas for energy importers it means diversification of energy supply.

Figure 5.2 Linking Central Asian Gas to MENA

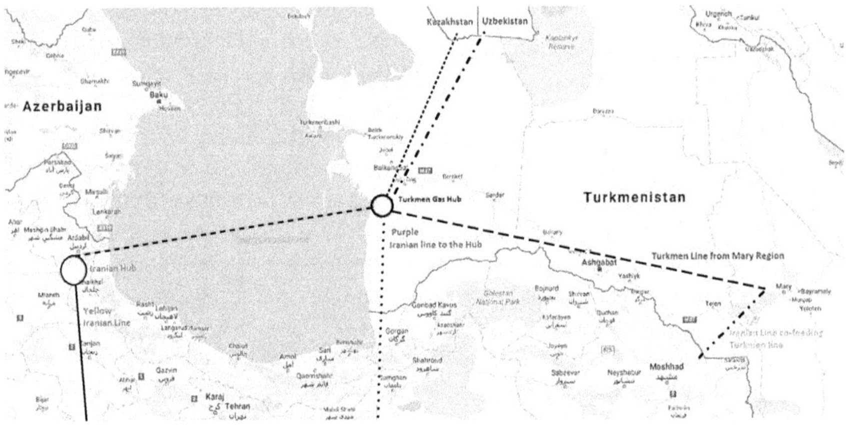

Note: Clockwise, the first two lines denote pipelines providing gas from Kazakhstan and Uzbekistan to the Turkmen gas hub (represented in a circle). The last line clockwise represents the Trans-Caspian gas pipeline project that would link the Turkmen and the Iranian gas hubs. Although the Trans-Caspian pipeline project would be the easiest option physically, it would not materialize without Russian participation as one of the Caspian countries. Clockwise, the third line represents a gas pipeline of Turkmen fields in the Mary region. Its extension to the southwest represents Iranian imports of Turkmen gas for Iran's Northern provinces. Iran could provide gas either from its own fields or/and from the South Pars field in the Persian Gulf, which it shares with Qatar. Clockwise, the fourth line represents Turkmen gas going to Iran, whereas the straight black line denotes a pipeline from Northern Iran toward Qatar.

Generally speaking, domestic supply in Algeria, Oman and the UAE is failing to keep pace with demand growth. Conflicts in Yemen are disrupting LNG production and resulting in power outages. Iran needs to renovate its ageing energy infrastructure. This state of affairs has led to a substantial amount liquefaction capacity being underutilized, especially in Algeria, Egypt, Oman and Yemen.[53]

In a nutshell, MENA countries are directing natural gas production for the domestic market while investing in import infrastructure and capacity. This is the easiest remedy to advance energy security under the prevailing circumstances. In view of the low spot LNG prices, several countries of the region have resorted to FRSUs as a

53. Arab Investment Petroleum Corporation, "Competition stiffens for MENA LNG exporters," p. 4, available at http://apicorp-arabia.com/Research/EnergyReseach/2016/APICORPEnergyResearch_V02_N01_2016.pdf.

quick fix before contemplating more expensive long-term solutions.[54] Moving forward, it appears that Iran (arguably in the medium or long term) and Qatar are the best-placed MENA countries to provide LNG supplies within the region. Pushing for a Pan-Arab Gas Market (PAGM) with these two nations (even if Iran is not an Arab country) on board would set a promising framework for regional gas trade. The promotion of multilateral ventures within the region might constitute a way to secure more Qatari LNG supplies. Furthermore, such projects could cater for the required infrastructure to unlock Iran's tremendous potential.

One can draw the conclusion that, in terms of feasibility (namely level of complexity) to develop major infrastructure projects for gas trade in the MENA region, it is credible and major clients such as Europe as well as other regions do need the gas, and therefore a regional market would be sizeable enough to justify the infrastructure costs.

2.2.2 Suggestions for the Arab World

The European Union (EU), the Association of South-east Asian Nations (ASEAN), and MERCOSUR in South America are revealing examples of how trade schemes can support peace in war-torn or conflict-ridden regions. The Arab region is currently at the epicenter of major ravages triggered by civil wars in Iraq, Libya, Syria, and Yemen. These, in turn, have led to the largest refugee exodus in recent history, thereby placing a strain on already struggling hosting countries such as Djibouti, Jordan, Lebanon and Tunisia. In addition, major oil and gas exporters such as Algeria, Iran and the GCC countries are confronted with low oil prices, youth unemployment as well as undiversified economies.[55] The magnitude of these circumstances needs to be acknowledged when providing a set of recommendations.

The following policy recommendations are geared toward making the MENA energy market operational. More specifically, these recommendations can be inscribed within the implementation of the PAEM in the electricity industry as well as within the creation of a suitable framework for the gas sector (i.e., the suggested PAGM).

1. Investment in trans-MENA energy networks should be encouraged by all available means, including Foreign Direct Investment (FDI).[56] Initiatives such as the Maghreb interconnection, EIJLLPST interconnection or the GCC interconnection are positive precedents. These projects blaze the way for MENA to create the necessary infrastructure to secure energy flows in the region. Such endeavors, may they be bilateral or even multilateral, can be the prelude to a

54. Arab Investment Petroleum Corporation, "MENA: LNG's top growth target," p. 3, available at http://apicorp-arabia.com/Research/EnergyReseach/2016/APICORPEnergyResearch_V01_N09_2016.pdf.
55. World Bank, "Middle East and North Africa region. Overview," available at http://www.worldbank.org/en/region/mena/overview#1.
56. Similar recommendations are made in the "Analytical Foundation for Increased Pan-Arab Regional Gas Trade Report," of the World Bank's Pan-Arab Regional Energy Trade Platform (P157314), May 1, 2017 (on file with author).

well-connected MENA energy market through the progressive fusion of these schemes between the nations of the region involved.

2. As illustrated by the case of the EU, small groups of countries within a region can start successful undertakings that should hopefully attract neighboring countries. These projects will take some time to come to fruition. However, some preliminary decisions should be taken in the meantime. In that vein, the World Bank has already identified the ullage of existing infrastructure; the promotion of bilateral connections; and the interconnections between transmissions systems as some of the quick wins that can be achieved under the prevailing market conditions.[57]
3. Measures to foster the liberalization and opening-up of electricity and gas markets should be adopted. In that sense, a degree of political and infrastructure convergence is essential to establish a PAGM and to reinforce the existing PAEM. Action in the gas sector is particularly urgent as there is currently no shared vision for regional gas trade. This is all the more regrettable considering that the gas reserves located in the area have the potential to address many of the region's needs: secure long-term fuel supply to promote affordable and reliable electricity, reduction of power outages, and diversification of the region's energy supply.
4. It should therefore be explored whether the political willingness to enhance regional gas trade exists for the sake of all countries involved. This should most likely be the case of states that are grappling with low oil prices and are in dire need of diversifying their economies. Therefore, countries such as Algeria, Iran and the GCC countries are the ones with more at stake to push for the establishment of a regional gas market. Indeed, gas is their clearest alternative to achieve economic diversification. Counting with nations endowed with major gas reserves or that are major LNG exporters (such as Iran and Qatar, respectively) is already a promising default setting.
5. The creation of specialized MENA bodies to facilitate cooperation between transmission system operators (TSOs) and energy regulators should be encouraged. European TSOs and the European Agency for the Cooperation of Energy Regulators in the case of the EU, or independent institutions that serve as think tanks for generic energy policy issues such as the ASEAN Centre for Energy, are specialized institutions that play an essential role in the energy policy of their respective regions. Such entities are necessary to regulate the various aspects of energy integration between Member States. Such specialized bodies would directly contribute to the liberalization of the electricity and gas markets.

For instance, TSOs are vital to address technical issues, but also to push policymakers to take the required successive decisions to implement and advance the

57. World Bank, "Opportunities for Gas Trade in the MENA Region," p. 5, available at https://www.ief.org/_resources/files/events/5th-ief-igu-ministerial-gas-forum/alexander-huurdeman--session-2.pdf.

nascent electricity market in the MENA region and an eventual MENA gas trade market. Furthermore, such entities are the best-suited avenues to centralize a specific aspect of energy between several participating Member States. An example of an initiative for energy integration is the South Asia Regional Initiative for Energy integration. South Asia could be an example of regional cooperation that MENA could emulate with its own specificities and characteristics. A way forward is to strengthen regional electricity and gas trade cooperation based on sharing the benefits for the region's economic growth, political and social stability, energy security and sustainability.

3 Conclusion

While it is trite knowledge that several countries within the MENA region have successfully exploited their energy resources, there is sufficient evidence that the global landscape for energy production, trade and consumption is changing—changes which necessitate a re-evaluation of the way in which energy is traded in the region. These countries need to change their approach toward energy trade, not only to maintain a position of relevance in global energy trade, but also to ensure that the socioeconomic benefits of energy trade are optimized at a regional level. It is also clear that a significant amount of cross-border cooperation is required for this, both from within and without the region.

The current high political climate and instability in the region has resulted in a situation where foreign investors are cautious about investing in the region. De-risking investment is imperative. Many investors would like to see their business model in the MENA region. New investments on power projects and cross-border transmission interconnections are necessary. Moreover, utilizing the existing interconnections is a must.

Furthermore, the PAEM would facilitate the greater utilization of existing power trade capacity before aiming at new investment on power projects. The benefits of such integration would be far-reaching and would immensely impact the socioeconomic situation in the region by assisting in the further diversification of what has been, for many countries within the region, an economy heavily reliant on energy trade. This change in paradigm would hopefully occasion a change in policy, which would permit the consideration of greener energy, where the region has enormous potential, and this would go a long way to significantly contribute toward the improvement of the energy security situation within the region.

The advocated investment opportunities for the MENA region will in themselves be insufficient to prompt effective Pan-Arab energy trade. Nevertheless, if the right investments and project choices are made, financial frameworks could well be the ideal catalysts to galvanize the attractive investment climate that the MENA countries crave for. However, a clear and predictable regulatory framework—one subjected to the market rather than governmental decisions—emerges as the *sine qua non* requisite to attract foreign investment. Furthermore, only if the political will of, at least, a few regional champions subscribes, can an eventual economic injection lead to further regional energy ventures in the MENA region. Such projects could represent the building blocks toward the Pan-Arab Regional Energy Trade Platform. The World Bank can contribute to this effort by engaging international and regional development partners (*e.g.*, African Development Bank, Arab Fund for Economic Development, Islamic Development Bank as well as the EU and its partners).

While acknowledging the current lack of proper regional institutions and political consensus, the objective of the MENA countries should be to reach a cohesive energy policy for the MENA. Under the prevailing circumstances, the progressive integration of regional initiatives (i.e., a bottom-up approach) seems more suitable than pressing for a MENA common position (i.e., a top-down approach) for the creation of an internal energy market. Yet, without regional institutions, it will be difficult to enhance energy trade in the region. Although a harmonized position by the MENA region would undoubtedly be preferable to an array of discordant national measures, pushing for a MENA common position in such a delicate and versatile matter as energy might prove counterproductive, especially if the MENA countries involved lack the political will to do so. Without establishing a proper institution such as a TSO or a regional regulatory institution, it may prove difficult to move the agenda forward.

Such a bottom-up approach may not lead to a truly internal energy market any time soon. Nonetheless, it offers the opportunity of securing reliable foundations for the project, given that a bottom-up approach is more likely to guarantee the effective commitment of the cluster of MENA countries willing to embark in such an undertaking. Although we propose intra-MENA cooperation to boost energy trade, the notion of an Energy Union, pioneered by the European Union, could be applied to other regions of the world, especially developing countries, which are in great need of achieving sustainable energy. For instance, in the same way that the EU is currently using an Energy Union to (1) decarbonize its economy by 2050, (2) to enhance its energy security, (3) to increase its energy efficiency, (4) to finalize its internal energy market, and (5) to invest in technology and innovation (these last two points being reasons to be hopeful), one could propose ways to create a MENA Energy Union, an African Energy Union, an ASEAN Energy Union, and a Mercosur Energy Union, to obtain similar benefits in the economies of those regions (namely MENA, Africa, ASEAN, and Mercosur). This idea is all the more imperative if we consider that two thirds of greenhouse gas emissions come from energy. This would be a rather long-term aspiration for the MENA region, as prior effective and progressive regional integration remains indispensable.

There is reason for optimism in the MENA region because, in addition to large amounts of fossil fuels in some MENA countries, MENA is blessed with an abundance of renewable natural resources and, therefore, has the potential to be the number one region in the world in renewables. Renewable energy and trade in (renewable) energy, in turn, could be the catalyzer for EU-MENA relations, for instance, by linking Morocco and Libya to the south of the EU. The MENA region could export electricity from renewables to the EU. Prices of renewables have gone down considerably in recent years and it is expected that the trend will continue in years to come. Thus, promoting renewable energy has the potential to result in regional cooperation. The region is endowed with copious solar irradiance and wind power also has significant potential.

The chronic need for economic diversification, coupled with the resolution to press for energy sustainability, are further grounds to engage in the promotion of renewable energy. Further optimism comes from Goldman Sachs's prediction that new wind turbines and solar panels worldwide will provide more energy over the next five years than US shale-oil production has over the past five years.[58]

The MENA region is bound to be at a crossroads shortly. The region may find itself in a similar juncture to that of Europe in the aftermath of World War II. Despite simmering with political turbulence, the Arab world has the basic tools to promote peace through trade, in this case via the anticipated Pan-Arab Regional Trade Platform. Such tools are the potential for the creation of a MENA single market based on cooperation, given their shared values, religion, language, culture and history. Indeed, energy cooperation marked the genesis of the EU, perhaps the most manifest success story in the world in terms of regional integration. If, prior to Brexit, 28 European countries with such disparity of languages and cultures could come together to achieve shared prosperity in the context of the EU, there is even greater reason to believe that the Arab world could and should do the same. All Arab countries share the same language, religion, and culture. The MENA region has already a promising default setting with the PAEM in place as well as other multilateral schemes not only in electricity but also in the gas industry.

Moving forward, action should primarily be targeted toward regional gas trade, which is arguably at an embryonic phase, due largely to politics and prices distortions. The task of creating a market for regional gas trade is even more imperative considering that almost half of the world's gas reserves are located in the MENA region.[59] Thus, a Pan-Arab Gas Market should follow suit to the establishment of the PAEM. These two initiatives would streamline smoother cooperation in the electricity and gas sectors. In addition, the region should focus on a bottom-up regional approach to overcome prevalent distrust in the region as a building block toward a vision of a PAGM. Given MENA's geographical proximity to the EU, especially the North African countries, it seems a good idea to create alliances with it, but also with the African Union, Central Asia, South Asia and Russia.

Finally, it appears that Iran (arguably in the medium or long term) and Qatar are the best-placed MENA countries to provide liquefied natural gas supplies within the region. Pushing for a PAGM (even if Iran is not an Arab country) with these two nations on board would set a promising framework for regional gas trade.

58. Martin, C. "Wind, Solar Power to Supply More Energy Than Shale, Goldman Says," Bloomberg, November 30, 2015, available at http://www.bloomberg.com/news/articles/2015-11-30/wind-solar-power-to-supply-more-energy-than-shale-goldman-says.
59. World Bank, "Pan-Arab Regional Energy Trade Platform," available at https://www.ief.org/_resources/files/events/ief15-ministerial/world-bank-ief15-parallel-roundtable-2-presentation.pdf.

III EU-RUSSIA ENERGY TRADE

1 Introduction[60]

The energy sector is a vital segment of any economy, be it a sovereign state, a political or an economic union of states. Hence, securing essential supply of energy, by internal or external means, is a crucial goal for any sovereign establishment. Complications come into play when several sovereigns act both in a united and individual capacity, depending on the context, as in the case of the European Union ("EU," the "Union").

This case study aims to provide an analysis of the EU external relations in the energy sector, focusing on the EU-Russia energy trade relations and the Energy Community. The hypothesis we have undertaken is that securing reliable flows of diverse sources of renewable energy would foster the EU's energy security and decrease its vulnerability toward major fossil fuel exporters. There are currently efforts toward much greater investment in renewables in the EU and major oil companies will need to learn from renewables companies if they wish to enter this new field. In fact, the pandemic created by COVID-19 has offered the world what a renewables future might look like.

For instance, Rome-based Enel (which is the EU's largest utility) intends to invest EUR 160 billion by 2030 to triple its renewable-energy capacity and transform its grids in Europe and Latin America in order to be ready for an electric future.[61] Spain's Iberdrola, another very large energy company, has pledged to invest EUR 75 billion in renewables and grids by 2025.[62] The same green trend is true in other parts of the world outside the EU. For instance, even if under President Trump, the US became the world's largest producer of crude oil[63]—which does not mean that the US is energy independent[64]—America's NextEra, a pioneering utility, has also decided to invest heavily in wind and solar energy.[65] China is also investing heavily in renewables.[66] The scope of this case study, however, is limited to the participation of the EU in the Energy Charter Treaty (ECT),[67] the Energy Community Treaty, and a general analysis of the EU-Russia energy trade relations.

After this short introduction, Section 2 provides an overview of pertinent data related to the EU's energy production and consumption, as well as current trends and challenges the EU faces in the energy realm. Section 3 covers related competence issues with regard to the EU treaties. Section 4 examines the energy-related treaty regimes that

---

60. Sections of this case study have been co-written with Iuliia Dragunova.
61. The Economist, "Schumpeter: The Climate Centurion," November 28, 2020, p. 64.
62. *Ibid.*
63. Yet, the US's oil boom is going down and coal-fired plants seem to be on their way out. *See* The Economist, "Donald Trump's Record: Pumped Up," October 24, 2020, pp. 34-35, at 34.
64. In 2019, the US imported 9.1 millin barrels a day of petroleum. *Ibid.*
65. The Economist, "Schumpeter: The Climate Centurion," November 28, 2020, p. 64.
66. Ideally, many Asian countries that are not yet ready to make the move to renewable energy should develop natural gas instead of coal in the coming years. This move may not be the best option for climate-change mitigation, but it is a step in the right direction.
67. R. Leal-Arcas (ed.), *Commentary on the Energy Charter Treaty*. Edward Elgar Publishing Ltd, 2018.

the EU is a part of, whereas section 5 analyzes the EU's participation in the Energy Community (EnC) and related issues. EU-Russia energy relations concerning trade in oil and gas are analyzed in section 6, which is followed by the conclusion with recommendations in section 7.

2 EU Energy Consumption and Production, Current Trends and Challenges

2.1 EU Consumption

The EU is one of the world's largest consumers of energy. In 2017,[68] the EU-27[69] held the third place in the total energy supply, the index which shows the amount of energy needed to satisfy total energy consumption demand of a particular geographical entity.[70] Petroleum products constituted 34.8% of the gross inland energy consumption, natural gas—23.8%, solid fossil fuels—13.6%. Hence, 72.2% of the EU-27 energy consumption was based on fossil fuels while allocating the remaining 12.6% and 13.9% of the gross inland energy consumption to nuclear heat and renewable energy sources respectively.[71]

Nevertheless, the distribution of shares of energy sources in gross energy consumption in different EU Members States depends on the economic structures, available natural resources, and national choices of particular countries.[72] It differs tremendously across the EU landscape.[73] For instance, the largest share of oil consumption was determined in Cyprus (92.5%) with it being a small island lacking natural resources and Luxembourg (63.5%) affected by "fuel tourism."[74] Renewable energy constituted more than 40% of energy sources in gross energy consumption only in Latvia (42.5%) and Sweden (41.5%), where in Luxembourg it amounted to merely

68. For all statistical data provided in this section, data of 2017 will be used, unless explicitly stated otherwise. In 2017, Eurostat issued the relevant data on the EU-27, the EU-28 and energy.
69. For the purposes of this chapter, the EU-27 excludes the United Kingdom, which formally left the EU in 2020.
70. Eurostat, *The EU in the World—2020 Edition* (Luxembourg: Publications Office of the European Union, 2020), https://ec.europa.eu/eurostat/documents/3217494/10934584/KS-EX-20-001-EN-N.pdf/8ac3b640-0c7e-65e2-9f79-d03f00169e17, 119, accessed August 1, 2020.
71. Eurostat, *Energy, Transport and Environment Statistics—2019 Edition* (Luxembourg: Publications Office of the European Union, 2019) https://ec.europa.eu/eurostat/documents/3217494/10165279/KS-DK-19-001-EN-N.pdf/76651a29-b817-eed4-f9f2-92bf692e1ed9, 20, accessed August 1, 2020.
72. For further details on EU energy law and policy, *see* R. Leal-Arcas (ed.), *EU Energy Law and Policy: The External Dimension*. Eliva Press, 2020; R. Leal-Arcas et al., *The Great Energy Transition in the European Union*, Volumes 1 and 2. Eliva Press, 2020; R. Leal-Arcas and J. Wouters (eds.), *Research Handbook on EU Energy Law and Policy*. Edward Elgar Publishing Ltd, 2017; R. Leal-Arcas, C. Grasso, and J. Alemany Rios, *Energy Security, Trade and the EU: Regional and International Perspectives*. Edward Elgar Publishing Ltd, 2016; R. Leal-Arcas, *The European Energy Union: The Quest for Secure, Affordable and Sustainable Energy*, Claeys & Casteels Publishing, 2016; R. Leal-Arcas, A. Filis, and E. Abu Gosh, *International Energy Governance: Selected Legal Issues*, Edward Elgar Publishing Ltd, 2014.
73. Eurostat, *Energy Statistics* 20.
74. *Ibid.* at 21.

6.3%.[75] Natural gas share was fluctuating from 39.5% in the Netherlands to less than 2% in Sweden and Cyprus.[76]

Renewable energy constituted more than 40% of the energy consumption mix only in two countries, Latvia (42.5%) and Sweden (41.5%), where in Luxembourg it amounted to only 6.3%.[77] In the evolution of the final (as opposed to gross inland index) energy consumption, the share of renewable energy sources has grown from 3.8% in 1990 to 9.7% in 2017.[78] The EU commitment to expand its share in its total energy consumption to 20% by 2020[79] seems to be an overestimation by now.

2.2 *EU Production and International Trade Supply*

In 2017, the EU's primary production of energy constituted only 44.9% of its total demand where the rest 55.1% have been secured from the external energy trade partners.[80] Renewables and waste energy sources constituted amounted to 29.9% of the primary energy production of the EU when around 27.8% was supplied by means of nuclear energy.[81] For earlier years, 66% of EU's primary production used to be attributed to just six member-states: the United Kingdom (UK), France, Germany, Poland and the Netherlands.[82] Over the decade of 2007-2017, the trend for fossil fuels primary energy production dropped tremendously while, on the opposite, a trend for renewable energy sources witnessed a 65.6% increase.[83] Nevertheless, although decreased, the 2017 EU total production amounted approximately to 4% share in petroleum and 13.6% share in natural gas, while being spread unequally among states.[84] It still required the gap between demand and supply to be closed by means of external trade relations.[85]

This data is illustrative of an idea of energy dependency where the shortfall of production has to be met by net imports. The EU-28 energy dependency increased throughout last two and a half decades with its net imports of energy being larger than its primary production (from 44% in 1990 to 55.1% in 2017).[86] Since 2004, the EU's total dependency rate surpassed 50% rising to its highest in 2017—55.1% (with highest

75. *Ibid.*
76. *Ibid.*
77. *Ibid.*
78. *Ibid.* at 19.
79. European Commission, "Communication from the Commission: Europe 2020. A European Strategy for Smart, Sustainable and Inclusive Growth," https://eur-lex.europa.eu/legal-content/EN/TXT/PDF/?uri=CELEX:52010DC2020&from=en 32 (accessed December 18, 2020). *See also* Rafael Leal-Arcas, "Smart Grids as a Way to Democratize Energy," *Yale Journal of International Affairs* (2020).
80. Eurostat, *Energy Statistics*, 19.
81. Eurostat, *The EU in the World*, 115.
82. *See* Rafael Leal-Arcas and Andrew Filis, "Conceptualizing EU Energy Security Through an EU Constitutional Law Perspective," 36 *Fordham International Law Journal* 5, 1239 (2013).
83. Eurostat, *Energy Statistics*, 10.
84. *Ibid.* 11-12.
85. Eurostat, *The EU in the World*, 115.
86. Eurostat, *Energy Statistics*, 19.

rates for oil (86.7%) and natural gas (74.3%)).[87] However, dependency rates with regard to various energy sources are also spread unevenly among the EU Member States. For instance, Denmark is entirely energy imports independent unlike Cyprus or Malta.[88] Although the EU has ambitious goals in switching to renewables (and is comparatively successful in doing so vis-à-vis other countries), this data reinforces the importance of the EU's external relations in the energy sectors, particularly those in energy trade, to fulfill its needs in fossil fuels demand. This raises another pertinent issue: the energy security of the Union.

2.3 *Trends and Challenges in the EU External Relations in the Energy Sector*

First and foremost, presented data delineates the disparity between the energy realities of EU Member States. Difference in energy mixes of Member States is mainly contingent on natural resources and industrial assets available, as well as political-economic relations with external energy suppliers. Such energy realities are crucial for lobbying the policies that separate members would aspire to promote at the EU level.[89] For instance, where Poland or the Baltic states might be opposed to intensifying the EU-Russia energy integration, Germany on Luxembourg could, on the contrary, pursue it. This leads to the lack of an always coherent stance with regard to the external energy relations of the Union from within of the establishment.

The EU is generally highly dependent on external energy supplies. The Commission recognizes that the more globalized the energy market becomes, the more the EU increases its vulnerability toward global trends, *e.g.*, price fluctuations or energy supply shocks.[90] Hence, pursuing strategic bilateral and regional external relations with energy suppliers reflect the EU's interests to ensure energy security and reliable supply flows.[91]

For decades, the European external energy policy has been primarily shaped by the legacies of single events, *e.g.*, the 1973 oil crisis, the 2006 and 2009 Ukraine gas crisis, the 2011 Fukushima nuclear incident, the 2014 Russia-Ukraine crisis.[92] These events led to new forms of trade and new energy mixes.[93] However, in the last decades, the EU admitted the need for common policy. For example, in 2011, the EU Commission stated that the development of a coherent external EU energy policy would increase the collective EU energy security as well as assist the EU in addressing attempts of external suppliers to influence markets or impose its energy leverage in

87. *Ibid.* 19.
88. *See* Leal-Arcas, "Conceptualizing EU Energy Security," 1237.
89. *Ibid.* 1241.
90. *See* Energy Roadmap 2050.
91. *See* Rafael Leal-Arcas, Juan Alemany Rios, and Costantino Grasso, "The European Union and Its Energy Security Challenges," 8 *Journal of World Energy Law and Business* 292 (2015).
92. *See* Luca Franza, Coby van der Linde, and Pier Stapersma, "The Internal and External Dynamics of EU Energy Relations," 72 *Clingendael Spectator* 2 (2018), https://spectator.clingendael.org/pub/2018/2/eu-energy-relations/.
93. *Ibid.*

political realm.[94] In 2015, the Foreign Affairs Council adopted Council Conclusions on EU Energy Diplomacy that contained the EU Energy Diplomacy Action Plan.[95] Decarbonization and diversification of the energy mix have been the central points of the EU energy policy that aims at promoting energy security, decreasing energy dependency and protecting the environment have been. Pursuant to the Title XXI of the Lisbon Treaty, the EU energy policy is clearly linked to the EU's environmental goals. However, this view has faced certain criticism of the conditionality of the EU energy policy.

To sum up, differences of the energy realities in the EU Member States bear considerable impact on the challenges that lay in front of the EU as a single Union to pursue common energy policy. The EU while being highly dependent on external energy supplies aspires to lower such dependence by means of pursuing energy security and "greening" of its total energy mix.

3 Competence Issues Regarding the EU Treaties

It has never been an easy task to achieve the common stance of the EU external action matters, even less so when they are related to energy. Discrepancies in geopolitical context, energy priorities and energy security capabilities make it more unrealistic to achieve common EU energy security, which is in the realm of the EU exclusive competence.[96] Moreover, energy is a multifaceted concept that lies across a variety of policy areas, such as foreign trade, human rights, foreign policy, national security, economics, etc.[97] This adds complications to the fact that the EU competences are unequally shared between the Union and its Member States depending on particular issue. Moreover, the EU competencies involve actions by a variety of agencies and institutions, with provisions enshrined across the Treaty on European Union (TEU) and Treaty on the Functioning of the European Union (TFEU).[98]

Considerable shift toward fulfilling energy policy and dividing competences has been achieved with the conclusion of the Treaty of Lisbon ("Lisbon"). Before Lisbon, energy was not a distinct area of foreign policy where Lisbon introduced a new energy section and expanded the foreign policy toolkit by establishing a European External Action Service.[99] The treaty enshrines a specific section on energy, which affirms most critical responsibilities and general energy-policy goals. Nevertheless, although the Treaty of Lisbon has reinforced the political value of energy, it has not assigned

94. European Commission, *An External Policy to Serve Europe's Energy Interests* (2006) Paper from Commission/SG/HR for the European Council, https://www.consilium.europa.eu/ueDocs/cms_Data/docs/pressData/en/reports/90082.pdf (accessed 01.08.2020).
95. European Union External Action, *EU Energy Diplomacy* (2016) https://eeas.europa.eu/topics/energy-diplomacy/406/eu-energy-diplomacy_en (accessed 02.08.2020).
96. *See* Leal-Arcas, "Conceptualizing EU Energy Security," 1254.
97. *Ibid.* 1255.
98. *Ibid.*
99. *See* Sascha Müller-Kraenner, *The External Relations of the EU in Energy Policy*. Heinrich Böll Foundation (2010) https://www.ecologic.eu/sites/files/publication/2015/external-relations-of-the-eu-in-energy-issues_0.pdf (accessed 01.08.2020), 6.

considerable additional decision-making power to the Union's institutions. This imposes limitations on the EU's ability to sustain coherent external energy policy.[100]

Although the EU has the institutional capacity to act as a single unit on behalf of all its Member States, it does not possess such capacity with regard to all policy areas, especially with regard to such complex one as energy.[101] The competences of the EU are shared between the Union and its Member States in the following ways: (1) the EU exclusive competences where the Union can exclusively act on behalf of member states to their exclusion; (2) shared competences where both the Union and its Member States have a degree of competence to act upon and usually relate to policy areas, for which treaties do not make express provisions; (3) exclusive competence of the EU Member States where the Union lacks any powers.[102]

As the EU is a customs union, questions related to external trade relations, common commercial policy, and competition lay in the realm of the EU exclusive competence. Hence, according to Article 3 of the TFEU, such energy-related matters as fostering competitive conditions of the energy trade in the internal market or addressing the issue of tariffs when energy commodities enter the EU, are within the exclusive competence of the EU.

Under Articles 4(2)(i) and 194 of the TFEU, energy *per se* (in its wide sense) lies within the realm of shared competencies between the EU and its members. Matters of shared competence may be handled to the EU pursuant to the subsidiarity and proportionality principles when the actions seem to be more efficient at the supranational level.[103] Hence, where the external energy-related policy matter falls within the first category, the EU would enjoy unrestrained powers to pursue it. When the latter category comes into play, it will be necessary to assess how the proposal should be decided at the EU level, unanimously or by majority for the policy to be adopted.[104]

Nevertheless, each EU member-state has the right to determine its own conditions for exploiting its energy resources and determining its energy mix.[105] The compelling energy security measures with distinct diplomatic nature remain within the powers of the EU Member States. This seems to be reasonable from the perspective of preserving certain of their sovereign prerogatives. In addition, implementation of energy-sector treaties concluded by the EU is also within the realm of exclusive competence of Member States.

The complications of the competence allocation are related to the lack of the clear-cut division of such competences. Moreover, different energy priorities and backgrounds of the EU Member States contribute to the difficulty to reach consensus. However, explicit inclusion of energy to the realm of competences along with further development of related policies delineate the steady development in the energy-related domain.

---

100. *See ibid.*
101. *See* Leal-Arcas, "Conceptualizing EU Energy Security," 1227.
102. Articles 3, 4, 6 of the TFEU.
103. *See* Leal-Arcas, "Conceptualizing EU Energy Security," 1252.
104. *Ibid.* 1245.
105. Article 194(2) of the Lisbon Treaty.

4 Participation in Treaty Regimes

The Treaty of Lisbon formalized the legal personality of the Union, so the EU became capable of concluding international agreements and joining international organizations as a single entity,[106] which is fundamental for engaging in external energy relations. Many EU Member States, separately and united, participate in various fora to promote their energy interests, *e.g.*, the ECT,[107] the EnC, the North Atlantic Treaty Organization (NATO), the International Energy Agency (IEA), the International Solar Alliance (ISA), the Organisation for Economic Co-operation and Development (OECD), the Paris Agreement framework, etc. The competence allocation also bears its impact on the conclusion of energy-related treaties.

The EU has bilateral and regional cooperation agreements with all of its neighboring countries. For instance, external energy relations with Norway, which is one of the most critical suppliers of oil and gas for the EU, are governed by the Agreement of the European Economic Area that makes Norway an integral part of the EU's internal market.[108] The European Energy Community Treaty was concluded with the EU's eastern and southern neighboring states to integrate them into the EU internal energy market (it will be analyzed in detail in the next section). The EU concluded the Eastern Partnership association agreement with some of the CIS States to reduce high dependence on Russian energy imports, diversify fossil fuels suppliers and integrate these States in the EU's energy infrastructure. As for the Arctic's region, although Denmark is part of the Arctic Council, the EU does not participate in it *sui juris*.[109]

As for the multilateral efforts, the United Nations (UN) Framework Convention on Climate Change and the Paris Agreement play a fundamental role in the EU securing the reduction of greenhouse gas (GHG) emissions and increasing the share of renewables in the EU's energy mix.[110] The World Trade Organization (WTO) instruments enable the EU to ensure energy transit and various energy-related matters on the supranational level. The EU participates in the International Solar Alliance that promotes efficient exploitation of solar energy to reduce the share of fossil fuels in the energy mix.[111] Moreover, the EU participates in the International Renewable Energy Agency (IRENA). Although these multilateral initiatives are crucial for the EU's

106. Article 216 of the TFEU.
107. Italy withdrew in 2016.
108. *See* Müller-Kraenner, 4.
109. *See* Leal-Arcas, "Conceptualizing EU Energy Security," 1275.
110. For further details, *see* Rafael Leal-Arcas et al., "Of International Trade, Climate Change, Investment and a Prosperous Future," 12:2 *Trade, Law and Development* (2020); R. Leal-Arcas, "Climate Clubs and International Trade Across the European and International Landscape," 29:3 *European Energy and Environmental Law Review* 72-88 (2020); R. Leal-Arcas et al., "The Contribution of Free Trade Agreements and Bilateral Investment Treaties to a Sustainable Future," 23:1 *Zeitschrift für Europarechtliche Studien—ZEuS* 3-76 (2020); R. Leal-Arcas, *Solutions for Sustainability: How the International Trade, Energy and Climate Change Regimes Can Help*. Springer International Publishing, 2019; W. Leal and R. Leal-Arcas (eds.), *University Initiatives in Climate Change Mitigation and Adaptation*, Springer, 2018, R. Leal-Arcas, *Climate Change and International Trade*, Edward Elgar Publishing Ltd, 2013.
111. https://thediplomat.com/2020/05/can-the-international-solar-alliance-truly-be-indias-gift-to-the-world/.

external policy relations, their analysis does not fall within the scope of the current paper. Therefore, we will focus in more detail on the ECT.

The ECT is the most frequently involved treaty in investor-state dispute settlement (ISDS).[112] This treaty was signed in 1994 and demonstrated an attempt to apply WTO's free-trade disciplines to the trans-European energy sector.[113] It came into effect in 1998 and was signed by 41 states including the EU. However, the treaty has not been ratified by the major energy moguls, such as Russia or Norway, with the former suspending the treaty in 2009 after facing the *Yukos* saga. The ECT contains disciplines on energy-sector foreign investment and its protection once such investment has been admitted to the host state territory. Nondiscriminatory treatment of investments (most-favored nation and national treatment standards) is the foundation of the Treaty.

The ECT has been continuously subject to criticism due to undermining the environmental protection initiatives, such as the Green Deal and the Paris Agreement, and allegedly being a vehicle for fossil fuel industries as it hinders states to issue green policies and raise public interest related defenses in ISDS.[114] Renewable energy generally received little attention when the ECT was drafted. For various reasons, the call for its modernization has been roaring throughout the last decade.[115]

In 2012, parties to the treaty agreed to commence the process of modernization during negotiations rounds called the Warsaw Process. However, the Energy Charter Conference, an intergovernmental decision-making body for the ECT process, made an official announcement of beginning of these renegotiation efforts only in 2019 while still lacking firm consensus among its members on the direction of the amendments, especially in relation to investment protection and dispute settlement.[116] The latest (third) round of negotiations took place in November 2020.[117] On December 2, 2020, the representative to the European Commission stated that "[i]f core EU objectives...are not attained within a reasonable timeframe, the Commission may consider proposing other options, including the withdrawal from the ECT."[118]

In light of the global commitment to climate action, the ECT needs a "fundamental overhaul" with the new version encouraging low-carbon investments and discouraging fossil fuels ones, supporting policy measures to accommodate transition to the green economy.[119] It is particularly critical for the EU while being a pioneer in the environmental march. In case such treaty modernization ceases to be feasible, the EU

112. *See* Martin Dietrich Brauch, "Modernizing the Energy Charter Treaty: A Make-or-Break Moment for Sustainable, Climate-Friendly Energy Policy," (2019) *International Institute for Sustainable Development Blog* https://www.iisd.org/articles/modernizing-energy-charter-treaty-make-or-break-moment-sustainable-climate-friendly-energy (accessed 10.08.2020).
113. *See* Müller-Kraenner, 4.
114. *See* Brauch.
115. *Ibid.*
116. *Ibid.*
117. European Commission, "Third negotiation round to modernise Energy Charter" (2020) *European Commission Website*, https://trade.ec.europa.eu/doclib/press/index.cfm?id=2206 (accessed December 18, 2020).
118. "Answer given by Executive Vice-President Dombrovskis on behalf of the European Commission," P-005555/2020, December 2, 2020, https://www.europarl.europa.eu/doceo/document/P-9-2020-005555-ASW_EN.pdf (accessed December 18, 2020).
119. *See* Brauch.

would indeed most probably have to withdraw from it, as the treaty as it is today opposes the EU's aspirations in the energy sector.

Generally, energy-related treaty regimes represent a critical instrument for the EU in achieving its goals related to decarbonization and energy security, decreasing energy dependency on major fossil fuels suppliers.

5 Participation in the Energy Community

The EnC was set up by the Energy Community Treaty ("EnC Treaty") between the EU and several neighboring states in 2005. In 2006, it entered into force and, in 2016, was extended until 2026.[120] It aims at expanding the EU internal energy market to its neighboring countries while creating a stable regulatory and market framework to attract investment, creating integrated regional market related to the EU and its regulations, ensuring security of supply, boosting competition and fostering environmental protection.[121] The EnC framework obliges its participants to adopt the key EU energy laws. Eventually, the EnC has established its own body of law known as EnC acquis that has to be transported in domestic legislations of member states.[122] Hence, its crucial role is to ensure regulatory convergence to achieve a more coherent and predictable investment environment and more integrated energy market for the EU and its neighbors.[123]

Initially, the treaty was concluded by the potential EU membership candidate countries (Albania, Bosnia-Herzegovina, Kosovo, Macedonia, Montenegro, Serbia). However, accession of Moldova (2010) and Ukraine (2011) expanded the geopolitical perspective of the treaty to non-candidates. Generally, membership is open to other states. Several parties to the treaty eventually joined the EU, *e.g.*, Bulgaria, Romania, Croatia. Non-EU states can participate as observers, *e.g.*, Armenia, Norway, Turkey. The European Commission primarily represents the EU in its interaction with the EnC parties. However, the EU Member States can still attain a status of a participant on its own behalf, which has been done by 19 EU members.

Although the EU strongly supports the EnC objectives, the shortcomings have been identified in the EnC framework, *e.g.*, issue of potential expansion, weak implementation of the energy acquis in several state parties, lack of adequate mechanisms for enforcement. These issues have been commissioned to be analyzed by the

120. European Parliament, *Energy Community: Prospects and Challenges* (2015) Briefing https://www.europarl.europa.eu/RegData/etudes/BRIE/2015/569011/EPRS_BRI(2015)569011_EN.pdf (accessed 10.08.2020) 2.
121. Article 2 of the EnC Treaty. *See* more generally R. Leal-Arcas, *International Trade and Investment Law: Multilateral, Regional and Bilateral Governance*. Edward Elgar Publishing Ltd, 2010; R. Leal-Arcas (ed.), *The Future of International Economic Law and the Rule of Law*. Eliva Press, 2020; R. Leal-Arcas (ed.), *International Trade, Investment and the Rule of Law*. Eliva Press, 2020.
122. European Parliament, *Energy Community* 2.
123. *See* Leal-Arcas, "Conceptualizing EU Energy Security," 1261-1262.

specially organized high-level group and addressed in the following Ministerial Councils.[124]

Though initially created to accommodate Balkan states as potential EU membership candidates, the EnC framework has expanded beyond such potential members participation. This draws to the question of whether the single membership is realistic within the EnC context when its parties are not adopting the acquis with the same rigor, which weakens the credibility of the establishment. Although widening the EnC membership might serve the geopolitical interests of the EU, the potential EnC members should possess a strong incentive to implement all necessary regulations, which might be too burdensome for states lacking a realistic prospect of the EU membership.[125]

The initial EU acquis package included provisions on matters related to gas and electricity markets, environment, renewable energy, obligations to sign the Kyoto Protocol, to apply the EU principles in the areas of competition, etc. The EnC includes a rather flexible mechanism allowing the acquis to expand to new areas of the EU law making it dynamic and reflective of changing reality. Although "dynamic" acquis allows the EnC to rapidly adapt to changing EU law, in many cases, parties to the treaty merely transplant the regulations into their domestic legislation without proper implementation.[126]

Albeit the fact that this treaty envisages tools for enforcement, *e.g.*, a case can be taken to the Secretariat by a public authority, legal entity or person, so far, with 19 enforcement cases commenced, binding penalties and financial sanctions have not been introduced. Two of these cases resulted in breach of the EnC Treaty, however, nothing sufficient has been done from the side of the state parties to remedy the situation and implement enforcement decisions. Hence, lack of binding financial sanctions leads to fundamental enforcement problems.

The EnC framework illustrated an adequate tool for the EU to strengthen its external energy policy and ensure energy security and diversification of energy sources. However, the regime requires fundamental revision and modernization. Certain recommendations on such changes have been made by the High-Level Reflection Group in its assessment of the efficiency of the EnC, *e.g.*, establishment of the independent Court of Justice, ensuring direct effect of the EnC Treaty in domestic legislations of parties to the treaty, providing a flexible two-tier membership. However, the following public consultations lacked the consensus on these recommendations. For now, many potent initiatives have been undertaken under the auspices of EnC, such as Energy Community Regulatory School increasing expertise of its members authorities or the dispute settlement center resolving ISDS disputes. Hence, the EU does not cease to grow its external energy policy leverage and influence among neighboring states.

124. European Parliament, *Energy Community*, 1.
125. *Ibid.*
126. European Parliament, *Energy Community*, 4-5.

6 EU-Russia Trade Relations in the Energy Sector[127]

As a result of globalization, real GDP per capita in emerging economies more than doubled between 1995 and 2019 in purchasing-power-parity terms.[128] In the case of developed economies, however, it grew only by 44%.[129] Between 1990 and 2008, international trade as part of global GDP rose from 39% to 61%.[130] Explosive trade growth, however, has come to an end,[131] the end of hyperglobalization seems to have arrived, trade has started to go regional, globalization has become slowbalization, and openness in global trade seems to have also come to an end, with the protection of domestic producers as a matter of national security. Moreover, global competition has placed much pressure on exporters to become more efficient and competitive.

In the specific case of energy trade relations, Russia is the largest supplier of oil, natural gas, and hard coal to the EU.[132] In 2017, 60% of total imports from Russia to the EU were in the energy sector. Considering the large gap between the primary energy production of the EU in the realm of fossil fuels and the pertinent demand, the relationship with Russia still remains as one of the most critical in the energy sector.

The EU-Russia energy relationship was strengthened during the aftermath of the Iraq War in 2003, when the EU turned closer to Russia to secure its energy supply while turning away from US leadership.[133] For Russia, with the global oil prices raising tremendously at that moment, the EU market was an ideal consumer. The 2008 Russia-Georgia conflict made the EU more vigorously engaged in diversifying its energy mix to shift from its compelling energy dependency on the oil-rich partner. The 2014 Russia-Ukraine crisis reiterated the EU's aspirations toward decreasing overreliance on Russia's energy resources, which the latter has been continuously using as a geopolitical leverage.[134]

The EU-Russia external energy relations are largely sustained by means of bilateral and (fewer) multilateral efforts. However, the institutionalized regimes between two entities, such as the ECT and the 1994 Partnership and Cooperation Agreement, for now, ceased to exist for various reasons. The latter agreement has governed the relations for a decade after its ratification in 1997 and, for that period, has been the main legal framework for EU-Russian energy trade and investment. Nevertheless, the fact that these frameworks ceased to exist has not discouraged mutual trade

127. For a detailed analysis of how EU trade law and policy function, *see* R. Leal-Arcas, *EU Trade Law*, Elgar European Law Series, Edward Elgar Publishing Ltd, 2019; R. Leal-Arcas, *Theory and Practice of EC External Trade Law and Policy*. Cameron May, 2008.
128. The Economist, "Free Exchange: Turning Inward," November 7, 2020, p. 64.
129. *Ibid.*
130. *Ibid.*
131. That said, trade has not done as poorly as initially thought. In April 2020, the World Trade Organization predicted that goods trade would fall by 13%-32% in 2020; in October 2020, it seemed to be only around 10%. *See* The Economist, "Special Report: The World Economy," October 10, 2020, pp. 1-14, at 8.
132. Eurostat, 2017.
133. *See* Isabelle Beegle-Levin, "EU-Russia Energy Relations: Considerations for United States Policymakers" (2018) *Master Thesis* https://www.researchgate.net/publication/326426407 (accessed 13.08.2020).
134. *Ibid.*

and investment flows.[135] The European Commission still perceives Russia as an important energy partner, regardless of the strong EU's commitment to decarbonization.[136]

The relationship with Russia is central to the external energy policy of the EU, being its *raison d'être*.[137] However, it would be deceitful to perceive energy as purely economic matter when entrusted to a supranational body, such as the EU, where its Member State can separately pursue its energy needs and interests.[138] The approach toward this relationship differs across the EU Member States. For instance, Germany's energy relations with Russia are more favorable than those with Hungary's (or any other state with difficult historic relations with Russia and its predecessors, the USSR and the Russian Empire). Hence, Germany will not only seek to ensure that its energy interests are well served on the national level but also attempt to secure a beneficial framework with Russia on the level of the EU. Construction of the Nord Stream 1 and 2 pipelines and circumvention of Russia's neighbors with strained relations along the way with direct supply to Germany represent an evident illustration of disparity in national interests of the EU Member States when it comes to the question of energy relations with Russia.[139]

However, Russia desires a good relationship with the EU energy market as much as the EU is reliant on Russia's reliable supplies of natural gas and oil. Russia's energy policy perceives the European market as vital in the energy sector.[140] Russia is interested in a stable demand for its energy resources, ensuring the economic efficiency of the activities of Russian energy companies, attracting investments, advanced technologies and experience.[141] For its part, Russia is trying to ensure guarantees of stability and reliability of supplies, flexibility in pricing the proposed energy resources, technological interaction with European energy systems. Therefore, the central aspect of the EU-Russian external energy relations is interdependency.

Nevertheless, Russia has used its ascendancy in the energy sector as a leverage in its geopolitical aspirations, *e.g.*, in the Ukrainian gas crisis of 2000s, which led the EU to pursue a more independent energy mix as well as more secure supply. Certain energy policy interests might differ substantially between Russia and the EU. The EU while concerned about supply of the required amount of goods to its internal market promotes energy market liberalization, decrease of energy subsidies, foreign investment protection. Russia, on the contrary, is interested in keeping as much control over

135. *See* Leal-Arcas, "Conceptualizing EU Energy Security," 1288-1289.
136. *Ibid.* 1266-1267.
137. Oliver Geden, Clémence Marcelis* and Andreas Maurer, "Perspectives for the European Union's External Energy Policy" (2006) *Working Paper FG 1, 2006/17 SWP Berlin*, https://www.swp-berlin.org/fileadmin/contents/products/arbeitspapiere/External_KS_Energy_Policy__Dez_OG_.pdf 16.
138. *See* Leal-Arcas, "Conceptualizing EU Energy Security," 1271.
139. *See* Franza et al.
140. Ivanov and Likhachev, "Relations of Russia and the EU in the Energy Sector" (2017) *Russian Council on International Affairs Publication* https://russiancouncil.ru/papers/Russia-EU-Energy-RIAC-DGAP-Report35ru.pdf 13.
141. *Ibid.*

its energy system and its influence as possible, securing constant flow of hard currency needed to upgrade its upstream and downstream infrastructure.[142]

To sum up, the EU is already successfully addressing mentioned issues by adopting strong policies toward low-carbon energy sources supply, diversifying its energy mix and energy suppliers.[143] The fact that some EU Member States perceive Russia as a "whimsical" partner, where others consider it to be a reliable ally,[144] complicates achieving the common external energy strategy toward Russia. What can be said without hesitation is that the EU and Russia energy relations will continue to develop with the EU's high energy demand and Russia's possibilities of supply, hopefully, along the common track of shifting to a greener economy.

7 Conclusion

Energy is a vital part of the existence of any social or economic system, be it a state or a state union. The EU energy structure is highly dependent on external supplies. This fact explains why external energy relations play a crucial role in foreign-policy development. The ability to accommodate an effective EU energy acquis among its neighboring states as well as ensure a stable relation with the biggest energy supplier are two of the major external goals of the Union. However, to do so efficiently while ensuring energy security requires certain policy solutions, which the EU has been implementing for some time: diversifying energy supplies, *e.g.*, by means of trade liberalization in green goods and services, the promotion of renewables, as well as the lobbying of the ECT modernization. Finally, securing reliable flows of diverse and sustainable energy sources would foster the EU's energy security.

The next chapter brings together the links between trade, climate change, investment and a prosperous future for citizens.

142. *See* Leal-Arcas, "Conceptualizing EU Energy Security," 1295-1296.
143. *See* Beegle-Levin, 82.
144. *See* Leal-Arcas, "Conceptualizing EU Energy Security," 1296.

ANNEX

*Figure A1 Energy Situation in the MENA Region**

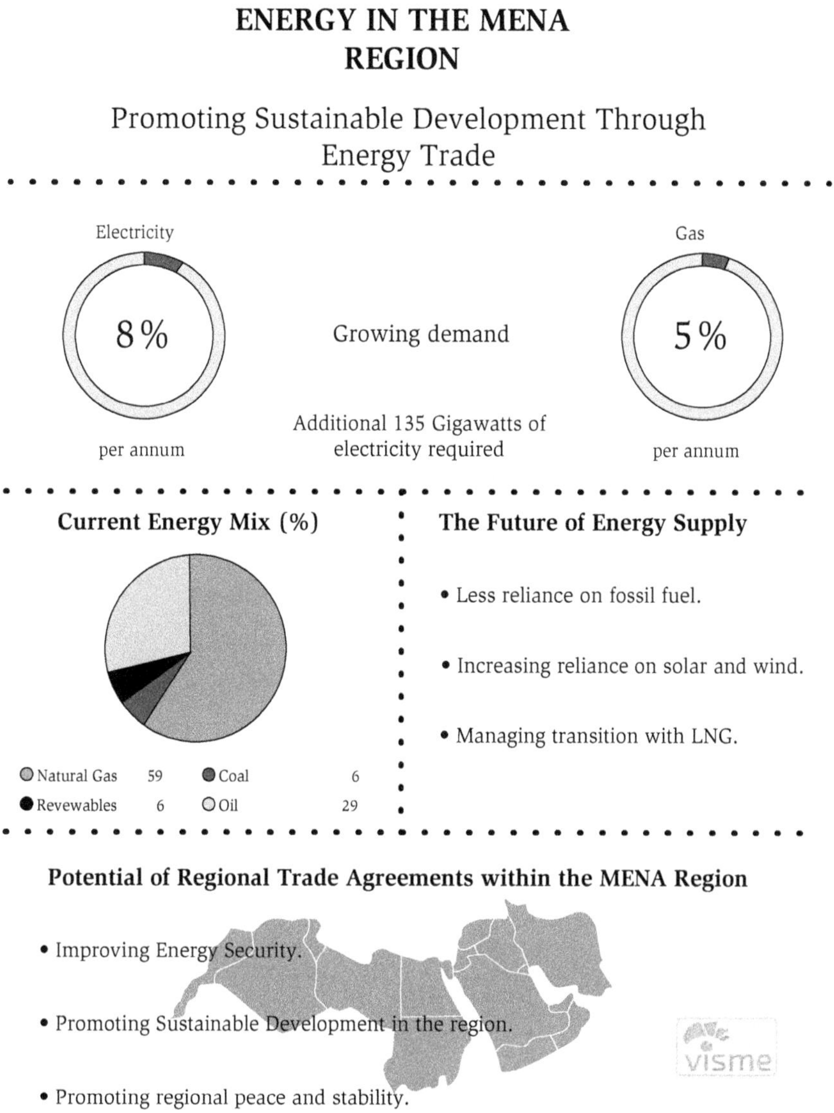

* Based on data provided by the World Bank and the International Energy Agency.

CHAPTER 6

Linking International Trade, Climate Change, Investment and a Prosperous Future

A transformative, integrated, and holistic approach to sustainability is necessary to reach a prosperous future for all. With growing inequality in the world, demographic change, rapid technological development, and 40% of the world's population with no access to digital technology, access to digital technology for all and, ultimately, having a prosperous future are a must today. Doing so will give a voice to the voiceless. For that, governments, companies, and transnational institutions should invest as much as possible to make it happen and provide solutions that are rules-based, inclusive, innovative, disruptive, and simple. We, the citizens, should push for this human right to materialize. That way, everyone will benefit and no one will fall behind. Access to energy, mitigating climate change, and benefiting from international trade and investment all can be achieved thanks to technological advancement. These global challenges are complex, interconnected, multidimensional, interdependent, and multicausal.

I INTRODUCTION[1]

The world is slowly realizing that something needs to change if we want to have a sustainable future where we can protect the environment and, at the same time, prosper economically.[2] COVID-19 has taught us that this can be done, to a large extent although it is not enough, if we close many shops, offices, and factories, as well as if we stop many people from travelling frequently. At the same time, this new reality may

1. Parts of this chapter were published as Rafael Leal-Arcas et al., "Of International Trade, Climate Change, Investment and a Prosperous Future," 12:2 *Trade, Law and Development* 405-455 (2020).
2. A. Freedman, "Global Warming to Push Billions Outside Climate Range That Has Sustained Society for 6,000 Years, Study Finds," *The Washington Post*, May 4, 2020, available at https://www.washingtonpost.com/weather/2020/05/04/human-climate-niche/.

make people think that it is not so urgent to move from fossil fuels to renewables, based on the new way of life resulting from COVID-19. In fact, by containing the propagation of COVID-19, demand for oil has fallen drastically in a very short period of time. Nonetheless, when putting together ecology, the economy, and geopolitics, in April 2020, people from 14 countries gave their views as to whether climate change was as serious an issue as COVID-19. In China, 87% agreed it was; in the US, the answer was 59%.[3]

Climate change has become a serious global issue because human activity is adding greenhouse gases (GHGs) on a daily basis to the atmosphere at such a rate that is impossible to predict. Carbon dioxide and other GHGs are produced when, for instance, fossil fuels are burned to generate energy.[4] In 1988, scientists, environmental activists, and policymakers got together for a World Conference on the Changing Atmosphere because they were alarmed by the concentration of carbon dioxide in the atmosphere. As a result, the Intergovernmental Panel on Climate Change was created in that same year. By 2019, the GHG emissions from human activity equaled 55 billion tons of carbon dioxide. The carbon dioxide from fossil-fuel emissions and industrial activity amounted to 37 billion tons.[5] In the next 50 years, it will be necessary that 90% of the energy currently produced be provided by renewables or nuclear power.

The world's natural capital (composed of animals, water, air, soil, plants, fossil fuels, and minerals) is expected to decline by 20% by 2040 and the change in CO_2 emissions (in a high-emissions scenario) to increase by 7% by 2040 based on the 2015 levels in the developed world and by 53% in the developing world.[6] It is well known that natural resources raise the level of GDP in a given nation, but slow its growth rate. If countries continue being energy-hungry and with high energy demand, the world will get even richer, but natural resources will get consumed rapidly.

So there is recognition that the time has come to make the transition from fossil fuels to clean energy—which is a global trend[7]—and that we must leave hydrocarbons in the ground and move toward a windy and sunny future. This recognition may materialize if the world's appetite for hydrocarbons decreases due to a change in habit, greener regulations, or cleaner technology that can be used as an accelerator and enabler of sustainable development.[8] Even then, the world would still have more than

3. The Economist, "Never Let a Crisis Go to Waste: A Trillion-Dollar Question," May 23, 2020, pp. 14-15, at 15.
4. Of the energy that people pay for, "34% comes from burning oil, 27% from coal and 24% from gas. [...] hydroelectric power and all other renewables combined provide just 15%." *See* The Economist, "Carbon Cycles: Where Nature Ends," May 9, 2020, pp. 47-48, at 47. Around 70% of the world's industrial CO2 comes from energy-related emissions. *See* The Economist, "Climate Change: Goodish News," February 15, 2020, p. 69.
5. The Economist, "Hotting Up," June 20, 2020, pp. 57-60, at 57.
6. New Zealand has more natural wealth per capita than oil-rich countries such as Saudi Arabia or Kuwait, and Gabon more than any country in the world. According to the Inclusive Wealth Report, 47% of the world's natural capital is fossil fuels and minerals. *See* The Economist, "Resource Economics: The World's Natural Wealth Is in Decline," July 18, 2020, p. 73.
7. In the case of the Middle East, renewable-energy capacity has doubled since 2010 and is expected to double again by 2024. In fact, between 2008 and 2018, investment in solar energy in the Middle East increased 12-fold. *See* The Economist, "Solar Power: Rays of Hope," May 9, 2020, p. 33.
8. The Economist, "Crude Oil: After the Fall," June 13, 2020, pp. 53-54, at 54.

90% of decarbonization left to do to reach the Paris Agreement's target of a climate not warmer than 1.5°C compared to the pre-industrial revolution temperature.[9]

The forecast is that more than 3 billion people could live in extreme heat by 2070.[10] Getting rid of GHG emissions seems to be the only sustainable way forward to stop global warming.[11] The Economist states that "cutting CO2 emissions could slow the rate of warming as early as 2033, but only if they are ended worldwide in 2020."[12] That would imply getting rid of 80-90% of the energy sources used today overnight, which is not a credible scenario.[13] In addition, CO2 emitted today will take decades and even centuries until it is reabsorbed by the oceans and the flora.[14]

Various policymakers are offering ways and means to decarbonize the economy. The President of the European Commission Ursula von der Leyen has proposed a WTO-consistent carbon border tax to penalize imports from countries that have looser environmental regulation than those of the European Union (EU). She has also proposed a European Green Deal to make the EU's economy sustainable.[15] The European Commission said in December 2019 that the Paris climate deal should from now on be included in the "essential elements" clauses of any trade agreements that the EU negotiated with other countries around the world.[16] The Prime Minister of Finland Sanna Marin aims to make her country carbon-neutral by 2035, making the point that, with proper planning, it is possible to have economic prosperity and equality at the same time as environmental sustainability.[17] The British government has a target of banning new diesel, hybrid, and petrol vehicles by 2035.[18] The German government agreed to cut GHG emissions to 55% of their 1990 levels by 2030 and stop using coal by 2038.

The US's switch from coal to natural gas has decreased the country's GHG emissions. In fact, natural gas, solar and wind are now slowly putting coal aside. This is largely due to the fact that solar farms are cheaper, faster and safer to build than oil and gas power plants.[19] In 2019, renewables produced more of the electricity consumed in Europe than coal.[20] In many parts of the world, renewables are now cheaper than fossil fuels. On the other hand, oversupply and the increasing competitiveness of renewable energy mean that the price of oil may remain low for years to come. Costs

9. The Economist, "Seize the Moment," May 23, 2020, p. 7.
10. BBC News, "Climate Change: More Than 3bn Could Live in Extreme Heat by 2070," May 5, 2020, available at https://www.bbc.com/news/science-environment-52543589.
11. In the Siberian Arctic, temperatures were 10°C hotter between January and June 2020 than the average between 1981 and 2010.
12. The Economist, "Delayed Cool," July 11, 2020, p. 66.
13. *Ibid.*
14. *Ibid.*
15. https://ec.europa.eu/info/strategy/priorities-2019-2024/european-green-deal_en.
16. J. Brunsden, "Brussels and Britain Clash over Climate Conditions in Trade Deal," *Financial Times*, May 6, 2020, available at https://www.ft.com/content/0f09f819-77b3-45d8-9ba3-76a3042c240c?segmentId=3f81fe28-ba5d-8a93-616e-4859191fabd8.
17. *See* speech by Sanna Marin, Prime Minister of the Republic of Finland, at Columbia University in the City of New York, March 6, 2020, available at https://worldleaders.columbia.edu/content/ms-sanna-marin-prime-minister-of-the-republic-of-finland.
18. The Economist, February 8, 2020, p. 8.
19. The Economist, "Solar Power: Rays of Hope," May 9, 2020, p. 33.
20. The Economist, "The Contentious and Correct optioO," May 23, 2020, pp. 56-59, at 57.

of photovoltaic solar energy have dropped, efficiency has tripled from 6% in the 1950s to 17%-20% in 2020.[21] As a result, many grid managers are preferring solar to fossil-fuel energy generation. Greece's Prime Minister wants renewable sources to cover 35% of Greece's energy needs by 2030, which means investing heavily in wind and solar energy, something foreign investors are already doing.[22] All of this makes the energy transition hopeful.[23]

Oil corporations also have a big role to play in decarbonizing the economy and are starting to commit to reducing their carbon footprint. BP aims to reach net zero carbon dioxide emissions by 2050. Repsol, a Spanish oil company, said it would reduce its net carbon footprint to zero within 30 years.[24] Other major energy companies, such as Spain's Iberdrola and Italy's Enel, have invested heavily in renewables around the world. Yet, according to The Economist, "of a whopping $80bn or so of capital expenditure by Europe's seven biggest listed energy firms last year, only 7.4%—less than $1bn each on average—went to clean energy."[25] Part of the reason is that oil executives know that, for now, oil generates more wealth than clean energy. Nevertheless, BP reportedly aims to invest beyond USD 5 billion in the next five years with the intention of providing more energy with lower greenhouse gas (GHG) emissions.[26]

How damaging the impacts of climate change will be to international trade and the economy at large depends greatly on how much global warming will take place and how people will react, which are unknowable at present. Equally, the trading system is experiencing serious tensions between China—the dominant economic force in Asia—and the US; the WTO's judicial branch is going through a crisis due to the discontinuation of its Appellate Body, and trade may fall by as much as a third due to lack of demand, not barriers to trade.[27] In the past, trade was about economic efficiency. Today, it seems to be about helping local farmers, resulting in protectionism. Is the trading system moving toward autarky?

The current situation in the trading system may lead to the rise of regionalism where, in the supply chain, suppliers may be replaced by intra-regional ones. This rise of regionalism, in itself largely due to COVID-19, may be beneficial for the idea of creating climate clubs[28] to mitigate climate change.[29] As a result, small countries in any

21. The Economist, "Photovoltaics: Solar's New Power," May 23, 2020, pp. 68-69, at 68.
22. The Economist, "Energy: Greening Greece," February 1, 2020, p. 30.
23. For a thorough analysis of energy transition, *see* R. Leal-Arcas et al., *The Great Energy Transition in the European Union*, Volume 1. Eliva Press, 2020; R. Leal-Arcas et al., *The Great Energy Transition in the European Union*, Volume 2. Eliva Press, 2020.
24. The Economist, "Blowin' in the Wind," January 18, 2020, p. 66.
25. *Ibid.*
26. *Ibid.*
27. The Economist, "Globalization: Torn Apart," May 16, 2020, pp. 57-59, at 57.
28. These clubs could be similar to existing ones in global governance such as the G7, the G20, the EU, the African Union, or the Association of South-East Asian Nations (ASEAN). They could be interconnected: for instance, the presidency of the G20 could be connected to that of the EU or the African Union to have a broader geographical reach.
29. For an analysis of regionalism in trade, *see* R. Leal-Arcas, *International Trade and Investment Law: Multilateral, Regional and Bilateral Governance*. Edward Elgar Publishing Ltd, 2010.

given region would benefit from neighboring larger countries. Equally, companies may opt for automating services instead of offshoring them. Despite the existing global health crisis created by COVID-19 with potentially serious economic consequences, we argue that, just as change through trade can eventually lead countries toward openness and democracy, trade can also be used as a powerful tool to mitigate climate change.

After this introduction, section II analyzes how globalization has affected the impacts of international trade[30] and climate change in major economies. Section III answers the question of what role the European Union (EU)—which is the world's largest exporter of services and the second largest exporter for goods only behind China—and the WTO have played and should play in climate change. It also examines the options to put a price on carbon emissions: in the case of the EU, it can introduce either border carbon adjustments or low-carbon standards. The aim is to meet economic and environmental objectives by connecting trade with the carbon issue. It further explores the option of climate clubs as a potentially effective way to mitigate climate change.[31] Both climate change and COVID-19 involve the entire world and are predicted by science. We recognize that climate change (a chronic and slower issue than COVID-19), like the current COVID-19 pandemic (an acute issue), is a global issue that requires various levels of cooperation; however, we argue that a more effective way to mitigate climate change and expand economic growth is via climate clubs.

Section IV explores the possibility of creating a climate club to enforce the commitments of the Paris Agreement on Climate Change and, at the same time, to create climate-resilient trade laws for the enhancement of free trade that is climate-friendly. Section V explores how EU bilateral investment treaties contribute to sustainable development. We make the argument that investing in infrastructure promotes the flow of goods, services, capital, and technology and, therefore, economic growth. Our view is that we can maximize the use of technology for a sustainable future, whether it is in social-justice issues, in north-south cooperation, or for renewable energy. A good example is sharing knowledge thanks to digital technology. We have witnessed this fact during the confinement of much of the world. Yet, people who did not have the financial means to attend conferences were able to do so via webinars. Equally, another advantage of digital technology is that students with limited financial resources can actually learn online from home. COVID-19 has taught us that it is actually possible to have a prosperous future if we invest in digital technology.

30. B. Davis and L. Wei, *Superpower Showdown: How the Battle Between Trump and Xi Threatens a New Cold War*. Harper Business, 2020; M. Klein and M. Pettis, *Trade Wars Are Class Wars*. Yale University Press, 2020.
31. For an analysis of the role of climate clubs in international trade, *see* R. Leal-Arcas, "Climate Clubs and International Trade Across the European and International Landscape," 29:3 *European Energy and Environmental Law Review* 72-88 (2020).

Section VI explores the role citizens can play in shaping sustainable trade. Sustainability constitutes a principal concern for various regimes that, because of their nature, are in constant interaction with climate change.[32] It is one of the three parts of the "energy trilemma" (i.e., the pursuit of the ideal balance between energy equity, energy security, and environmental sustainability), and it has also been a subject of considerable controversy in the field of international trade. More specifically, it has been argued that international trade, as a practice that makes people dependent on products produced far from their bioregion, often contributes to the destruction of ecosystems and the continuation of social injustice, therefore threatening the very idea of sustainability.[33] On the other hand, an argument has been developed that trade can, in fact, go hand in hand with sustainable development and even promote it, if certain standards are met adequately.[34]

Governments and intergovernmental organizations (IGOs) such as the United Nations and the European Union have favored the latter approach (i.e., that trade can go hand in hand with sustainable development), an admittedly sensible choice in today's globalized world—although the concept of "sustainable trade" is not necessarily interpreted in one single way. An area that has until now remained rather unexplored is whether, and to what extent, sustainable trade could also be shaped and supported by the help of citizens. Citizens have gained an important role in the aforementioned regimes closely connected to trade (namely climate change and energy) in ways that will be examined below. Therefore, these regimes could possibly set an example to be replicated—to the level that this is feasible—in the trade sector. This hypothesis will be examined further in this chapter. If it succeeds, it would mean that the traditionally very technocratic field of trade could be significantly democratized. Section VII concludes.

---

32. For a policy history on climate change, *see* Cass R. Sunstein. *Worst-Case Scenarios*, "A Tale of Two Protocols," Harvard University Press, 2007, 71-117, ; Ian Parry and William Pizer, "Emissions Trading Versus CO2 Taxes Versus Standards" Resources for the Future, 2007. Link: https://www.rff.org/publications/issue-briefs/emissions-trading-versus-co2-taxes-versus-standards/; Kelly Levin, Benjamin Cashore, Steven Bernstein, and Graeme Auld, "Overcoming the Tragedy of Super Wicked Problems: Constraining Our Future Selves to Ameliorate Global Climate Change," *Policy Sci* 45 (2012); Nathaniel Rich, "Losing Earth: The Decade We Almost Stopped Climate Change," *The New York Times Magazine*, August 1, 2018, available at https://www.nytimes.com/interactive/2018/08/01/magazine/climate-change-losing-earth.html; David Roberts, "The Once and Future Democratic Consensus on Climate Change," Vox (2017), available at https://www.vox.com/energy-and-environment/2017/11/18/16669094/democratic-consensus-on-climate-change; Justin Gundlach, "How Much Does the Existing Regulatory Patchwork Reduce U.S. Greenhouse Gas Emissions?," Columbia Law School, 2015, available at https://doi.org/10.7916/D8959HDJ; Yale Environment 360, "The Waxman-Markey Bill: A Good Start, or a Non-Starter?" (6/18/09), available at https://e360.yale.edu/features/the_waxman-markey_bill_a_good_start_or_a_non-starter; Amanda Reilly and Kevin Bogardus, "7 Years Later, Failed Waxman-Markey Bill Still Makes Waves," *E&E News* (June 27, 2016), available at https://www.eenews.net/stories/1060039422.
33. Patricia Perkins, "What Is Sustainable Trade?," in Satya Dev Gupta and Nanda K. Choudhry (eds.) *Globalization, Growth and Sustainability*, Recent Economic Thought Series, Volume 58. Springer, 1987, 273, 276.
34. *Ibid.* 280.

II HOW HAS GLOBALIZATION AFFECTED INTERNATIONAL TRADE AND CLIMATE CHANGE IMPACTS IN THE EU AND OTHER MAJOR ECONOMIES?

1 Introduction

Globalization became an academic phenomenon in the 1980s,[35] when the world's greatest economies began expending their powers to become stronger and increasingly prosperous in their domains. This section discusses the positive and negative effects that globalization has had both on international trade and climate change intertwined. Due to the increase of public concern regarding climate change and the preservation of our planet, global powers such as the European Union, the United States (which in 2018 became the largest oil producer in the world due in part to its shale-oil revolution), and China (which has become rather repressive at home and increasingly assertive abroad, bullying other countries) have begun to attempt to find answers for the outcry for drastic progress and change within all capacities, including in the area of international trade, mainly through the creation of international organizations and agreements. Granted, the relationship between these three powers (the EU, the US, and China) is not currently optimal: Europeans see China's rise as an economic threat that is building a financial system to avoid US dollar-based payment mechanisms, whereas the Americans see it as an issue of national security. China, in turn, sees a great chance to expand its global leadership and influence in the absence of American leadership.[36] In fact, in 2017 President Xi of China proclaimed that China has now taken a driving seat in global affairs to make contributions to mankind.[37]

One of several prevalent debates is whether or not trade has any effect on the climate whatsoever,[38] and what the consequences of stringency of regulation may envisage, specifically in terms of economic prosperity of States. We conclude that, although the effect of international trade directly may not be significant on the climate on the outset, there are deeper issues that may arise over the years such as climatic disasters. The section will conclude that the various environmental agreements in place are entirely toothless due to non-enforceability and little accountability.

The suggestion of this section is for world leaders of major economies to commit financially and intellectually alongside their technology and create an agreement with full liability measures in place. This is the most realistic solution to ensure the most

35. P. Dicken, *Global Shift: Industrial Change in a Turbulent World*. Harper and Row, 1986. (This was arguably one of the earliest academic writings that spiked the interest in globalization—originally read in R. Lee and J. Wills, *Geographies of Economies*. Taylor & Francis Ltd, 1997, p. 134.)
36. Between 1992 and 2012, China's GDP grew sevenfold. Its CO2 emissions more than tripled. By 2006, China had already become the largest GHG emitter in the world. *See* The Economist, "A History of Action and Inaction: The Challenge Without Precedent," April 25, 2020, pp. 52-53, at p. 53.
37. The Economist, "China's Post-COVID Propaganda Push," April 16, 2020, pp. 29-31, at 31.
38. C. McAusland and D.L. Millimet, "Do National Borders Matter? Intranational Trade, International Trade, and the Environment," 65:3 *Journal of Environmental Economics and Management* 411-437 (2013).

adequate and agile outcomes for a very imminent threat to our planet. Technology can provide a better predictive analysis for climate change, it can bridge the development divides between rich and poor countries, and it can get rid of technical barriers to trade in technology transfer for the benefit of developing countries. Competition and innovation can make technology cheaper.

2 Globalization

Although a frequently voiced term, the globalization phenomenon has not been occurring for an extensive period of time, nor has the matter been taken sincerely.[39] Globalization can be defined as "...the opening of international borders to increasingly fast flows of goods, services, finance, people and ideas; and the changes in institutions and policies at national and international levels that facilitate or promote such flows."[40] This lexicon conveys that there has been a clear universal shift toward globalization generating a vast increase in the amalgamation of state systems, resulting in positive and negative effects. One of the ways in which the increase in this modern change began was the increase in trade throughout the European continent as a result of enhanced transport links, which then gradually engulfed other states, including those across waters. Prosperity is one of its most paramount effects. The process can have very generous impacts on the state economy, both through imports and exports, as prices can be decreased as a consequence of negotiation.[41]

Negotiation is another positive effect of globalization. Since 1979,[42] there have been worldwide efforts to understand the impacts and possible solutions that mankind has on the planet. Although the agenda has now moved beyond just definitions and awareness, it is clear that a global effort continues to be necessary to have an impact on the situation of the climate in 2020. Nations have been forced to consult on all matters, with trade being perhaps one of the most discussed topics, as evidenced by the creation of world organizations for convenience and transparency such as the WTO.

All states constantly compete in the contemporary globalized domain, as it has never been easier to find replacements for virtually anything or negotiate a better deal elsewhere. With increased movement of business, there has been an upsurge in the number of international firms across the globe, with different parts of their operations often being in a handful of states. Since trade is such an enormous part of the economy of a country, it is essential that research and development on combatting climate change be a priority for international organizations and governments to combat some of the processes involved that may contribute to greenhouse gas emissions.

39. R. Lee and J. Wills, *Geographies of Economies*. Taylor & Francis Ltd, 1997, p. 134.
40. World Health Organization, "Health Topics: Globalisation," https://www.who.int/topics/globalization/en/.
41. K.H. O'Rourke and J.G. Williamson, "When Did Globalisation Begin?," *European Review of Economic History*, (2002), 6, 23-50, p. 36.
42. 1979 World Climate Conference, World Meteorological Organization, Geneva, February 12-23, 1979.

The following part of this section will discuss the various agreements that have been negotiated as a result of globalization and their strategy to make a positive impact on climate change beside trade.

3 International Trade and the Climate

One of the positive effects that globalization has contributed upon both international trade and climate change has been the creation of world organizations and various agreements. Major and developing economies ordinarily participate in the negotiations, as can be seen in the WTO context. The WTO Agreement[43] features one of the earliest modern concerns and needs to negotiate trade in light of preserving the planet.[44] This is contained in the Marrakesh Agreement that created the WTO, stating "expanding the production of and trade in goods and services, while allowing for the optimal use of the world's resources in accordance with the objective of sustainable development, seeking both to protect and preserve the environment … .[45]" Albeit a promising move forward with so many nations involved, what must be questioned is the enforceability and the effectiveness of such statements.

As of December 2019, the WTO no longer has a functioning Appellate Body due to political disagreements.[46] The Appellate Body is one of the most essential elements of this organization, allowing the WTO's member states to use it as a form of dispute resolution and to hold states accountable for their actions, which is not possible to reach via multilateral trade negotiations. There is currently no viable option for states to be able to uphold the values of one another through the Organization, nulling its enforceability. This is one of the problems with upholding such statements about the climate to any extent as mentioned above. Arguably, this is not the biggest problem in the WTO. As has been witnessed in recent history, it has become more and more complicated for states to agree on terms of new Agreements during negotiations, particularly the Doha Round.[47]

Although the WTO has a great incentive to uphold environmental measures in all negotiations, there is a struggle to come to any agreement at all. The complication comes from the shift toward globalizing trade and attempting to negotiate agreements on a multilateral scale. This brings about the WTO's own challenges due to the variety of cultures, morals, local practices, and underlying political issues between members

43. Marrakesh Agreement Establishing the World Trade Organization, April 15, 1994.
44. D. Esty and S. Biniaz, "Introduction," in D. Esty and S. Biniaz (eds.), *Cool Heads in a Warming World: How Trade Policy Can Help Fight Climate Change*, Yale Center for Environmental Law & Policy, 2020, p. iv, available at https://envirocenter.yale.edu/sites/default/files/files/CoolHeads_Introduction(3).pdf.
45. *Ibid.*
46. A. Rathore and A. Bajpai, "The WTO Appellate Body Crisis: How We Got Here and What Lies Ahead?" (Jurist, April 14, 2020) https://www.jurist.org/commentary/2020/04/rathore-bajpai-wto-appellate-body-crisis/.
47. C. Ehlermann, "Decision Making in the World Trade Organisation: Is the Consensus Practice of the World Trade Organisation Adequate for Making Revising and Implementing Rules on International Trade?," WilmerHale, 23, September 2005, 1-24, p. 14.

that are present in the current 164 WTO members.[48] It would therefore be naïve to argue that multilateral trade negotiations in light of climate change are at the forefront of the WTO's priorities. It is more likely that the priority of this time is to resolve the riddle of the Doha Round. The WTO also does not offer any further explanation as to what this statement may involve in practice and therefore is a mere proclamation. The WTO would need to re-evaluate its approach broadly, "ensur[ing] that the trading system gives wider leeway for policy interventions aimed at climate change mitigation and adaptation efforts.[49]"

Evidently, the issue of enforceability seems to be somewhat of a trend in international agreements on trade and the climate. The Paris Climate Agreement is another example of a consequence of globalization; yet it is necessary to analyze its capabilities. What is seen from the outset as a constructive initiative slowly begins to show cracks in its core. Nonetheless, the Paris Agreement does have a structural approach, with states being required to submit nationally determined contributions to the United Nations Framework Convention on Climate Change (UNFCC) to track progress of each nation. Yet, these are nationally, rather than internationally, determined, which can cause inconsistencies.[50] Since the major economies of the world are also the world's biggest emitters of greenhouses gasses (GHG), some accountability is present.

However, is this truly *accountability*? Unfortunately, there are effectively no legal consequences for states if they do not reduce their GHG emissions. One consequence, as witnessed recently, can be political pressure and the withdrawal of countries from the Paris Agreement on Climate Change, trying to protect their own economic agenda.[51] This showcases a negative effect of globalization—it can sometimes be seen by nations as an intrusion on their sovereignty and freedom to decide their own fate. Since there are no major consequences and no apparent enforceability measures, apart from political and social pressures, it is simpler for leaders of states to exit such efforts to combat climate change or to not comply with the requirements.

In this case, globalization presents a threat to some states and, in turn, compromises the global efforts toward the savior of the planet. It also changes the dynamic of international trade in major economies, as can be seen with the reaction of France and China as a result of the United States's intention to withdraw from the Paris Climate

48. World Trade Organization, "Members and Observers," https://www.wto.org/english/thewto_e/whatis_e/tif_e/org6_e.htm.
49. D. Esty and S. Biniaz, "Introduction," in D. Esty and S. Biniaz (eds.), *Cool Heads in a Warming World: How Trade Policy Can Help Fight Climate Change*, Yale Center for Environmental Law & Policy, 2020, p. vi.
50. *Ibid.*, p. vii.
51. On the notification of the United States's withdrawal from the UNFCC Paris Climate Agreement, *see* M. Pompeo, "On the U.S. Withdrawal from the Paris Agreement" (U.S. Department of State, November 4, 2019), https://www.state.gov/on-the-u-s-withdrawal-from-the-paris-agreement/; BBC News, "Paris Climate Accords: US Notifies UN of Intention to Withdraw" (BBC News, November 5, 2019) https://www.bbc.co.uk/news/world-us-canada-50297029.

Agreement.[52] Since the US's notification to withdraw from the Paris Agreement, several US states have been active in the implementation of the Agreement.[53] Moreover, the Paris Agreement demands movement toward state-of-the-art technologies, which requires substantial financing promptly. Necessitating both economic and social transition will be difficult even in the major economies in the EU and elsewhere,[54] especially in unprecedented times such as the coronavirus pandemic[55] since 2020.[56] This is a clear example that the fruits of globalization come with obstacles, prominently major political impacts upon society as a whole.

A similar climate initiative is to be found in the UN Sustainable Development Goals (UN SDGs), specifically Goal 13 on "Climate Action." Moreover, it seems that the entirety of the work is centered around climate action since almost every UN SDG mentions "sustainability."[57] The creation of the United Nations itself occurred as a result of globalization, demonstrating the innovation of new ways to encourage and urge states to combat climate change. Similar to the previous trends in other climate change agreements, there is no enforceability to ensure that countries meet the standards set by the UN. However, due to the progress reports that are published globally and can be accessed by the public, there is now also pressure from individuals onto states to obey the targets and do better on climate action. In democratic systems, the public are able to outvote the individuals for not doing enough, which, in turn, influences international trade policies.

The amount of agreements created to combat climate change should be questioned. These may not solely focus on climate change. However, they are plenty in number multilaterally, bilaterally and regionally that pledge to negate greenhouse gas emissions in some way. Why is there a need for so many avenues? Unquestionably, this approach only contributes to complexity and confusion.[58] It is clear that it is

52. Pennetiere M., "China, France Reaffirm Support of Paris Agreement, Call It 'Irreversible'" (*Reuters*, November 6, 2019) https://www.reuters.com/article/us-china-france-paris-agreement/china-france-reaffirm-support-of-paris-climate-agreement-call-it-irreversible-idUSKBN1XG0QJ.
53. Michael Fullilove suggests a coalition of the responsible among like-minded countries who are serious about mitigating climate change. *See* The Economist, "Special Report: The New World Disorder," June 20, 2020, pp. 1-12, at 6.
54. A. Cosbey, "Climate Policies, Economic Diversification and Trade" (United Nations Conferences on Trade and Development, October 3, 2017, Geneva), https://unctad.org/en/PublicationsLibrary/ditcted2018d4_en.pdf, p. 1.
55. One interesting climate-change observation of the COVID-19 pandemic is that, in early April 2020, daily GHG emissions worldwide were 17% lower than they were the year before. This fact is due to less energy use and therefore less burning of fossil fuels.
56. The Trump administration gave formal notice to the World Health Organization (WHO) that the US will withdraw from it on July 6, 2021, largely due to differences of opinion between President Trump and the WHO on how to handle the COVID-19 pandemic.
57. UN, "Sustainable Development Goals," (United Nations Sustainable Development Knowledge Platform) https://sustainabledevelopment.un.org/?menu=1300.
58. *See* generally R. Leal-Arcas, *International Trade and Investment Law: Multilateral, Regional and Bilateral Governance*. Edward Elgar Publishing Ltd, 2010; R. Leal-Arcas, "The Fragmentation of International Trade Law: Is Now the Time for Variable Geometry?," 12:2 *The Journal of World Investment and Trade* 145-195 (2011).

necessary worldwide to combat negative effects of national regulation;[59] yet is there a prerequisite for this type of repetition? For example, what is the difference between signing the Paris Climate Agreement and abiding by the UN Sustainable Development Goals, or already being a member of the WTO that is continuing very similar if not the same efforts? The ultimate purpose is mirrored.

Arguably, there is no need for several agreements promising to accomplish an identical goal. This is why it is contended that carbon clubs may not be a satisfactory solution to climate change mitigation, since they would essentially be attempting to do the same as what is currently being done on a multilateral scale, yet with fewer members. The creation of carbon clubs would also likely lead to inconsistencies and highly likely have very low to no enforceability mechanisms. What is currently absent is an agreement that has stricter legal and economic consequences as well as enforceability such as economic sanctions in the form of increased tariffs, for example.[60] The price of higher tariffs is paid by the consumer and the exporting country may need to devalue its currency.

Similarly, there could also be an introduction of economic incentives for businesses if hard policy is perceived as damaging. The consequence of such an action would actually be threefold: disapproval by the public of that country, the political pressure of other nations, and economic consequences for businesses. According to a meeting held in 2017, there were only 12 years left to combat climate change to a degree of mitigating irreversible damage.[61] It is clear from that meeting that leaders understand the consequences of delay in this scenario; yet, there still does not seem to be a combined rigorous effort to do something about it. At the time of writing, the time left is only 9 years and nations, organizations, and governments in major economies are still not committing enough to mitigate such damage.

4 Other "Side" Effects

A huge negative impact of globalization on trade and the climate can be seen in the distribution of work forces around the world by big corporations. A new mechanism used by some parts of the world has seen some effects on the way companies behave and the decisions they make. Tradable pollution permits have seen an increase in popularity around the world,[62] allowing companies to "offset" a certain amount of greenhouses gases and are able to trade the rest for whatever they wish. This gives the

59. C. McAusland and D.L. Millimet, "Do National Borders Matter? Intranational Trade, International Trade, and the Environment," 65:3 *Journal of Environmental Economics and Management* 441 (2013).
60. S.R. Milliman and R. Prince, "Firm Incentives to Promote Technological Change in Pollution Control," 17 *Journal of Environmental Economics and Management* 247-265, p. 248 (1989).
61. UN, "Only 11 Years Left to Prevent Irreversible Damage from Climate Change, Speakers Warn during General Assembly High-Level Meeting" (United Nations, Meetings Coverage and Press Releases, GA/12131) https://www.un.org/press/en/2019/ga12131.doc.htm.
62. For example, the EU Emission Trading System (which includes Iceland, Lichtenstein and Norway), or that of the United States, New Zealand, and Australia among others.

initiative to pollute less and be able to trade the rest of the permits for money to those companies who may need more.

Perhaps the most sophisticated system can be found in the EU, with the European Union Emission Trading System (EU ETS). However, there is a loophole in these types of systems, since not all companies have factories and manufacturers located in the same state as their head offices. This allows them to potentially continue to exploit the policies of less developed nations although the majority of their operations and decisions are made within the EU. The burden of the additional costs of permits is also likely to fall onto the consumer, which can impact market competition.[63] Often, large corporations have many more widespread operations and they often take advantage of less developed countries, which allows them to benefit from cheap labor, less regulation, and, most importantly, the ability to free-ride the less obstructive systems for the protection of the planet.[64]

For example, a global athletic apparel brand such as Nike has 533 factories in 41 different countries around the world.[65] Although some of the countries in which Nike manufactures have considerate national and international climate policies, others do not. Among others, the most prevalent would include Turkey, Bulgaria, Romania, and Japan.[66] This is not to state that Nike does not implement strategies to combat climate change. However, it is clear that the company is able to spread manufacturing around the world, whatever the reasoning may be. A giant conglomerate like Nike has the ability to bring jobs to developing economies and is likely to have an implied (or in some cases direct) impact onto the policies of that country for the continuation of a newfound increase in economic prosperity. By achieving international cooperation among nations on climate change policies, we could have a more balanced market competition level among competitors in many industries.[67]

Regardless of any motives corporations such as Nike may have, there will be far more leniency in these nations on GHG emissions and other climate-related matters. This is due to limited budget for research, technology and overall less education on this type of matter.[68] Nevertheless, it is clear that there is a push by international organizations[69] to continue the development of the economic market for pollution

63. A. Dechezlepretre and M. Sato, "The Impacts of Environmental Regulations on Competitiveness," 11:2 *Review of Environmental Economics and Policy* pp. 183-206, at p. 185 (Summer 2017).
64. A. Cosbey, "Climate Policies, Economic Diversification and Trade" (United Nations Conferences on Trade and Development, October 3, 2017, Geneva), https://unctad.org/en/PublicationsLibrary/ditcted2018d4_en.pdf, p. 3.
65. Nike Inc., "Nike Manufacturing Map" (November 2019), http://manufacturingmap.nikeinc.com/.
66. Climate Change Performance Index, "Category Results: Climate Policy—Rating Table for All Counties" (CCPI 2020) https://www.climate-change-performance-index.org/category-results#climate-policy.
67. F. Branger and P. Quirion, "Climate Policy and the 'Carbon Haven' Effect," WIREs Clim Change 2014, 5:53-71. DOI: 10.1002/wcc.245, p. 58.
68. A. Cosbey, "Climate Policies, Economic Diversification and Trade" (United Nations Conferences on Trade and Development, October 3, 2017, Geneva), https://unctad.org/en/PublicationsLibrary/ditcted2018d4_en.pdf, p. 5.
69. Such as the World Bank, the International Civil Aviation Organization, and local development banks, among others. *See* D. Esty and S. Biniaz, "Introduction," in D. Esty and S. Biniaz (eds.),

permits and emission trading systems due to it being inexpensive and contributing to reducing GHG emissions,[70] alongside other methods to maximize impact that trade may have on the climate. One way in which globalization does contribute positively here is that, although some developing economies may not have the knowledge, they may be able to acquire it from neighboring countries or by being a part of a major international organization, through knowledge spillovers.[71]

Nonetheless, although greenhouse gas emissions may be lower in major economies as a result of more stringent policies or any type of regulation,[72] it may be that those emissions are simply reallocated in another country across the world, causing carbon leakage.[73] What is mostly feared is the competition between nations to attract trade and grounding of companies in their territory and, as a result, lowering the standards of their climate policies in major economies.[74] What is encouraging is the scope of power of the millennial generation, who are shifting the conversation purely with their choices and ultimately creating pressure on major companies to create "greener" choices for their outdated alternatives, such as the giant fashion conglomerate H&M.[75] Nevertheless, this is not a matter of which border has the least GHG emissions, or which enormous company creates a sustainable line of product for the purpose of a higher political standing or economic gain; what truly matters is the amount of GHG emissions on the planet overall.

Correspondingly, another trend of globalization is its ability to merge forces of nations for the benefit of public welfare. Conversely, this is not as tranquil as it may seem. Perhaps one of the foremost issues of globalization, both in respect of climate change and international trade, but also more generally, is the bliss ignorance of all the differentiating values, morals, cultures, priorities, and approaches of all the various countries involved. Not every state is pro-globalization since this ideology may not align with the principles and wishes of its citizens.[76] The globalization movement is being led by major economies of the world; their ideas and priorities are often outweighed by those of the smaller fish in an enormous ocean. Major economies have the resources to be able to invest in innovation and technology to fight the impacts of

Cool Heads in a Warming World: How Trade Policy Can Help Fight Climate Change. Yale Center for Environmental Law & Policy, 2020, p. ii.

70. OECD, "Emission Trading Systems," https://www.oecd.org/environment/tools-evaluation/emissiontradingsystems.htm.
71. A. Cosbey, "Climate Policies, Economic Diversification and Trade" (United Nations Conferences on Trade and Development, October 3, 2017, Geneva), https://unctad.org/en/PublicationsLibrary/ditcted2018d4_en.pdf, p. 4.
72. *Ibid.* p. 10.
73. F. Branger and P. Quirion, "Climate Policy and the 'Carbon Haven' Effect," WIREs Clim Change 2014, 5:53-71. DOI: 10.1002/wcc.245, p. 54.
74. A. Dechezlepretre and M. Sato, "The Impacts of Environmental Regulations on Competitiveness," 11:2 *Review of Environmental Economics and Policy* 183-206, at p. 183 (Summer 2017).
75. B. Morgan, "11 Fashion Companies Leading the Way in Sustainability" (Forbes, February 24, 2020), https://www.forbes.com/sites/blakemorgan/2020/02/24/11-fashion-companies-leading-the-way-in-sustainability/#3dd687b86dba.
76. An example here would be the exit of the United Kingdom from the European Union.

climate change;[77] yet many do not. The EU and other major economies are the largest GHG emitters; however, this does not withstand the fact that they do contribute to GHG emissions globally, not simply locally.

Foreign direct investment (FDI) is another factor to consider. All states compete for FDI on a daily basis to be able to build on their existing infrastructures. The imposition of green policies, which may not align with investors, can cause lack of FDI, especially in developing nations. Similarly, investment and expansion of companies in individual states can also play a big role in the development of climate-change policies.

Perhaps one of the best examples of this scenario is the United States. The fossil-fuel industry is still highly influential on policies, with the US government continually subsidizing the extraction of fossil fuels and the respective companies that do so.[78] It is no secret that giant energy companies are often funding political campaigns in America, and perhaps elsewhere in the world, alongside campaigns to block any pro-environment changes.[79] A way forward would be getting rid of fossil-fuel subsidies.

China is an interesting example of FDI inflows in recent years.[80] FDI in China has been increasing significantly year upon year, creating the world's second largest economy after the United States. China is, by far, the world's largest GHG emitter;[81] yet its climate policies are still considered "insufficient."[82] Therefore, some academics argue that the amount of relaxation in green policies determines the amount of FDI inflows.[83] This seems to suggest that, should there be *real* changes in climate policy around the world, not only will the individual governments of each nation contribute financially, they may also encounter the likely loss of FDI as a result.[84] The world leaders are highly likely aware of this, knowingly avoiding the true momentum that is necessary and scientifically suggested. As mentioned previously, the lack of enforceability is allowing such practices. It is necessary for strict mechanisms to be put in place for greater answerability.

77. A. Cosbey, "Climate Policies, Economic Diversification and Trade" (United Nations Conferences on Trade and Development, October 3, 2017, Geneva), https://unctad.org/en/PublicationsLibrary/ditcted2018d4_en.pdf, p. 7.
78. J. Elsmoor, "United States Spend Ten Times More on Fossil Fuels Subsidies Than Education," (Forbes, June 15, 2019), available at https://www.forbes.com/sites/jamesellsmoor/2019/06/15/united-states-spend-ten-times-more-on-fossil-fuel-subsidies-than-education/#4dff17684473.
79. S. Laville, "Top Oil Firms Spending Millions Lobbying to Block Climate Policies, says Report" (The Guardian, March 22, 2019), https://www.theguardian.com/business/2019/mar/22/top-oil-firms-spending-millions-lobbying-to-block-climate-change-policies-says-report.
80. China is equally interesting in the trade field in that it has chosen, as has also Russia, to do its trade deals in its own currency, as opposed to US dollars. For instance, Russia's trade with China was nearly all in dollars in 2013; by 2020, it was less than half. *See* The Economist, "Dethroning the Dollar," January 18, 2020, pp. 68-70, at 69.
81. In 2017, China was the world's biggest importer of oil, and the second biggest importer of liquefied natural gas, after Japan, making China heavily dependent on foreign fuels. *See* The Economist, "Commodities: Customs of the Country," May 9, 2020, pp. 57-58, at 57.
82. Climate Action Tracker: https://climateactiontracker.org/countries/china/.
83. K.L. O'Brien and R.M. Leichenko, *Double Exposure: Assessing the Impacts of Climate Change within the Context of Economic Globalization*, Global Environmental Change, 2000, 10, 221-232.
84. A. Dechezlepretre and M. Sato, "The Impacts of Environmental Regulations on Competitiveness," 11:2 *Review of Environmental Economics and Policy* 183-206, p. 192 (Summer 2017.

Lastly, it is essential to examine the argument that the flow of international trade itself does have an impact on climate change,[85] as suggested throughout this section. Some studies have concluded that, in some cases, international trade actually has lower GHG-emission value in comparison to intra-national trade.[86] Some academics argue, however, that globalization has had a "double exposure" effect on certain areas of the world.[87] For example, it has been argued that, although China may have been exposed to FDI, which has helped it grow economically, it has also endangered the lives of some communities along its coastline.[88] This is due to disastrous climatic events that are likely to occur if ocean levels continue to rise, for example. One is not suggesting that climate action ought to be prioritized over FDI; nonetheless, there should be careful consideration of impacts that can occur as a result of increased pollution.

III WHAT ROLE HAVE THE EU AND THE WTO PLAYED IN CLIMATE CHANGE? WHAT ROLE SHOULD THEY PLAY?

1 Introduction

Global climate change is an issue. Actually, it is certainly the major issue the world will have to face in the coming decades. While the priority is not to point out the responsibilities of particular countries or organizations regarding to what led us to face such a dramatic situation, it appears interesting to question ourselves on the role of certain international organizations. Indeed, international organizations have a particular power in the area of environment regulations.

United States National Space Agency (NASA) defines climate change as follows:

> Climate change is a change in the usual weather found in a place. This could be a change in how much rain a place usually gets in a year. Or it could be a change in a place's usual temperature for a month or season.
>
> Climate change is also a change in Earth's climate. This could be a change in Earth's usual temperature. Or it could be a change in where rain and snow usually fall on Earth.[89]

This definition helps to establish the foundations of our reflexion on the subject. When referring to climate change, we intend to consider the way economical activities

85. *See* generally R. Leal-Arcas, *Climate Change And International Trade*. Edward Elgar Publishing Ltd, 2013.
86. C. McAusland and D.L. Millimet, "Do National Borders Matter? Intranational Trade, International Trade, and the Environment," 65:3 *Journal of Environmental Economics and Management* 411-437, at p. 413 (2013).
87. K.L. O'Brien and R.M. Leichenko, *Double Exposure: Assessing the Impacts of Climate Change within the Context of Economic Globalization*, Global Environmental Change, (2000), 10, 221-232.
88. *Ibid.* p. 230.
89. D. Stillman and J.C. Green, "What Is Climate Change ?," NASA Web site, at https://www.nasa.gov/audience/forstudents/k-4/stories/nasa-knows/what-is-climate-change-k4.html

have been able to influence the climate, in such a way that jeopardizes our civilizations.

Though this very abstract definition has to be exposed in order to understand precisely what climate change is, for the purposes of our section, we will take the introduction of climate change in the United Nations (UN) website as a starting point:

> Climate Change is the defining issue of our time and we are at a defining moment. From shifting weather patterns that threaten food production, to rising sea levels that increase the risk of catastrophic flooding, the impacts of climate change are global in scope and unprecedented in scale. Without drastic action today, adapting to these impacts in the future will be more difficult and costly.[90]

Acknowledging these elements as facts, several questions arise. We will focus on two of them. What role have the European Union (EU) and the WTO played in global climate change? What role should they play in a globalized world?

These questions need to be considered in a particular context. The WTO and the EU, though powerful, are dependent on the will of their members and thus, on the trend of international relationships between countries. In fact, the recent trend in international trade is the formation of regional and bilateral trade arrangements, instead of multilateral arrangements, such as the pan-African free trade agreement. Therefore, it is necessary, in order to understand their action to take into consideration the actions of other major international actors such as the UN, very active on environmental issues. Moreover, it is essential to analyze the consequences of their actions as leading actors of globalization and trade.

To answer these questions, we will first consider the role played by major international organizations such as EU and WTO. It appears necessary to draw up the assessment of what have been the role of globalization, and more precisely of trade expansion, encouraged by these two organizations, in climate change. However, it appears that their role changed, and we will discuss their new role as well. This section will then aim to explore a suitable solution in order to tackle efficiently climate change. We will discuss the issue of free-riding in international environmental agreements and then explore the option offered by climate clubs.

2 The Role of the EU and the WTO

Since the 1970s and the adoption of the first regulations in matter of environment, climate change is a preoccupation for policymakers, especially at an international level. Though the subject is nowadays a major issue almost all economic actors decided to tackle, it is interesting to see that this awareness is very recent.

Major international organizations, particularly WTO and EU, since their creation have as the principal goal to encourage exchanges between their members, to

90. UN Website, Climate Change, at https://www.un.org/en/sections/issues-depth/climate-change/.

encourage globalization.[91] From the correlation between the growth of economic exchanges worldwide in these last 50 years and the aggravation of global warming,[92] we argue that globalization have deeply damaged environment:

> Globalisation, which is partly synonymous with rising international trade, has fostered the rapid production, trade and consumption of material goods in unprecedented quantities. This has weighted the ecological footprint of human activities around the world.[93]

Through a certain number of mechanisms, globalization increases economic activity and trade resulting in an increase of damages for the environment. In 2014, exports were more than six times larger than in 1970.[94] In the meantime, CO_2 emissions have increased by about 90%.[95]

It would be naive to think that these two major growths are unrelated. As a matter of fact, the sole growth of the volume of goods transportation (by road, sea or air) represents an important volume of GHG emissions.

Every year, sea transport represents between 2% and 4% of GHG emissions worldwide; to this figure, it is necessary to add the emissions caused by route and air transport (in 14 years, between 1990 and 2004, GHG emissions from air transportation increased by 86%).[96]

If the increase of GHG emissions coming from transportations can be seen as a direct consequence of globalization, the growth of trade, largely encouraged by WTO and EU, caused major damages to environment, worsening global warming. Without going into details, it appears important to voice that the development of trade has directly impacted the development of consumption and of industrial production. According to the United States Environmental Protection Agency (EPA), 22% of US GHG emissions in 2018 were coming from industry.[97] The part of industry in GHG is much more important in developing countries were industry is still the first economic sector.

Finally, it appears that the opening of markets, induced by globalization and development of trade within the WTO, creates, for a certain number of countries, a need for space in order to be able to reach the demand, particularly in agriculture sector. This space is often found using forest.

91. "The WTO shall provide the common institutional framework for the conduct of trade relations among its Members," Article II of Marrakesh Agreement Establishing the WTO.
92. That said, mild winters due to global warming do have some benefits: less use of heating and therefore less money spent on energy bills, flu seasons are shorter and therefore fewer people die.
93. J.Y. Huwart and L. Verdier, "What Is the Impact of Globalisation on the Environment?," in *Economic Globalisation: Origins and Consequences*, OECD Publishing, 2013.
94. F. Tena-Junguito, "A Tale of Two Globalizations: Gains from Trade and Openness 1800-2010." London, Centre for Economic Policy Research, 2016 b. (CEPR WP.11128).
95. T.A. Boden, G. Marland, and R.J. Andres, "Global, Regional, and National Fossil-Fuel CO2 Emissions." Carbon Dioxide Information Analysis Center, Oak Ridge National Laboratory, U.S. Department of Energy, Oak Ridge, Tenn., U.S.A. R.J.(2017). doi 10.3334/CDIAC/00001_V2017.
96. *Ibid*. 4.
97. EPA, *Inventory of U.S. Greenhouse Gas Emissions and Sinks: 1990-2018* (published 2020).

According to a satellite-based study, the earth has lost 1.3 millions square kilometers of forest since 1990. Between 1990 and 2015, every hour, the equivalent of 1,000 football fields of forest has been lost in deforestation.[98]

The reasons that lead to all these phenomena are not, in our opinion, difficult to understand. It would be hypocritical to judge, *a posteriori*, countries for having invested in the development of their economies, playing in a competitive globalized world with their own assets. China developed its industry leaning on its enormous workforce; Brazil invested in agriculture taking advantage of a favorable climate and of huge cultivable space taken to rainforest.

We will not argue here that the solution to climate change issue is the "degrowth" theory mainly inspired by the work of economist Nicholas Georgescu-Roegen.[99] This would lead to a reduction of trade and exchanges between countries all over the world, and this section defends the idea that only a world-based solution, aiming to gather the larger possible number of countries in a common effort, will be able to tackle the issue of climate change.

Taking into account the observations made above, international organizations such as EU and WTO have changed their paradigms of functioning these last years. The awareness of climate change and the absolute need to fight it forced international organizations as well as most of the countries to adapt their economic policies in order to take this new goal into account.

In this process, the WTO and the EU are, of course, accompanied by the United Nations. In this logic, the United Nations Framework Convention on Climate Change (UNFCC) was adopted on May 1992. The objective of this convention is clearly set up in article 2:

> The ultimate objective of this Convention (...) is to achieve, in accordance with the relevant provisions of the Convention, stabilization of greenhouse gas concentrations in the atmosphere at a level that would prevent dangerous anthropogenic interference with the climate system.[100]

Every year, the parties to the UNFCCC gather in a Conference of Parties (COP) in order to assess the progress made in the final goal of reducing GHG emissions. At the occasion of these COP, countries concluded international agreements creating binding obligations (mainly for developed countries) to reduce their GHG emissions. In 1997, in Japan, the Kyoto Protocol was adopted and became an extension of UNFCCC. 192 countries are parties to the Protocol, which entered into force in 2005.

The Kyoto Protocol relied on the principle of common but differentiated responsibilities. This principle is stated directly in Article 3 of UNFCCC:

> The Parties should protect the climate system for the benefit of present and future generations of humankind, on the basis of equity and in accordance with their

98. T. Crowther et al. "Mapping Tree Density at a Global Scale." 525 *Nature* 201-205 (2015).
99. A. Missemer, "Nicholas Georgescu-Roegen and Degrowth." 24:3 *European Journal of the History of Economic Thought* 493-506 (2017).
100. UN General Assembly, United Nations Framework Convention on Climate Change: resolution adopted by the General Assembly, January 20, 1994, A/RES/48/189.

> common but differentiated responsibilities and respective capabilities. Accordingly, the developed country Parties should take the lead in combating climate change and the adverse effects thereof.[101]

According to this article, not every country shall bear the same responsibility in the fight against global warming. Developed countries, considering the strength of their economy but also their historic responsibility in the level of GHG in atmosphere, are made responsible for the largest effort of reduction.

The Kyoto Protocol is widely considered as a failure. However, it appears that among the 36 countries involved in the application of the protocol, only nine did emit higher level of GHG than the Kyoto set target. Even if some good results need to be highlighted, a certain number of them was the consequence of the dissolution of Soviet Union in 1991 and of the 2007 economic crisis.[102]

Though important in the history of development of environment regulations at a global scale, the Kyoto Protocol lacked ambitions. These considerations led to the adoption in 2015 of the Paris Agreement. In total, 189 counties have become parties to it, as of February 2020. We will discuss in depth the issue of the Paris Agreement later in our development.

EU countries are fully involved in the fight against climate change. Though economic considerations are still a break preventing from radical and more efficient actions, environment preoccupation have become a central object of EU policies. In this matter, the EU adopted in 2019 the European Green Deal. The purpose of this new ambitious policy is to gather all European efforts aiming to tackle global warming issue. The stated goal of the European Green Deal is:

> (...) to transform the EU into a fair and prosperous society, with a modern, resource-efficient and competitive economy where there are no net emissions of greenhouse gases in 2050 and where economic growth is decoupled from resource use.[103]

Moreover, it appears that the EU in its international-trade-treaty negotiations has adopted a more environment-oriented logic, imposing on its contracting parties ambitious environment-related rules as a condition for the conclusion of its treaties.

It has been particularly true in the context of the negotiation of EU-Mercosur free-trade agreement, where French President Emmanuel Macron expressed his opposition to the Agreement, considering that the engagement of Brazil was not sufficient in order to defend the rainforest. He considered that a withdrawal of Brazil from the Paris Agreement would make any agreement impossible.[104]

101. *Ibid.* 11.
102. I. Shishlov, R. Morel and V. Bellassen, "Compliance of the Parties to the Kyoto Protocol in the First Commitment Period," 16:6 *Climate Policy* 768-782 (2016), DOI: 10.1080/14693062.2016.1164658.
103. Communication from the Commission to the European Parliament, The European Council, The Council, the European Economic and Social Committee and the Committee of the Regions—The European Green Deal—COM/2019/640 final.
104. Independent (The), *Emmanuel Macron says he will block EU trade deal with Brazil over Amazon forest fires.*

Implementing climate change mitigation policies directly in bilateral free trade agreement is considered by observers as a good and efficient manner to use free trade and globalization as a tool in fighting climate change.[105]

The WTO has a much more ambiguous attitude concerning environment regulations. For example, the then Director-General of the organization, Roberto Azevedo, expressed his doubt about the link between international trade and climate change during a discussion at the World Economic Forum's annual meeting in Davos: "This whole argument that there is an impact (of intention trade) on the environment, I stand to be corrected, but I haven't seen one, single credible study that shows that."[106]

Many studies affirm that international trade and economic growth, allowing economies to become mature and developed, have *in fine* virtuous effects on the environment.[107] Others have shown that in the process, intentional trade have a tangible effect on the environment, particularly on deforestation.[108]

3 Climate Clubs as a Possible Solution

Considering the previous considerations, global environment agreements are of great importance. It seems interesting to analyze the issue of free-riding within these agreements. Facing this issue, some may think that forming climate clubs would be the best solution in order to tackle the issue of free-riders and, finally, in order to mitigate climate change.

As mentioned above, after the "failure" of the Kyoto Protocol, the adoption of the Paris Agreement in 2015 has been perceived as a game changer in the fight against global warming. The purpose of the Paris Agreement is set up in article 2: "This Agreement (...) aims to strengthen the global response to the threat of climate change, in the context of sustainable development and efforts to eradicate poverty."[109]

Countries efforts, in the same vein of what was decided for the Kyoto Protocol, are based on the principle of "common but differentiated responsibilities and respective capabilities."[110] The objective of this agreement is clearly to install an efficient mitigation of climate change while preserving equity between countries. More precisely, the Agreement aims to hold the increase of global average temperature below 2°C above preindustrials levels.

If the Paris Agreement is considered as ambitious by most of the observers, some consider that it is still not enough. Though the agreement appears to be a great

105. *See* for instance, R. Leal-Arcas, *Climate Change and International Trade*. Edward Elgar Publishing, 2013.
106. Azevedo Sewing, Malmstrom on Global Trade: Davos Panel, accessible at https://www.bloomberg.com/news/videos/2019-01-24/sewing-azevedo-malmstrom-on-global-trade-davos-panel-video [accessed May 5, 2020].
107. *See* for example, J. Frankel, "Environmental Effects of International Trade," January 2009 (RWP09-006).
108. Defries, Rudel, Uriarte and Hansen, Deforestation Drive by Urban Population Growth and Agricultural Trade in the Twenty-First Century. 3 *Nature Geoscience* 178-181 (2010). 10.1038/ngeo756.
109. Paris Agreement. Paris, December 12, 2015, Article 2.
110. *Ibid.* 20.

achievement from the perspective of its scope (no previous agreement have never gathered such a number of countries around environmental subject), the lack of enforcement mechanisms and of strict obligations seems to hurt its announced ambitions.[111]

As a matter of fact, the Paris Agreement is based on the good will of countries to reach their commitments. The Agreement is based on a bottom-up approach, where countries determine freely their contributions to the objectives set up in the Agreement. These contributions are called Nationally Determined Contributions (NDCs). This system poses the risk of facing the same issue than previous environment agreements: the problem of free-riding.

NDCs have the advantage of being more incentive for poor countries that could opt for low cost mitigation measures than what could be decided at a global level. Primary NDCs to fight GHG emissions are related to carbon pricing with two types of measures: the implementation of a carbon tax or of an Emission Trading System (ETS).

In the eventuality of a carbon tax, countries will set up a price for the GHG ton that will be paid by the emitter business. One should also remember that the externalities of GHG emissions are not properly priced in the real cost of energy. Given how low energy prices are, it is easier to introduce a tax on carbon.

In the ETS, "governments instead put a cap on overall emissions, which is tightened over time, allocate allowances that emitters have to submit for every tonne of GHGs, and allow participants to trade these emission units among themselves."[112]

The ETS seems more effective as to incite private-sector actors to actively participate to the aim of reducing GHG emissions. By allowing small emitters to sell a part of their allowance to bigger ones, it will incite the former to reduce their emissions in order to make a real profit selling their surplus to the latest. Bigger emitters will be incited to reduce their emissions too in order to reduce the cost associated to their emissions of GHG.

However, the risk of unequal fight against GHG emission is still real and so is the risk of carbon leakage. The European Commission defines carbon leakage as "the situation that may occurs if, for reasons of costs related to climate policies, businesses were to transfer production to other countries with laxer emission constraints."[113]

This situation will occur if a country or a group of countries implements rigorous NDCs in order to reach the goals set up in the Paris Agreement, while other countries do not implement NDCs or at least implement less rigorous ones. Companies from the first group of counties will be tempted to move their polluter businesses in less strict countries.

At the same time, less rigorous countries will enjoy the benefits of ambitious mitigation policies adopted by the other counties; this situation is called free-riding. William Nordhaus, the author of reference on matters of climate clubs, defines

111. Rogelj J., et al., "Paris Agreement Climate Proposals Need a Boost to Keep Warming Well Below 2 °C, *Nature* (2016).
112. ICTSD, *Carbon Market Clubs under the Paris Climate Regime: Climate and Trade Policy Considerations*, 2016.
113. European Commission Website, Carbon leakage, accessible at https://ec.europa.eu/clima/policies/ets/allowances/leakage_en [accessed May 5, 2020].

free-riding as follows: "Free-riding occurs when a party receives the benefits of a public good without contributing to the costs."[114]

Nordhaus identified two main free-riding dimensions. The geographical free-riding, when countries enjoy the reduction of GHG emissions from other countries without taking any climate change mitigation policies, and the temporal free-riding: "the present generation benefits from enjoying the consumption benefits of high carbon emissions, while future generations pay for those emissions in lower consumption or a degraded environment." He argues that this free-riding phenomenon is the consequence of the Westphalian Dilemma. International relations are essentially based on the consent of countries in joining and applying international treaties.

Nordhaus was one of the first to identify a solution to free-riding issue. He explained that the formation of clubs of countries could be a suitable solution. He identifies four main elements necessary for a climate club:

> (i) that there is a public-good-type resource that can be shared (whether the benefits from a military alliance or the enjoyment of a golf course); (ii) that the cooperative arrangement, including the dues, is beneficial for each of the members; (iii) that non- members can be excluded or penalized at relatively low cost to members; and (iv) that the membership is stable in the sense that no one wants to leave.[115]

Since then, Nordhaus has been joined in his analysis by many scholars and observers. Studies analyzing the solution of climate clubs, and more specifically of carbon market clubs, in the context of the Paris Agreement flourish. Most of the proposals are oriented toward the creation of a carbon market club.[116] The result would be a harmonization of carbon tax and of Emission Trade System that would be effective internationally.

The most interesting in these proposals is the possibility offered to spread the Emission Trade System at an international level rather than at a purely national level. This system could be put in place between the members of a club and only between them. A condition to the effectiveness of the system would be a harmonization of accounting standards as well as on maximum GHG emissions caps. Without it, the carbon market club could result in an increase of GHG emissions.

Carbon markets already exist. The European Union have an ETS in place for several years. The purpose of a club would be to standardize carbon price between countries. That could be a solution to tackle the two major issues of carbon-based mitigation policies: competitiveness and carbon leakage.

If members of the climate club fix a unique price for carbon, it will reduce possibilities (and interest) for companies to seek lower carbon prices abroad. Then, the

114. William Nordhaus, "Climate Clubs: Overcoming Free-Riding in International Climate Policy." 105:4 *American Economic Review* 1339-1370 (2015).
115. *Ibid.* 25.
116. *See* ICTSD. 2016. Carbon Market Clubs under the Paris Climate Regime: Climate and Trade Policy Considerations. Geneva: International Centre for Trade and Sustainable Development (ICTSD).

members of the club would lose less competitiveness compared to other counties (it will of course depend on the number of club members).

Some issues have been raised related to these types of carbon market club. The issue here is to know if the creation of a carbon market club would be in accordance with WTO rules. Some scholars argue that carbon is neither a "good" nor a "service" and should not follow General Agreement on Tariffs and Trade (GATT) or General Agreement on Trade in Services (GATS) rules.[117]

Other options could be considered such as using the general exceptions of Article XX of GATT and XIV of GATS, both relating to "the conservation of exhaustible natural resources if such measures are made effective in conjunction with restrictions on domestic production or consumption." In the *Shrimp—Turtle* case,[118] WTO panel accepted an exception to GATT rules in order to preserve environment.

There are no existing rules within WTO agreements that provide a harmonization of environment rules as a prerequisite before entering WTO. Maybe such a provision should be considered as a tool to reduce the risk of carbon leakage and free-riding. This could be viewed as a first step to the formation of a climate club combining strong climate change mitigation efforts with a more sustainable international trade system.[119]

Further options that should be considered would be the creation of a plurilateral WTO Agreement—to deal with coalitions of the willing—or a preferential trade agreement. Agreements concluded in this form would be in accordance with WTO rules as long as they provide benefits for the members of the agreements without sanctioning non-members. The best incentive of such an agreement would certainly be the creation of free-trade zones between the members of the climate club. The respect of strict and ambitious mitigation policies would offer the access to important markets. Nonetheless, it appears that the great progress made against climate change since 2015 is not enough to consider the fight against climate change a success.

IV A CLIMATE CLUB TO ENFORCE THE COMMITMENTS OF THE PARIS AGREEMENT AND ENHANCE FREE TRADE

1 Introduction

One of the current major global problems that are extremely important is global climate change. Since the severity of the problem has increased dramatically, higher temperatures increase the frequency of many types of disasters, including storms, floods, heatwaves, and droughts. Moreover, this situation can create the devastating and

117. A. Petsonk and N. Keohane, *Creating a Club of Carbon Markets: Implications of the Trade System*. E15Initiative. Geneva: International Centre for Trade and Sustainable Development (ICTSD) and World Economic Forum, 2015.
118. United States—Import prohibition of certain shrimp and shrimp products, accessible at https://www.wto.org/english/tratop_e/envir_e/edis08_e.htm.
119. The pledge by Jeff Bezos of Amazon in February 2020 to give USD 10 billion to scientists, activists and NGOs working to mitigate climate change is also welcome. By doing so, he created the Jeff Bezos Earth Fund.

costly consequences, jeopardizing access to clean drinking water, fueling out-of-control wildfires, damaging property, creating hazardous-material spills, polluting the air,[120] and leading to loss of life.[121] Therefore, from what has been overlooked for a long time, it became the hot topic which was raised to consider more seriously. In 2015, the world joined together, trying to solve the problem before it is too late.

The 2015 United Nations Climate Change Conference, COP 21 or CMP 11 held in Paris, allowed countries to negotiate, and hoping they will reach a new agreement that will replace the Kyoto Protocol the former international environment accord, which is seen by many critics as a complete failure. After discussion, the results that the world waited for appeared, it is called "The Paris Agreement." It is an accord within the United Nations Framework Convention on Climate Change (UNFCCC), dealing with greenhouse-gas-emissions mitigation. Many scholars and related important people such as François Hollande, former president of France, and David Cameron, former prime minister of the United Kingdom both had a very positive view to this agreement. World leaders hail Paris climate deal as "major leap for mankind."[122] The Paris Agreement's status is comparable to the light at the end of the tunnel that brings hope to humankind. However, despite being highly appreciated, from analytical studies appeared to show its important weaknesses, which might be the obstacles to the implementation of this accord. The result of the worst-case scenario will lead to the failure of this agreement like many other international agreements.

If the Paris Agreement is unlikely to be successful, keeping the expectations on the solution that deal solely on traditional international law seems to be a very foolish mindset. Therefore, to deal with the climate issues, many countries, especially the countries that are members of the EU, have developed a new approach by creating a friendly group called a "climate club." Considering the results of the implementation of this integration, it is found that the climate club is effective in dealing with the climate issues and seems to be able for coping the climate change commitments of the Paris Agreement better than the Paris Agreement itself, while the climate club also enhances free trade. This section will discuss the inefficiency of the Paris Agreement and will discuss the potential of the climate-club model for the enforcement of the climate change commitments of the 2015 Paris Agreement. It will then focus on EU policies related to climate-change mitigation. The section has been divided into three parts: (1) the limitations of the Paris Agreement, (2) the advantages of the climate-club model compared to the Paris Agreement, and (3) the EU policies that relate to the climate-club notion.

---

120. One benefit of the COVID-19 confinement in China is that there was cleaner air, which "may have saved 17 times more lives than the (official) number lost to the virus." *See* The Economist, "India on lockdown: Impossible sums," April 25, 2020, p. 30.
121. Melissa Denchak "Global Climate Change: What You Need to Know" (2017). https://www.nrdc.org/stories/global-climate-change-what-you-need-know accessed April 21, 2020.
122. The Guardian, "World Leaders Hail Paris Climate Deal as 'Major Leap for Mankind'" https://www.theguardian.com/environment/2015/dec/13/world-leaders-hail-paris-climate-deal, accessed May 5, 2020.

2 The Paris Agreement

By considering a landmark environmental accord that was created for dealing with global climate change, the Paris Agreement has the key aim of holding the increase in the global average temperature to well below 2 Celsius above preindustrial levels and to pursue efforts to limit the temperature increase to 1.5 Celsius above preindustrial levels, recognizing that this would significantly reduce the risks and impacts of climate change.[123] To overcome the failure of the Kyoto Protocol, this accord launched the new way of cooperation called Nationally Determined Contributions (NDCs). NDCs have received a lot of international attention. With ideas that allow countries to go back and consider their action plans for climate change management. This can be considered a very smart idea, as the result is that this agreement has been signed by many countries. In total, 188 states and the European Union have ratified the Agreement, meaning more than 87% of global greenhouse gas emissions are ready to be in the Paris Agreement. Moreover, China, the United States, and India, the countries that are the major greenhouse gas emitters of the UNFCCC members, have a positive opinion on this accord. Despite the great success of the Agreement in terms of the number of members that have ratified it, the Paris Agreement still has problems in terms of features and characteristics similar to other international agreements. Since negotiating and ratifying agreements takes years, by the time their rules are ready, the world has moved on.

The experts' analysis clearly demonstrates that, although the Paris Agreement is able to demonstrate the unity of international cooperation, the content and methods of implementing the Agreement cannot be accessed and resolved in global climate change issues. The standard set by academics is called "The people's test," which specifies the requirements of an agreement that is effective enough to deal with this problem. This test is achieved by following the criteria below:

1. catalyze immediate, urgent and drastic emission reductions;
2. provide adequate support for transformation;
3. deliver justice for impacted people; and
4. focus on genuine, effective action, rather than false solutions.[124]

The Paris Agreement cannot meet these conditions, nor can it make people have confidence in it.

As for the NDCs model that is highly flexible, it was seen as being too open to the point that there was no concrete action plan to hold. This system may be able to operate if every country works together and sacrifices their own interests for the benefit of the public. This fanciful idea is almost impossible in reality. Importantly, we must not forget that each country has a different living status; developing countries that have not

123. "The Paris Agreement," https://unfccc.int/files/meetings/paris_nov_2015/application/pdf/paris_agreement_english_.pdf.
124. John Foran, "The First Draft of History Thirty-Four of the Best Pieces on the Paris Agreement at COP 21" (2015), http://ejcj.orfaleacenter.ucsb.edu/wp-content/uploads/2018/02/2016.-John-Foran-The-First-Draft-of-History-January-4-2016-Thirty-Four-of-the-Best-Pieces-on-the-Paris-Agreement-at-COP-21.pdf, accessed April 21, 2020.

yet utilized the industry to develop their own countries to become more prosperous and have shifted their status to developed countries would not accept to be exploited by developed countries such as the United States, and the United Kingdom, which already use coal or other fossil fuels, and heavily released GHG emissions into the atmosphere in the past. Conversely, developed countries may consider developing countries are using this status as an advantage for pushing the burden of solving global climate change on them like the conflict that already occurs between the United States and China. Therefore, the expectation to believe that each country will go back and write NDCs that slash their bodies does not sound reasonable.

In addition, not only failing to meet the standard and the ambiguity of the implementation plan, the goals of the Paris Agreement with the NDCs model may not be able to be achieved, because, in reality, the current global climate change crisis is worse than that. It will result in the increase of global average temperature in the rate that the commitments of the agreement cannot follow.[125]

Second, the commitments under the Paris Agreement have no legally binding.[126] With the concept of NDCs model, it creates the bottom-up architecture of the Paris Agreement instead of a top-down structure like the most international treaties, which is great; however, without the binding there are no such things to guarantee that each country will follow the beautifully drawn NDCs plan. The final result will lead to the failure of achieving global temperature control to the desired level. The NDCs can be compared to the promise that opening for twisting at every opportunity unconditionally. The lack of binding provisions and the motivation to comply with this accord is considered a very serious weakness in every type of laws and regulations because when there is no one to take it seriously, those laws or regulations are considered as though it did not exist at all. From this weakness, combined with the differences in the status of each country, the implementation of the NDCs, or even the drafting of the NDCs that every country be in line with the goals of the Paris Agreement, is almost impossible; poor countries still have to rely on the carbon economy to develop their future.[127] This prisoner's dilemma—the game theory of decision-making—is one of the challenges that international cooperation must face.

Besides, the absence of binding and the motivation for implementing NDCs will naturally open the room for the free-riding to continue. When there is no need or benefit from any adverse actions or noncompliance with NDCs, countries that saw this fact could be able to take advantage of this weakness by not taking steps to help reduce GHG emissions and take advantages on the committed countries that have taken serious action in addressing the Global Climate change. If the problem of free-riders cannot be resolved, the Paris Agreement is likely to fail like the Kyoto Protocol.

125. Joeri Rogelj, "Paris Agreement Climate Proposals Need a Boost to Keep Warming Well Below 2°C," 534:7609 *Nature* 631-639.
126. Antero Ollila, "Challenging the Scientific Basis of the Paris Climate Agreement," 11:1 *International Journal of Climate Change Strategies and Management* 18-34 (2019).
127. Tucker Davey "Developing Countries Can't Afford Climate Change" (2016) https://futureoflife.org/2016/08/05/developing-countries-cant-afford-climate-change/?cn-reloaded=1, accessed May 5, 2020.

3 A Climate Club

Considering the weaknesses of the Paris Agreement discussed above, it is possible to predict the success rate of the Paris Agreement. To push the Paris Agreement to being a more complete resolution to solve the global climate change problem, it would take a very long time and human resources, including many other opportunity costs, to get an agreement that every country agrees on, and still not considering the fact that it may never occur at all. Accordingly, looking at the problem from different perspectives seeking to create new innovations is necessary. The climate club is one such innovation.

By the facts that, international environmental agreements in the world level mostly end up at failure, does not mean that establishing an international partnership should not be done. Conversely, this is something that is the core concept to solve the global climate change, because this is a big problem and very complicated, and affect every country. Therefore, it is a problem that a single country cannot handle it alone.

From the previous topic, the international environmental agreements such as the Paris Agreement or the Kyoto Protocol failed, not because they are the concept of international cooperation, but instead, it is because of an ineffective architecture of them. Then to make the international cooperation that has the potential to be successful, the change in its architecture will be the priority. The climate club comes with the new idea to deal with the Paris Agreements weaknesses, namely changing from a voluntary agreement to one with strong incentives to participate.

Overall, a very important factor that made the climate club, which is characterized as a club of nations success, is the mutual benefits of the club's members. Although the climate club might look like a voluntary group similar to the actions taken by NDCs, by pushing the club to the successful point will result in a clear, tangible benefit than the implementation of the Paris Agreement that members will receive from supporting the climate club, whether it is for international trade benefits, such as receiving tax privileges, or even the benefits of having a military alliance. Therefore, to be bound by paying club membership fees or to comply with the club's rules is interesting and potentially appealing.

According to William Nordhaus, "Individual countries are assumed to adopt climate policies that maximize their national economic welfare."[128] This statement clearly highlights the root idea that develops to be the climate club model. The key features of the climate club that make it different from international treaties are the rules for membership and the incentives/penalties system. First, the climate club has clear rules for membership, such as setting a limit for GHG emissions, which they must agree to undertake the will be able to participate in the community and shared the common interests of club members. This point can be compared to a sports club or gym membership, if people want to exercise by using the gym equipment or using the tennis court, they will have to pay fees, and must also comply with the rules and regulations

128. William Nordhaus, "Climate Clubs: Designing a Mechanism to Overcome Free-riding in International Climate Policy," 105:4 *American Economic Review* 1339-1370 (2015).

set forth as well. This idea makes the club become stronger compared to the cooperation that bound by the promises.

Second, the climate club model demonstrates the benefits and disadvantages of joining or not joining the club, which can be said to be a form of reward or punishment that is clearer and more concrete than the Paris Agreement as it will both attach and indirectly force countries to participate with the climate club. Also, in regards to the punishment, there are many forms of it; however, the most effective seems to be the administration of tariffs on the imports of non-cooperative countries. This will create the conditions for those countries to choose to cooperate with the climate club or to accept import disadvantages that create more burdens.

The presence of a mechanic in the form of incentives and penalties can be said to be indirect coercion for each country to take serious environmental problems. However, this does not mean that it can be enforced only by countries that are members of the club because when creating a strong community, the negotiation power of that climate club is also stronger. Countries that are not members of the club that wish to conduct international activities with the club or the countries under the club especially activities in international trade and investment must naturally comply with various regulations resulting from the determination of such climate club, otherwise, it may not be able to continue the activity or the activities may not go smoothly. With this core concept, by using the power of international trade and investment, the climate club can narrow down the way that free-riders can easily slip away, and drag them to take the responsibility like other countries.

However, the climate club's potential is relying on the strength of the founder of the club; with the weak or poor leader the club will not find the benefits to lure other countries to participate and also cannot create the sanction that force other to accept the deal. In the next paragraph, the policies and strategy of the European Union, one of the strong international pacts, which is considered the world leader in term of environmental protection, will be the main discussion.

4 EU Policies

Due to being a leader in the field of environmental protection, the European Union (EU) has set the goal of dealing with the problems of global climate change and has a new ambition called "the European Green Deal." The deal can be summarized in short that the EU will work with all parties to make the EU become the world's first climate-neutral continent by 2050. The European Green Deal will tackle many challenges. In the next section, some of the challenges will be raised.

The EU Emissions Trading System

As being the world's first international emissions trading system since 2005, the EU set the "Cap-and-Trade" principle, by setting a cap to limit the total amount of GHGs that the unit can emit, and this quota can be traded with other companies. This method has proved itself by making the amount of GHGs emissions fall. Under the European Green Deal, the EU plans to lower GHG emissions more, which in 2020 it seems to be sure that the GHGs emissions in the system will be lower from 2005 by around 21%, and on track to drop down about 43% from 2005 in 2030. However, the cap-and-trade system is not perfect; it still does not lead to the social awareness of the emitters. To change the habit, the suggestion of developing the carbon taxes in parallel is put on the table, because it might affect more in this angle. By setting the price for carbon emissions the behavior of entrepreneur would change.[129]

Forests

There is the fact that the forest is important when dealing with carbon emissions. It absorbs the carbon dioxide to use in its system. This is one of the most functional ways to deal with GHGs emissions because it is very effective and sustainable. However, this approach is a long-term carbon removal, and it requires vast trees. The deforestation to expand some of the human activities mostly for agriculture is the obstacle to this method. With this sense of awareness, the EU has developed the EU forest strategy and the EU Forest Policy.[130] These will cover sustainable forest management, optimizing the use of the forest and promoting social awareness, which will create a sense of global forest responsibility.

Transportation

Transport accounts for about 25% of the EU's GHGs emissions and is also the cause of air pollution in cities.[131] Therefore, the EU is pushing a strategy called "the Commission's low-emission mobility strategy," which consists of three main concepts:

1. Increasing the efficiency of the transport system.
2. Speeding up the deployment of low-emission alternative energy for transport.
3. Moving toward zero-emission vehicles.[132]

However, this strategy needs participation from cities and local authorities to help to create green cities.

Trade policies

In 2020, the EU is seeking more for sustainable development. They tend to focus on the reservation of the resources for future generations, whether the economic, social or environmental. Therefore, the policies in the field of international trade and investment of the EU must be in the same lane. In this day, signing a trade agreement with the EU is more complicated; the agreement must meet the requirements, which is considered

129. EU "EU Emissions Trading System (EU ETS)" https://ec.europa.eu/clima/policies/ets, accessed May 5, 2020.
130. EU "The European Union and forests" https://www.europarl.europa.eu/factsheets/en/sheet/105/the-european-union-and-forests, accessed May 5, 2020.
131. Environment Protection UK "Air Pollution and Transport" https://www.environmental-protection.org.uk/policy-areas/air-quality/air-pollution-and-transport/, accessed May 5, 2020.
132. EU "A European Strategy for Low-Emission Mobility" https://ec.europa.eu/clima/policies/transport, accessed May 6, 2020.

the standard of the EU sustainable development.[133] For instance, this idea can be seen in the Comprehensive and Economic Trade Agreement (CETA); this is an agreement between the EU and Canada that has the concept of environment protection in many chapters, especially in Chapter 24 of the Agreement, which is about environmental protection of each party and also encourages conservation and sustainable management of forests,[134] or in Chapter 16 of the EU-Japan Economic Partnership Agreement,[135] etc. In order to create the standard for cooperation, this would show the stance of the EU on environmental issues. It will greatly benefit the EU on the world stage, and because being a group of countries that are important to the world economy, the EU has a very high bargaining advantage. Therefore, this makes the EU one of the leaders in the fight against global climate change.

V EU BILATERAL INVESTMENT TREATIES IN THE CONTEXT OF SUSTAINABLE DEVELOPMENT

1 Introduction

International Investment Agreements (IIAs) have the main purpose of protecting foreign investors and their covered investments in the host country.[136] However, IIAs have also been used for a myriad of other purposes such as attracting foreign direct investment (FDI),[137] levelling the playing field,[138] and facilitating with domestic reforms.[139] While the EU and/or its Member States are parties to several international treaties that protect social aspects (such as human rights[140] and labor rights[141]) as well

133. EU "Sustainable Development," https://ec.europa.eu/trade/policy/policy-making/sustainable-development/, accessed May 6, 2020.
134. "The EU-Canada Comprehensive Economic and Trade Agreement" (2017).
135. "The EU-Japan Economic Partnership Agreement" (2013).
136. R. Dolzer and C. Schreuer, *Principles of International Investment Law*, 2nd edition. Oxford University Press, 2012, Chapter 2.b, p. 13: "BITs provide guarantees for the investment of investors from one of the contracting states in the other contracting state."
137. United Nations, UNCTAD series on international investment policies for Development, "the role of international investment agreements in attracting foreign direct investment to developing countries," 2002, executive summary: "Since they [IIAs] are a key instruments in the strategy of most countries ... to attract foreign investment," https://unctad.org/en/Docs/diaeia20095_en.pdf.
138. J. Bonnitcha, International Institute for Sustainable Development, "Assessing the Impact of Investment Treaties: Overview of the Evidence" Chapter 2.2: "Both the EU and the United States cite 'levelling the playing field' between domestic and foreign investors as a core justification for the investment chapter in the proposed ... TTIP. This rationale is equally relevant to investment treaties involving developing countries," https://www.iisd.org/sites/default/files/publications/assessing-impacts-investment-treaties.pdf.
139. R. Dolzer and C. Schreuer, *Principles of International Investment Law*, 2nd edition. Oxford University Press, 2012, Chapter 3.F p. 25: "Thus, investment treaties provide for external constraints and disciplines which foster and reinforce values similar to the principle of good governance with its emphasis on domestic institutions and policies."
140. E.g.: The Convention for the Protection of Human Rights and Fundamental Freedoms (European Convention on Human Rights, as amended) and The United Nations Universal Declaration of Human Rights.

as treaties dealing with specific environmental issues (such as nuclear damage[142] and oil pollution[143]), European Investment Protection Agreements (IPAs) have not been able to effectively address issues of sustainable development.

Leaving aside the discussion whether IIAs are the proper fora to address issues related to sustainable development, this chapter proposes that EU IPAs have, till date, not sufficiently addressed the issues of sustainable development and in particular those related to climate action and sustainable energy. While recognizing the latter, this chapter argues that EU IPAs could, however, be used as a vehicle to contribute to climate action and sustainable energy as IIAs have, although arguably, been contributing to market-orientated reforms and domestic standards, such as transparency.[144]

In assessing the latter, we shall look into the most recent IPA of the EU, which was concluded with Vietnam on June 30, 2019. The reason being that, apart from it being the most recent European IPA, the parties to the latter have had access to more information regarding climate change and the need for sustainable development, as well as opportunities to ascertain more (nonbinding) legal obligations regarding such.

In order to assess the (lack of) contribution and the opportunity to use the future EU IPAs as a vehicle to contribute to climate action and sustainable energy, this section shall follow the following structure. First, we shall briefly look into the notions of climate action and sustainable energy, after which the investment relationship between the EU and Vietnam shall be looked at. Third, we shall look into sustainable development in the broader context of the EU-Vietnam Free Trade Agreement (EVFTA), therefore analyzing not only the IPA but also the chapter on trade and sustainable development and the schedules of commitments. Lastly, we shall look into the opportunities to use EU IPAs as a vehicle to contribute to climate action and sustainable energy, before presenting some conclusions.

2 Sustainable Development

Sustainable development can be defined in a number of ways, taking into consideration various aspects. The Oxford Dictionary has defined sustainable development as "concept used to describe community and economic development in terms of meeting the needs of the present without compromising the ability of future generations to meet

---

141. E.g.: ILO Convention C105: Abolition of Forced Labor Convention (Convention concerning the Abolition of Forced Labor) (40th Conference session, Geneva, June 25, 1957).
142. E.g.: EC directive Laying down basic safety standards for protection against the dangers arising from exposure to ionizing radiation, and repealing Directives 89/618/Euratom, 90/641/Euratom, 96/29/Euratom, 97/43/Euratom and 2003/122/Euratom [2013] OJ 2 13/1.
143. E.g.: Authorizing the Member States to sign, ratify or accede to, in the interest of the European Community, the Protocol of 2003 to the International Convention on the Establishment of an International Fund for Compensation for Oil Pollution Damage, 1992, and authorizing Austria and Luxembourg, in the interest of the European Community, to accede to the underlying instruments [2004] OJ 2 78/1.
144. J. Bonnitcha, International Institute for Sustainable Development, "Assessing the Impact of Investment Treaties: Overview of the Evidence," p. 7: "investment treaties could facilitate domestic reforms in the countries that sign them. However, to date there is little evidence to support this supposed benefit."

their needs."[145] The United Nations (UN), however, sets 17 so called sustainable development goals, addressing global challenges to achieve a sustainable future. These include no poverty, gender equality, clean energy, and climate action.[146] Considering the scope of this chapter, I shall now briefly look at the notions of climate action and sustainable energy in the broader context of sustainable development.

Climate action has been defined as stepped up efforts to reduce greenhouse gas emissions and resilience and adaptive capacity to climate-induced impact.[147] Sustainable energy can be defined as energy that is produced using the sun, wind, etc. rather than using fuels which cannot be replaced.[148] Sustainable energy is, therefore, seen as central to climate action, as it reduces, *inter alia*, the emission of greenhouse gasses.[149] There has been an array of efforts to enhance climate action and sustainable energy such as the Paris Agreement[150] and the efforts by the G7.[151]

It has been argued that sustainable development has become the guiding paradigm in the continuing reform of international investment law, as several IIAs have included sustainable development at the center of these legal instruments, providing a greater balance between public policy and investor rights.[152] However, as the main purpose remains the protection of foreign investors in host countries, the significance of the latter has been debated.[153]

145. Oxford Reference, Oxford University Press, sustainable development, https://www.oxfordreference.com/view/10.1093/oi/authority.20110803100544392.
146. Sustainable development goals, United Nations, about the sustainable development goals, https://www.un.org/sustainabledevelopment/sustainable-development-goals/.
147. UNDP, financing solutions for sustainable development, Goal 13: Climate action, https://www.sdfinance.undp.org/content/sdfinance/en/home/sdg/goal-13--climate-action.html.
148. Cambridge Dictionary, Sustainable energy, https://dictionary.cambridge.org/dictionary/english/sustainable-energy.
149. Union of concerned scientists, "Benefits of Renewable Energy Use" (*Union of Concerned Scientists*, July 14, 2008, Updated December 20, 2017) "Increasing the supply of renewable energy would allow us to replace carbon-intensive energy sources and significantly reduce US global warming emissions," https://www.ucsusa.org/resources/benefits-renewable-energy-use.
150. The Paris Agreement, Article 14(3): "... as well as in enhancing international cooperation for climate action," https://unfccc.int/sites/default/files/english_paris_agreement.pdf.
151. L. Mead, "G7 Ministerial Addresses Climate Change, Oceans and Clean Energy" (*IISD—SDG knowledge hub*, September 25, 2018) "21 September 2018: The ... G7 environment, oceans and energy ministers met on the theme, "Working Together on Climate Change, Oceans and Clean Energy." The meeting marked the first time a G7 ministerial meeting linked together the three issues ..." https://sdg.iisd.org/news/g7-ministerial-addresses-climate-change-oceans-and-clean-energy/.
152. Stefanie Schacherer, International institute for sustainable development, October 2018, international investment law and sustainable development: key cases from the 2010s, introduction, https://www.iisd.org/sites/default/files/publications/investment-law-sustainable-development-ten-cases-2010s.pdf.
153. E.g.: Dissenting Opinion of Professor Donald McRae in William Ralph Clayton, William Richard Clayton, Douglas Clayton, Daniel Clayton and Bilcon of Delaware, Inc. v. Canada, para. 51, p. 19: Once again, a chill will be imposed on environmental review panels which will be concerned not to give too much weight to socio-economic considerations or other considerations of the human environment in case the result is a claim for damages under NAFTA Chapter 11," https://www.italaw.com/sites/default/files/case-documents/italaw4213.pdf.

3 Investment Relations Between the EU and Vietnam

The investment relationships between the EU and Vietnam have grown firmly since the first formal diplomatic ties in 1996.[154] The EU is ranked fifth out of Vietnams 80 FDI partners, having invested more than 23.927 billion US dollars at the end of 2018.[155] The most important sectors for FDI are manufacturing, electricity production and transmission, and, real estate, with the Netherlands having the highest number of investors.[156]

The economy of Vietnam is predominantly based on industries which are state-owned, with agriculture employing most of its workforce.[157] While the industrial and service sectors are also significant, the energy sector has notably grown.[158] This is evident in Vietnam, although being fairly new in the oil industry, becoming the third-largest South-Asian oil producer.[159]

While this is true, efforts have been made to rely on other energy sources, such as coal and hydropower.[160] Regarding the former, the presumption is that coal plants will not be affordable in the medium term, which is why some countries might be thinking of increasingly investing in wind and solar energy. The latter is specifically important as the rapid economic growth is expected to result in an energy demand that is eightfold in 2030.[161] Keeping the latter in mind, Vietnam has focused on investments in clean and renewably energy, the diversification of energy sources and energy efficiency.[162] The steps taken are significant and produce opportunities for EU investors

154. Delegation of the European Union to Vietnam, “guide to the EU-Vietnam trade and investment agreements,” updated March 2019, p. 12: “Bilateral trade and investment links between the EU and Vietnam have steadily strengthened since the two sides established formal diplomatic realties in 1996,” https://trade.ec.europa.eu/doclib/docs/2016/june/tradoc_154622.pdf.
155. *Ibid.* p. 18.
156. *Ibid.*
157. Banco Santander, SA, "Vietnamese Economic Outline" (*Santander Trade Markets*, Latest Update: April 2020): Vietnam’s economy is based on large state-owned industries such as ...”; “Agriculture represents 14.7% of GDP and employs 39.4% of the total workforce,” https://santandertrade.com/en/portal/analyse-markets/vietnam/economic-outline.
158. *Ibid.* “Industry contributes 34.2% of GDP and employs 25.8% of the total workforce”; Services represent 45.5% of GDP and employ 34.7% of the total workforce.”
159. *Ibid.* “Despite being a ‘newcomer’ in the oil industry, Vietnam has become the third-largest Southeast Asian producer.”
160. Dutch Ministry of Foreign Affairs, Renewable energy in Vietnam, Embassy of the Netherlands in Hanoi and Consulate General in Ho Chi Minh City, January 2018, p. 2: “The Vietnamese government is committed to the promotion of renewable energy and energy efficiency has taken several measures (including mechanisms, policies, incentives and supporting schemes) to improve the environment for expansion over the last few years,” https://www.netherlandsworldwide.nl/documents/publications/2017/09/13/factsheet-renewable-energy-in-vietnam.
161. *Ibid.* p. 1: “different estimations of energy demand in Vietnam vary from increasing threefold to eightfold from 2015 to 2030.”
162. Kingdom of the Netherlands, "Energy in Vietnam" (*Kingdom of The Netherlands*, 2019) “To ensure that the supply of energy meets the rise in consumption, Vietnamese policy emphasizes the need to diversify the country’s energy mix, whilst maximizing the use of local energy reserves. There has been an increased focus in recent years on increased investment in the clean-technology sector and particularly energy efficiency, renewable energy technologies and waste management” https://www.netherlandsworldwide.nl/countries/vietnam/doing-business/key-sectors/energy.

in a myriad of sectors, however, it has been noted that more changes are needed to attract FDI and expand its scope.[163]

4 Textual Analysis of the EU-Vietnam FTA

This part of the chapter shall look at how the notions sustainable development, and in particular climate action and sustainable energy, are addressed in the EU-Vietnam Free trade and Investment Agreement.[164] In doing so, we shall look into: (1) the IPA,[165] (2) the chapter on sustainability,[166] and (3) the schedules of commitments under the EVFTA.[167]

1) The IPA

There are three articles related to sustainable development. Each of them shall be discussed below.

First, there are two preambular references to sustainable development.[168] It should, however, be kept in mind that preambular references are not binding and can only assist in interpreting the obligations included in the agreement itself.[169] Furthermore, it should be noted that the latter references contain broad, undefined wording from which no obligations stem.

Second, Article 2.2 contains an exception that "reaffirms the right to regulate to achieve legitimate policy objectives, such as the protection of the environment."[170] This exception clarifies that the protection of the environment, which is linked to climate action and therefore sustainable development, may be used. However, such

163. UNCTAD, Investment policy review in Vietnam, 2008, pp. 88-91, https://unctad.org/en/Docs/iteipc200710_en.pdf.
164. Free trade agreement between the socialist republic of Viet Nam and European Union, http://trade.ec.europa.eu/doclib/press/index.cfm?id = 1437; Investment Protection Agreement: The European Union and the EU party (consisting of all the EU Member States) on the one part and the socialist republic of Viet Nam of the other part, http://trade.ec.europa.eu/doclib/press/index.cfm?id = 1437.
165. Investment Protection Agreement The European Union and the EU party (consisting of all the EU Member States) on the one part and the socialist republic of Viet Nam of the other part, http://trade.ec.europa.eu/doclib/press/index.cfm?id = 1437.
166. Free trade agreement between the socialist republic of Viet Nam and European Union, Chapter 13: Trade and Sustainable Development, https://trade.ec.europa.eu/doclib/docs/2018/september/tradoc_157373.pdf.
167. *Ibid.* Chapter 8.
168. *Ibid.* at 3, DETERMINED to strengthen their economic, trade and investment relationship in accordance with the objective of sustainable development, in its economic, social and environmental dimensions, and to promote trade and investment under this Agreement in a manner mindful of high levels of environmental and labour protection and relevant internationally recognised standards and agreements"; "REAFFIRMING their commitments to the principles of sustainable development in the Free trade agreement."
169. United Nations, Vienna Convention on the Law of Treaties, Treaty Series, vol. 1155, Article 31.1: "A treaty shall be interpreted in good faith in accordance with the ordinary meaning to be given to the terms of the treaty and in the light of its object and purpose," https://treaties.un.org/doc/Publication/UNTS/Volume%201155/volume-1155-I-18232-English.pdf.
170. Council Decision 2019/753 of March 30, 2020, Free Trade Agreement between the European Union and the Socialist Republic of Viet Nam, Article. 13, 2020 O.J. (L 186), Chapter 2, Article 2.2.

may not negatively affect the operations of investments or the investor's expectations of profits.[171]

Third, as to the obligation of national treatment (NT),[172] Vietnam is exempted from the latter.[173] In particular, NT shall not apply to areas crucial to sustainable energy such as oil and gas, hydroelectricity and power transmission and/or distribution.[174]

To conclude, the IPA contains a nonbinding preambular reference, an exception in favor of sustainable development, which may only be applied under stringent conditions, and a derogation from NT in sectors crucial to sustainable energy.

2) The chapter on sustainable development

Chapter 13 of the EVTA has the objective of promoting sustainable development by fostering the contribution of trade and investment-related aspects of environmental issues.[175] The articles of this chapter can be distinguished into two categories: (1) where no or very weak obligation is present and (2) where a weak obligation is present. After discussing the latter, I shall briefly look into the (one) exception and the enforcement mechanism of Chapter 13.

As to the first category, it contains articles which do not create any or very weak enforceable obligations upon the parties. It includes the reaffirmation obligations under a myriad of multilateral environmental agreements and recognizing their inter-dependency,[176] as well as the endeavor to facilitate trade and investment relevant to climate mitigation.[177] These obligations form little to no enforceable obligations, due to the broad, ambiguous wording and standards such as "endeavor" and "recognition." What the latter words hold for obligations is not included, nor what would lead to a violation of such.

As to the second category, such contains articles which contain a weak, however, enforceable obligation. Article 13.3 contains the obligations of parties not to waive or derogate or offer to do so, from its environmental or labor laws, in a manner affecting trade and investment between the parties.[178] Reading this article, it should be noted that the scope is very narrow (namely when it affects investment between the parties). It is also unclear what would amount to the latter, and whether there is a de minimis situation.[179] Second, there is an obligation to take into account scientific information

171. *Ibid.*, Article 2.2.4.
172. *Ibid.*, Article 2.3: "Each party shall accord to investors of the other Party and to covered investments, with respect to the operation of the covered investments, treatment no less favorable than it accords, in like situations, to its own investors and to their investments."
173. *Ibid.*, Annex 2: Exemption for Vietnam on National Treatment.
174. *Ibid.*, Annex 2, Article 1 (g) and (h).
175. *Ibid.*, Article 13.1 para. 1.
176. *Ibid.*, Article 13.1 paras. 2-3; Article 13.5 para. 2; Article 13.6 para. 1; Article 13.10 para. 1.
177. *Ibid.*, Article 13.10 para. 2 (c).
178. *Ibid.*, Article 13.3.
179. A legal term meaning too small to be meaningful or taken into consideration, Thomson Reuters, "Glossary: De Minimis" (*THOMSON REUTERS: practical law*) https://uk.practicallaw.thomsonreuters.com/1-382-3382?transitionType=Default&contextData=(sc.Default)&firstPage=true&bhcp=1.

when preparing/implementing measures for the protection of the environment.[180] This obligation, like the previous one, lacks specificity and the standard (to take into account) is ambiguous, as it does not set a floor or ceiling. Lastly, there is an obligation of consultations and cooperation of trade issues related to the environment (where there is mutual interest), as well as the sharing of information, experience and progress to ratifications or amendments to multilateral environmental agreements.[181] While these obligations, although narrow, are a little more specific, its implications to the contribution of sustainable energy are little.

Turning to the exception, Chapter 13 contains only one: not to apply environmental law in a way that it results in arbitrary or unjustifiable discrimination.[182] This, while containing some specificity, does not provide a de minimis or a test to assess the discrimination.

Having discussed that, attention should be drawn to the ability to enforce the obligations set out above. The dispute settlement mechanism and the mediation mechanism included in the EVTA do not apply to the obligations included in this chapter.[183] The only way of enforcing the latter is by consultations followed by a report of recommendations by the committee of trade and sustainable development or a panel of experts, who shall monitor the follow up of such.[184] Needless to say, the enforceability of the already weak obligations is limited and inadequate to effectively contribute to sustainable development.

To conclude, Chapter 13 contains weak obligations for sustainable development, which cannot be effectively enforced.

3) The Schedule of Commitments

This part shall look at the schedules of commitment of the EU and Vietnam.

The schedule of commitments of Vietnam contains two exceptions.

First, measures seeking to ensure the protection of natural resources and the environment may be adopted or maintained and shall not be seen as a violation of market access, whether or not included in the schedule.[185] This gives a broad exemption in favor of the protection of the environment.

Second, Vietnam significantly limits the liberalization in the field of energy, as can be seen in Figure 6.1.[186] This means that there is no obligation in of market access or NT in areas crucial to sustainable energy,[187] thereby limiting new investments in the latter.

180. Council Decision 2019/753 of March 30, 2020, Free Trade Agreement between the European Union and the Socialist Republic of Viet Nam, Article 13, 2020 O.J. (L 186), Article 13.11.
181. *Ibid.*, Article 13.5 paras. 1-3; Article 13.6 paras. 1-2.
182. *Ibid.*, Article 13.1 para. 4; Article 13.5 para. 4.
183. *Ibid.*, Chapter 15 and Annex 15-C.
184. *Ibid.*, Article 13.16 paras. 2-3; Article 13.17.
185. *Ibid.*, Chapter 8: Liberalization of Investment, trade in services and electronic commerce, Annex 8-B: Vietnam's schedule of specific commitments, Appendix 8-B-1: specific commitments on Cross-Border Supply of Services and liberalization of investments, para. 6.
186. *Ibid.*, p. 83.
187. *Ibid.*

Figure 6.1 Vietnam's Liberalization in the Field of Energy

Sector or sub-sector	Description of reservations
5. PRODUCTION; TRANSMISSION AND DISTRIBUTION ON OWN ACCOUNT OF ELECTRICITY, GAS, STEAM AND HOT WATER[58] (excluding Nuclear Based Electricity Generation)	
A. Production of electricity; transmission and distribution of electricity on own account (part of ISIC rev 3.1: 4010)[59]	Unbound.
B. Manufacture of gas; distribution of gaseous fuels through mains on own account (part of ISIC rev 3.1: 4020)[60]	Unbound.
C. Production of steam and hot water; distribution of steam and hot water on own account (part of ISIC rev 3.1: 4030)[61]	Unbound.

58. The horizontal limitation on public utilities applies.

59. Does not include operation of electricity transmission and distribution systems on a fee or contract basis, which are energy services

60. Does not include transportation of natural gas and gaseous fuels via pipelines, transmission and distribution of gas on a fee or a fee or contract basis, and sales of natural gas and gaseous fuels, which are energy services.

61. Does not include transmission and distribution of steam and hot water on a fee or contract basis and sales of steam and hot water, which are energy services.

When discussing the second exemption above, it is important to note that most of the energy production is state-owned, as discussed earlier. Therefore, Chapter 11 of the EVTA[188] should be taken into account. The latter makes clear that the principle of nondiscrimination and transparency shall not apply to a number of areas related to sustainable energy,[189] such as oil and gas exploration or services related to such,[190] hydropower, and, the transmission or distribution of all types of energy, power and alternatives or substitutes for electricity.[191]

Therefore, while the one exception can be interpreted as being in favor of climate action, the second exception, as discussed in the part of the IPA, significantly restricts the liberalization necessary to achieve sustainability of energy.

188. Free trade agreement between the socialist republic of Viet Nam and European Union, Chapter 11: Specific rules for Vietnam on State-owned enterprises, enterprises granted special rights or privileges, and designated monopolies, Annex 11A, https://trade.ec.europa.eu/doclib/docs/2018/september/tradoc_157369.pdf.
189. *Ibid.*, para. 4: "Articles 11.4 (Non-Discrimination and Commercial Considerations) and 11.6 (Transparency) do not apply to the following enterprises, their subsidiaries and successors pursuing the same public mandate, engaged in, and limited to, the activities described below ... [para. 5]."
190. *Ibid.*, para. 5.1: "Vietnam Oil and Gas Group: activities: ... exploration of oil and gas and flight operation, services for oil and gas activities."
191. *Ibid.*, para. 5.2: "Vietnam Electricity and any enterprise: activities: power generations by hydropower, nuclear power ..., transmission; distribution of all types of electricity, power and alternatives or substitutes for electricity."

The EU, on the other hand, contains very little exceptions compared to Vietnam. When analyzing this, it should be kept in mind that the EU consists of Member States, which can have differences in the extent of liberalization.[192]

First, it should be noted that the schedule makes clear that there are no reservations as to environmental services.[193] By doing so, the EU liberalizes the EU market of environmental services, thereby supporting sustainable development.

However, the EU is not bound when it comes to the production of electricity and the transmission and distribution of the latter.[194] This is also true for energy services where the EU has mostly opted-out,[195] apart from a few exceptions in certain EU Member States.[196] By restricting the liberalization of the latter sectors, the EU limits investments which are crucial to sustainable energy.

Thus, while the EU liberalizes environmental services, the liberalization in the energy sector is highly restricted.

The textual analysis of the FTA has, therefore, shown that: (1) the IPA contains nonbinding obligations, derogations from NT and stringent conditions to apply sustainable development rules; (2) Chapter 13 contains weak obligations for sustainable development, which cannot be effectively enforced; and (3) both Vietnam and the EU, while containing some commitments in favor of sustainable development, restrict the liberalization particularly necessary to achieve sustainable energy.

5 The Way Forward

Taking into consideration the aforementioned analysis and recalling the object and purpose of EU BITs, I would like to propose three steps due to which such could contribute more effectively and efficiently to climate action and sustainable energy in the broader context of sustainable development.

First, weak and very weak obligations (as discussed under the textual analysis of Chapter 13) should be made less ambiguous as to form enforceable obligations. Most notably, the weak and unenforceable standard of not lowering one's environmental laws, or offering to do so, is a negative obligation that can have a great impact on the contribution of sustainable development if enforced.

As to enforceability, these standards could be subject to enforceability as it is clear and precise enough and contains no de minimis. In this regard, it should be noted that human rights, a much more controversial aspect related to international investment law, has been included in IIA and are subject to enforceability.[197] This form of

192. *Ibid.*, Chapter 8: Liberalization of Investment, trade in services and electronic commerce, Annex 8-A: the Union's schedule of specific commitments, Appendix 8-A-2, p. 3, para. 9: "The Union takes commitments differentiated by its Member States, where applicable."
193. *Ibid.*, pp. 50-51: environmental services.
194. *Ibid.*, p. 22: production; transmission and distribution of electricity on own account.
195. *Ibid.*, pp. 69-71: 19: energy services.
196. *Ibid.*, p. 70: e.g., sector B: "pipeline transportation of fuels: in CY unbound except for electricity distribution … ."
197. Reciprocal investment promotion and protection agreement between the government of the kingdom of Morocco and the government of the federal republic of Nigeria, Articles 18.2

enforceability, if amended to the particularities of environmental protection while balancing the liberalization of investment, could contribute significantly to climate action and sustainable energy. For example, the very weak obligations in Chapter 13 combined with the proposal above, could be amended in a way in which the interdependent international environmental agreements are to be upheld. This could be done by not lowering the steps taken in pursuit of the latter agreements, and holding investors taking advantage of the lowering of such accountable in the home state.[198]

Third, derogations from standards, such as nondiscrimination and transparency, should be prohibited in sectors crucial to sustainable development. Vietnam is especially desirable for investors in the sectors relating to sustainable energy;[199] however, the derogation from such substantial standards are undesirable and form a hurdle to investments. Taking into consideration the schedule of commitments of Vietnam, the negotiation of such and the energy market, abiding by internationally recognized standards would be a first, but nonetheless important, step for the liberalization, promotion and development of sustainable energy.

VI HOW CAN CITIZENS PLAY A GREATER ROLE IN SHAPING SUSTAINABLE TRADE?

1 Introduction[200]

One of the biggest 21st century challenges for our planet and its habitants is the combat against climate change.[201] As the potentially disastrous effects of global warming seem

("investments shall uphold human rights in the host state") and 20 ("investors shall be subject to civil action for liability ... where such acts or decisions lead to significant damage, personal injuries or loss of life in the host state"), https://investmentpolicy.unctad.org/international-investment-agreements/treaty-files/5409/download.

198. As is done by in the field of human rights in the Reciprocal investment promotion and protection agreement between the government of the kingdom of Morocco and the government of the federal republic of Nigeria (*ibid.*).

199. Dutch Ministry of Foreign Affairs, Renewable energy in Vietnam, Embassy of the Netherlands in Hanoi and Consulate General in Ho Chi Minh City, January 2018, p. 1: "... the Vietnamese government actively supports renewable energy, and the sector is thereby becoming increasingly interesting for Dutch companies and organizations. While realizing challenges, business opportunities can be found in biomass, wind and solar and solar energy."

200. This section draws upon R. Leal-Arcas et al., "Aiming at Sustainable Trade in the Context of the Rule of Law: What Role for Citizens and How International Trade Can Help Reduce Fossil-Fuel Consumption," 11:2 *Trade, Law and Development* 193-238 (2019).

201. Johan Rockström et al., "A Safe Operating Space for Humanity." *Nature* 461, 472-475 (2009), available at https://www.nature.com/articles/461472a; G. Hardin, "The Tragedy of the Commons." 162 *Science* 1243-1248 (1968); D. Wallace-Wells, "The Uninhabitable Earth, Annotated Edition." *New York Magazine* (July 14, 2017), available at http://nymag.com/intelligencer/2017/07/climate-change-earth-too-hot-for-humans-annotated.html; E. Kolbert, *Field Notes from a Catastrophe*. Bloomsbury, 2006 (Updated 2015); K. Emanuel, *What We Know about Climate Change* (MIT Press, 2007); Markowitz and Shariff, "Climate Change and Moral Judgment." 2 *Nature Climate Change* (April 2012); W. Steffen et al., "Planetary Boundaries: Guiding Human Development on a Changing Planet." *Science* (2015); Intergovernmental Panel on Climate Change. *Report on Global Warming of 1.5 Degrees C: Summary for Policymakers* (2018), available at https://www.ipcc.ch/sr15/chapter/summary-for-policy-makers/; U.S.

to appear increasingly imminent, a key concept that has gained more attention in the last few decades is that of sustainable development, i.e., the development that covers present needs sufficiently, without putting in jeopardy the needs of the future generations. Sustainable development entered the world of international law in 1987, through a report named "Our Common Future," produced by the authority formerly known as the World Commission on Environment and Development (today operating as the Brundtland Commission).[202]

Nearly three decades later, in 2015, the United Nations General Assembly adopted the 2030 Agenda for Sustainable Development,[203] a document crystallizing the main targets that have to be met so that this type of development is possible: 17 Sustainable Development Goals (abbreviated as "SDGs"). The G20 also has produced an Action Plan on the 2030 Agenda for Sustainable Development.[204]

2 Defining Sustainable Trade

To achieve the maximum possible amount of clarity, the characteristics that render trade sustainable should be defined. According to the United Nations Conference on Trade and Development (UNCTAD), the SDG with the closest relation to trade is Goal 17—"Strengthen the means of implementation and revitalize the Global Partnership for Sustainable Development."[205] In particular, there are three of its by-targets that have been categorized by the 2030 Agenda as "trade-related," on which the UNCTAD, in collaboration with the International Trade Centre (ITC) and the World Trade Organization (WTO), has an obligation to provide and update indicators:[206]

- Target 17.10: Promote a universal, rules-based, open, non-discriminatory and equitable multilateral trading system under the World Trade Organization, including through the conclusion of negotiations under its Doha Development Agenda.
  - Indicator 17.10.1: Worldwide weighted tariff-average.

Global Change Research Program. Executive Summary, Climate Science Special Report: 4th National Climate Assessment (NCA4), vol. 1, available at https://science2017.globalchange.gov/downloads/CSSR_Executive_Summary.pdf; Scripps CO2 Program, "Keeling Curve Lessons," available at http://scrippsco2.ucsd.edu/history_legacy/keeling_curve_lessons; C. Mooney, "The Hockey Stick: The Most Controversial Chart in Science, Explained." *The Atlantic* (May 10, 2013), https://www.theatlantic.com/technology/archive/2013/05/the-hockey-stick-the-most-controversial-chart-in-science-explained/275753/; Zachos et al., "Trends, Rhythms, and Aberrations in Global Climate 65 Ma to Present." *Science* 292, 686 (2001).

202. World Commission on Environment & Development, "Our Common Future," Annex 2, U.N. Doc. A/42/427 (1987), available at http://www.un-documents.net/our-common-future.pdf.
203. UN General Assembly, "Transforming our world: the 2030 Agenda for Sustainable Development," October 21, 2015, A/RES/70/1, available at http://www.un.org/ga/search/view_doc.asp?symbol=A/RES/70/1&Lang=E.
204. https://www.b20germany.org/fileadmin/user_upload/G20_Action_Plan_on_the_2030_Agenda_for_Sustainable_Development.pdf.
205. UNGA, "Transforming our world: the 2030 Agenda for Sustainable Development," p. 14.
206. United Nations Conference in Trade and Development (UNCTAD), "Trade and the Sustainable Development Goals (SDGs)," https://unctad.org/en/Pages/DITC/Trade-Analysis/TAB-Trade-and-SDGs.aspx.

- Target 17.11: Significantly increase the exports of developing countries, in particular with a view to doubling the least developed countries' share of global exports by 2020.
  - Indicator 17.11.1: Developing countries' and least developed countries' share of global exports.
- Target 17.12: Realize timely implementation of duty-free and quota-free market access on a lasting basis for all least developed countries, consistent with World Trade Organization decisions, including by ensuring that preferential rules of origin applicable to imports from least developed countries are transparent and simple, and contribute to facilitating market access.
  - Indicator 17.12.1: Average tariffs faced by developing countries, least developed countries and small island developing States.[207]

Moreover, the 2030 Agenda for Sustainable Development further stresses the worth of international trade as an instrument toward the fortification of sustainability in paragraph 68:

> International trade is an engine for inclusive economic growth and poverty reduction, and contributes to the promotion of sustainable development. We will continue to promote a universal, rules-based, open, transparent, predictable, inclusive, non-discriminatory and equitable multilateral trading system under the World Trade Organization, as well as meaningful trade liberalization. We call upon all members of the World Trade Organization to redouble their efforts to promptly conclude the negotiations on the Doha Development Agenda. 19 We attach great importance to providing trade-related capacity-building for developing countries, including African countries, least developed countries, landlocked developing countries, small island developing States and middle-income countries, including for the promotion of regional economic integration and interconnectivity.[208]

Another IGO that has been a staunch supporter of trade being a pillar of sustainable development is the European Union (EU). The EU has not only adopted the UN's 2030 Agenda and its SDGs in its community policies, but also contributed in its very creation. Further than that, its trade policy recognizes the value of "social justice, respect for human rights, high labour standards and high environmental standards,"[209] all fundamental elements for sustainability. This is evident especially in the inclusion of sustainable development clauses in the trade agreements it has signed in the last decade, and the conduction of sustainability impact assessments preceding their signing.

207. UN "SDG Indicators—Metadata Repository," https://unstats.un.org/sdgs/metadata/?Text=&Goal=17&Target=.
208. UNGA, "Transforming our world: the 2030 Agenda for Sustainable Development," p. 29.
209. EU trade policy and sustainable development," http://ec.europa.eu/trade/policy/policy-making/sustainable-development/.

3 The Example of the Climate-Change and Energy Regimes

The climate change regime, although originally structured in a top-down approach, did acknowledge the importance of citizen awareness and participation from its early stages, namely in the Principle 10 of the Rio Declaration (1992)[210] and in Articles 4 and 6 of the United Nations Framework Convention on Climate Change, or UNFCCC (1992).[211] Furthermore, in the closely knit to climate change field of environmental law, the Aarhus Convention of 1998 also recognizes public participation as a significant procedural environmental right.[212]

The major step to achieve a relevant democratization in climate change, however, took place much more recently, with the signing of the Paris Agreement[213] by the parties of the UNFCCC in 2015.[214] The preamble and Article 12 reaffirm the duty of

210. UN General Assembly, "Rio Declaration on Environment and Development," available at https://www.un.org/documents/ga/conf151/aconf15126-1annex1.htm.
211. UN General Assembly, "United Nations Framework Convention on Climate Change" (May 9, 1992), S. Treaty Doc No. 102-38 (1992), 1771 UNTS 107, available at https://unfccc.int/resource/docs/convkp/conveng.pdf.
212. "Convention on Access to Information, Public Participation in Decision-Making and Access to Justice in Environmental Matters" ("Aarhus Convention") 2161 UNTS 447 (1999), available at https://www.unece.org/fileadmin/DAM/env/pp/documents/cep43e.pdf.
213. For an analysis of the Paris Agreement and climate change in general, *see* William Nordhaus, *Climate Casino*, Yale University Press, 2013; ES. "History of UN Climate Change Talks, available at https://www.c2es.org/content/history-of-un-climate-talks/; IPCC, *Fifth Climate Change Assessment Report: Synthesis Report*. 2014, available at http://www.ipcc.ch/pdf/assessment-report/ar5/syr/AR5_SYR_FINAL_SPM.pdf; Daniel Bodansky and Lavanya Rajamani, "Evolution and Governance Architecture of the Climate Change Regime," in Detlef Sprinz and Urs Luterbacher (eds.), *International Relations and Global Climate Change: New Perspectives*, 2nd edition, MIT Press, 2016; Daniel Esty and Peter Boyd, "To Move Paris Accord Forward, Bring Cities and Companies on Board," *Yale Environment 360*, 2018, available at https://e360.yale.edu/features/to-move-paris-accord-forward-bring-cities-and-companies-on-board; Daniel Esty and Dena Adler, "Changing International Law for a Changing Climate," *American Journal of International Law*, 2018, available at https://www.cambridge.org/core/services/aop-cambridge-core/content/view/DB50564B97DEE8620D818BFFCB9BBC94/S2398772318000764a.pdf/changing_international_law_for_a_changing_climate.pdf; Decision of the Parties to the 2015 Paris Conference of the Parties (COP21), available at https://unfccc.int/sites/default/files/english_paris_agreement.pdf; UN Environment, "Emissions Gap Report." 2018, available at https://www.unenvironment.org/resources/emissions-gap-report-2018; "Essential Elements of the Paris 'Rulebook'," Center for Climate and Energy Solutions, November 2018, available at https://www.c2es.org/site/assets/uploads/2018/11/essential-elements-paris-rulebook.pdf; Ted Nordhaus, "The Two-Degree Delusion: The Dangers of an Unrealistic Climate Change Target," *Foreign Affairs*, February 8, 2018, available at https://www.foreignaffairs.com/articles/world/2018-02-08/two-degree-delusion; Jules Kortenhorst, "The Paris Agreement: 10 Key Takeaways For the Global Energy Landscape," Rocky Mountain Institute, 2015, available at https://rmi.org/news/blog_2015_12_18_the_paris_agreement_10_key_takeaways/; "Adapt, Curb, Engage: 21 Solutions to Protect Our Shared Planet," Cities4 Climate, 2015, available at https://envirocenter.yale.edu/sites/default/files/files/21%20Solutions%20to%20Protect%20Our%20Shared%20Planet.pdf.
214. "Paris Agreement" (December 13, 2015), in UNFCCC, Report of the Conference of the Parties on its Twenty-First Session, Addendum, at 21, UN Doc. FCCC/CP/2015/10/Add. 1 (January 29, 2016), available at https://unfccc.int/sites/default/files/english_paris_agreement.pdf (text of the Paris Agreement) or https://www.un.org/en/development/desa/population/migration/generalassembly/docs/globalcompact/FCCC_CP_2015_10_Add.1.pdf (text of the COP21 Report).

states to guarantee public awareness, participation and access to information; while in Article 16§8 nongovernmental bodies and agencies are provided the opportunity to gain observer status in the meetings of the Conference of the Parties or COP, i.e., the supreme decision-making body of the UNFCCC.

What is more, the Paris Agreement adopted a structure for states' response to climate change that was drastically different from the legal instruments that preceded it: it introduced a bottom-up approach, in addition to the until-then-dominant top-down. As its name suggests, the bottom-up approach (named "the conceptual breakthrough" of the Agreement, although this idea of a harmonious, truly global framework was conceived earlier, in the rather-less-popular Copenhagen Conference of 2009) is translated into the ability of the states to define their climate change targets and actions by their own national policy, instead of them being dictated from an international authority.[215] What is more, customers, shareholders, employees, and other forms of bottom-up governance are asking companies to take bolder action in the fight against climate change. Ordinary citizens are also demonstrating on the streets, asking their local governments to take the fight against climate change more seriously.[216] Equally, Europe's carmakers are now taking seriously their plans for electric cars[217] to comply with stricter EU rules on GHG emissions from cars.[218]

Further to that, the COP21 Decision to which the Paris Agreement is annexed urges "nonparty stakeholders," a term that encompasses, according to its preamble, "civil society, the private sector, financial institutions, cities and other subnational authorities, local communities and indigenous peoples," to also contribute to the

215. Daniel Bodansky, "The Paris Climate Change Agreement: A New Hope," 110 *American Journal of International Law*. 288, 289-292 (2016).

216. *See* "Protest in Madrid as conservatives suspend ban on most polluting cars," *Reuters*, June 29, 2019, available at https://www.reuters.com/article/us-spain-environment-protests/protest-in-madrid-as-conservatives-suspend-ban-on-most-polluting-cars-idUSKCN1TU0VS?utm_source=EURACTIV&utm_campaign=c51561b51f-EMAIL_CAMPAIGN_2019_06_05_08_04_COPY_01&utm_medium=email&utm_term=0_c59e2fd7a9-c51561b51f-114690739.

217. On electric cars and electrification more generally, *see* Paige Jadun et al. "Electrification Futures Study," NREL, 2018, available at https://www.nrel.gov/docs/fy18osti/70485.pdf; Ravi Manghani and Rory McCarthy, "Global Energy Storage: 2017 Year-in-Review and 2018-2022 Outlook," GTM Research, 2018, available at https://www.greentechmedia.com/research/report/global-energy-storage-2017-year-in-review-and-2018-2022-outlook#gs.sgLYFwQ; Chris Nelder, James Newcomb, and Garrett Fitzgerald, "Electric Vehicles as Distributed Energy Resources," Rocky Mountain Institute, 2016, available at https://www.rmi.org/wp-content/uploads/2017/04/RMI_Electric_Vehicles_as_DERs_Final_V2.pdf; Robert Walton, "Short term pain for long term gain: EPRI assesses expanding US electrification," Utility Dive, 2018, available at https://www.utilitydive.com/news/short-term-pain-for-long-term-gain-epri-assesses-expanding-us-electrificat/521496/; Bill Chappell, "Electric Cars Hit Record In Norway, Making Up Nearly 60 Percent Of Sales In March," NPR, April 2019, available at https://www.npr.org/2019/04/02/709131281/electric-cars-hit-record-in-norway-making-up-nearly-60-of-sales-in-march; Kenneth Gillingham and James Sweeney, "Barriers to Implementing Low Carbon Technologies," *Climate Change Economics* (2012); Julia Pyper, "Utilities See Distributed Generation as a Challenge—And Owning It as the Solution," Greentech Media, 2016, available at https://www.greentechmedia.com/articles/read/utilities-see-distributed-generation-as-a-challenge-and-owning-it-as-the-so#gs.KlLKMj0; Eric Wanless and Kelly Carlin, "Growing the Minigrid Market in Sub-Saharan Africa," RMI, 2017, available at https://rmi.org/news/growing-minigrid-market-sub-saharan-africa/.

218. The Economist, June 29, 2019, p. 10.

efforts of the states to effectively tackle climate change, and invites them to join, for optimum coordination, the Non-state Actor Zone for Climate Action platform (NAZCA).[219] In 2018, this platform had a total number of 12,396 stakeholders, comprised of 9,378 cities, 126 regions, 2,431 companies, 363 investors and 98 civil society organizations.[220]

In parallel, but prompted by the Paris Agreement and the process around it, there has been happening a remarkable mobilization of citizens, citizen groups and citizens' representatives that takes place beyond the UN climate regime.[221] A special mention should be made for the various mayors' initiatives that have taken place in some of the world's largest cities, such as the Compact of Mayors and the Covenant of Mayors, merged, as of 2016, into the Global Covenant of Mayors for Climate and Energy, which represents hundreds of millions of citizens.[222]

Of course, those actions, along with the initiatives of private actors, such as businesses of various sizes, do not have the binding character that the actions of national and international authorities have. However, one could argue that, because of the way authorities are elected, citizens usually are more directly represented by their regional authorities than their national ones, and therefore are more likely to comply with possible commitments to combat climate change if they come from the former. It should also be noted that, due to the way the market functions, the pressure of the public opinion and fellow companies is often as efficient as legislation in making businesses adopt ecological standards.[223] And, although the reputation of big businesses has been damaged by the 2008 financial crisis, inequality, and environmental harm,[224] and that rapid economic expansion is "squeezing the planet dry,"[225] there is

219. COP21 Report, preamble and §117, 118, 133, 134, available at https://www.un.org/en/development/desa/population/migration/generalassembly/docs/globalcompact/FCCC_CP_2015_10_Add.1.pdf.
220. Global Climate Action—NAZCA homepage, http://climateaction.unfccc.int/.
221. All of this raises the question of global environmental governance. For further discussion, *see* Katherine Michonski and Michael A. Levi, "Harnessing International Institutions to Address Climate Change," CFR, 2010, available at http://www.cfr.org/climate-change/harnessing-international-institutions-address-climate-change/p21609; Robert Keohane and David Victor, "Cooperation and discord in global climate policy," *Nature Climate Change* (2016); Michele M. Betsill and Desirée Fiske, "International Climate Change Policy: Complex Multilevel Governance," in Regina S. Axelrod and Stacy D. VanDeveer (eds.) *The Global Environment: Institutions, Law, and Policy*, Sage Publishing, August, 2019; "Explainer: Dealing with the 'loss and damage' caused by climate change." Carbon Brief, 2017, available at https://www.carbonbrief.org/explainer-dealing-with-the-loss-and-damage-caused-by-climate-change; "Building Toward Breakthrough: Energizing the Paris 2015 Climate Negotiations and Post-Paris Action Agenda Through Broader Engagement." Yale Center for Environmental Law and Policy, (2015), available at https://envirocenter.yale.edu/sites/default/files/yale_climate_change_dialogue_white_paper.pdf; Daniel Esty and Maria Ivanova, "Revitalizing Global Environmental Governance: A Function-Driven Approach," in Daniel Esty and Maria Ivanova (eds.), *Global Environmental Governance: Options and Opportunities*, 2002.
222. Global Covenant of Mayors for Climate & Energy, https://www.globalcovenantofmayors.org/.
223. Daniel Bodansky, Jutta Brunée and Lavanya Rajamani, *International Climate Change Law*, OUP 2017, 264-265.
224. The Economist, "Popenomics," September 7, 2019, p. 62.
225. *Ibid.*

empirical evidence that it is possible to grow economically and greenly. A case in point is China's economy, whose annual GDP has been growing exponentially since the late 1970s, but its CO_2 emissions have not grown at the same pace, which might be due to China's recent green growth and investment in renewable energy.[226]

In contrast to the above, in the energy regime, the empowerment of citizens is driven not by international legal instruments, but mostly by the progress of technology, and more specifically the technology related to the production of renewable energy, which can be exported from the West to the rest. Thanks to it, a new trend has emerged: the energy prosumers. A term coming from the words producer and consumer, prosumers are defined as "self-generating energy providers, whether households or energy communities. Individuals contribute to the energy supply in their vicinity via their own installed renewable energy capacity, more often than not solar roofing, wind energy, or combined heat and power."[227] They may be functioning independently from each other, within their own households, or they can be organized in a communal way, forming energy cooperatives.[228]

This development in a traditionally centralized sector like energy is unprecedented—it means an empowerment of citizens and, at the same time, a movement further away from the dependence to fossil fuels, enhancing energy security.[229] Especially in the solar energy field, the technological advances in photovoltaic (PV) panels and in batteries design and production have rendered the use of PV

226. *See* the chart in The Economist, "Climate-Change Targets: From Smog to Slog," September 21, 2019, pp. 47-48, at 48.

227. Rafael Leal-Arcas, Feja Leniewska and Filippos Proedrou, "Prosumers as New Energy Actors," in M. Mpholo, D. Steuerwald and T. Kukeera (eds.), *Africa-EU Renewable Energy Research and Innovation Symposium* 2018 (RERIS 2018), Springer Proceedings in Energy, Springer, Chapter 12, 2018, 1.

228. Rafael Leal-Arcas, Feja Leniewska and Filippos Proedrou, "Prosumers: New Actors in EU Energy Security" (2017) Queen Mary University of London School of Law Legal Studies Research Paper No. 257/2017, 14.

229. On the links between national (energy) security, geopolitics, and climate change, *see* Jennifer Perron, "Fatih Birol and David Victor on the geopolitics of energy," *Brookings Institution* (July 2, 2018), available at https://www.brookings.edu/blog/planetpolicy/2018/07/02/fatih-birol-and-david-victor-on-the-geopolitics-of-energy/; Umair Irfan, "106 Lawmakers-Including 11 Republicans-Tell Trump Climate Change Is a National Security Threat," *Vox* (January 12, 2018), available at https://www.vox.com/energy-and-environment/2017/12/18/16791106/white-house-climate-change-national-security-strategy-threat-military; Tony Capaccio, Jennifer Dloughy, and Ari Natter, "Defense Department Warns about Climate Change Impacts to Armed Forces and Bases," *Time* (January 18, 2019), available at http://time.com/5507465/climate-change-impact-armed-forces-bases/; "Climate Change and Security: Strengthening Resilience to Climate-Fragility Risks," *UN Environment Programme*, available at https://postconflict.unep.ch/publications/ClimateChange_Security_twopager.pdf; Richard Morningstar, "Energy and Geopolitics: A Diplomat's View," *Atlantic Council (speech)* (October 12, 2018), available at https://www.atlanticcouncil.org/news/transcripts/energy-and-geopolitics-a-diplomat-s-view; Varun Sivaram, "Clean Energy Might Reduce Global Warming, But What Will It Do to Geopolitics?," *Council on Foreign Relations*, 2018, available at https://www.cfr.org/blog/clean-energy-might-reduce-global-warming-what-will-it-do-geopolitics; Ernest Moniz, "Shifting Dynamics Across the Energy Landscape" (August 6, 2018), available at http://energypolicy.columbia.edu/dr-ernest-moniz-shifting-dynamics-across-energy-landscape; Sarah Light, "Valuing National Security: Climate Change, the Military, and Society," 61 *UCLA Law Review* (2014), available at https://www.uclalawreview.org/pdf/61-6-3.pdf; James Woolsey, "Energy and National Security," 2010. Link: https://www.youtube.com/watch?v=12kXsu35PWA;

panels in households more accessible than ever.[230] Moreover, the past decade has seen the cost of wind energy go down by 50% and that of solar energy has dropped by more than 80%.[231] So there is hope for renewable energy.

The European Union, proving its commitments to sustainable development, recognized the importance of the trend, dedicating two articles of the currently in force Directive on renewables-derived energy to renewables self-consumers and renewable energy communities, and thus embracing their bottom-up logic as part of its energy policy.[232] In the same directive, it is stressed how instrumental it is for energy consumers to be provided with information for optimizing their energy performance. As observed above regarding the climate change regime, this is an absolutely fundamental element in order to achieve true citizen empowerment and democratization.

4 Citizen Empowerment and Sustainable Trade

If in the aforementioned regimes of climate change and energy, that have historically been operating in a top-down fashion, decentralization and citizen participation are gaining increasingly more ground, rendering them more robust, one could wonder if that could become the new reality in the trade regime as well, as part of its movement toward sustainability. After all, every single person's daily reality is affected in innumerable ways by the trade of goods and services, yet it remains, to a large extent, defined by diplomats.

That feature on its own is not damning, since a certain degree of expertise is necessary for building efficient trade agreements; however, viewing trade through a solely technocratic and maybe even elitist prism is admittedly possible to deprive it of its connection to the very people it directly affects, and therefore its democratic character. It is this absence of citizens from the negotiations that is deemed to be the reason behind the Doha Round's inability to reach an agreement, almost two decades after its beginning in 2001.[233]

Robinson Meyer, "Does Climate Change Cause More War?," *The Atlantic*, 2018, available at https://www.theatlantic.com/science/archive/2018/02/does-climate-change-cause-more-war/553040/; Amy Larkin, "Energy Independence, National Security, and Another Reason for Sustainability," *The Guardian*, 2014, available at https://www.theguardian.com/sustainable-business/2014/jun/04/energy-independence-national-security-sustainability; "The Economist Special Report: The Geopolitics of Energy," *The Economist* (March 17, 2018), pp. 312 ff; Daniel Yergin, *The Quest*, Penguin Books, 2011, Chapters 13-14, 33; Henry Tricks, "The New Power Superpowers," *The Economist* (March 17, 2018), available at https://www.economist.com/special-report/2018/03/15/clean-power-is-shaking-up-the-global-geopolitics-of-energy.

230. Wolf-Peter Schill, Alexander Zerrahn, and Friedrich Kunz, "Prosumage of Solar Electricity: Pros, Cons, and the System Perspective" (January 2017), DIW Berlin Discussion Paper No. 1637, 4.
231. The Economist, "Environmental Policy: The Great Divide," June 29, 2019, pp. 34-35, at 35.
232. Directive (EU) 2018/2001 of the European Parliament and of the Council of 11 December 2018 on the promotion of the use of energy from renewable sources, available at https://eur-lex.europa.eu/legal-content/en/TXT/?uri=CELEX:32018L2001, Articles 21, 22.
233. Rafael Leal-Arcas, "Empowering citizens for common concerns: Sustainable energy, trade and climate change," 6:1 *GSTF Journal of Law and Social Sciences* 31-32 (2017).

First and foremost, being pragmatic, it is impossible to demand an exact replication of the conditions under which the reshaping of climate change and energy happened, as, although those regimes are often closely intertwined with trade, it remains very different from them, mainly because of the WTO's *modus operandi* and its heavy reliance on a number of efficiency-restrictive procedural rules (e.g., the "consensus" decision-making). What actually could be possible would be to pursue a proportional adoption of some of the ideas that have rendered the other processes successful, as far as citizen awareness, mobilization and participation is concerned.

In a similar way to how nongovernmental citizen groups were legally given observer status and were recognized as stakeholders in the Paris process, citizen groups could be given at least observer roles in trade agreements negotiations, and have the ability, if not to participate, then to at least lobby with policy makers and communicate their interests to them. Since those procedures are often criticized for their information asymmetry between citizens and authorities, as well as their lack of transparency, it would be a most welcome offer.

This would, of course, mean a modification of the procedural rules of the WTO, if the negotiations involve multilateral agreements, while if the agreements are regional trade agreements (RTAs) or free trade agreements (FTAs), and therefore their provisions not subject to its authority, it could be realized through rather simpler procedures. A recent positive example of an RTA that took a step toward this direction with its legal provisions is the Comprehensive Economic and Trade Agreement (CETA) between EU and Canada, which entered provisionally into force on September 21, 2017. The CETA contains specific provisions on trade and sustainable development (Chapter 22), trade and labor (Chapter 23) and trade and environment (Chapter 24), promoting in each of them transparency of information, public debate and public awareness on the relevant existing standards (Chapter 22§2, 23§6, 24§7).[234]

As NAZCA was created for digitally cataloguing the various civil society stakeholders for climate action, an online platform under the auspices of the WTO could undertake this function on behalf of citizen groups in pursue of sustainable trade. Public information campaigns, conducted by either the WTO or countries entering into RTAs or FTAs, would also be particularly helpful to those citizens who are not necessarily organized in groups, for various reasons (e.g. living within considerable distance from the headquarters of the closest one available). It should not be neglected to carefully process the way the messages of those campaigns are communicated to their recipients: there should not be an abundance of technical terms, that would impair their comprehensibility, but the language should not be overly simplistic either.

Furthermore, the Internet could be utilized in order to achieve the maximum possible mobilization from the group of citizens most likely to be affected by the adverse effects of climate change: the younger members of the society. Since a significant amount of social interaction takes place via various social media platforms, and the younger citizens have always been the ones more familiarized with technology, it would be of no trivial importance if the WTO or the state authorities that are

234. European Commission, "CETA chapter by chapter," available at http://ec.europa.eu/trade/policy/in-focus/ceta/ceta-chapter-by-chapter/.

involved in the construction of a trade agreement could make use of those platforms. Their presence could extend from merely uploading facts that could be of interest regarding upcoming negotiations and agreements to conducting online surveys and Q&A or debate sessions that can keep the audiences, especially the young ones, thought-provoked, involved and aware.

It is an approach that has until now served well the UN and its agencies (active in as many as eleven different social media platforms[235]), the World Bank (six different platforms[236]), and NGOs such as Client Earth, WWF and the Climate Group (active in at least three different social media platforms[237]). The EU, for example, has shown willingness to engage in this activity, by creating a designated Twitter account within its online platform regarding the proposed Transatlantic Trade and Investment Partnership (TTIP) between it and the US, to facilitate the citizens' information and interaction with the decision-making bodies.[238]

Sustainability in trade can be further supported via citizens through their role as consumers. When a choice of purchasing a good or a service is made, the consumer supports financially its provider; which is why environmentally and socially aware citizens are more frequently and pressingly trying to trace how sustainable those providers are.[239] However, given the plethora of goods necessary for a modern household or workspace, this effort seems understandably daunting.

This is why eco-labelling, i.e., the practice of labelling products and services based on their environmental impact, is of paramount importance, so that consumers can show their preferences and support sustainable businesses more efficiently. Further to that, there has been a general interest in labelling goods according to the processes and production methods (PPMs), e.g., if they were produced in an energy-efficient way, or if genetically modified organisms (GMOs) were involved in the process.[240]

A number of such labels have been introduced by various countries and countries' associations, such as the EU Ecolabel, the US Energy Star, the Nordic Swan and others.[241] However, they should be popularized even more if consumers are to exercise their role more actively. Of course, it is not only the authorities in this case that have to be involved for the benefit of the citizens empowerment, but also the private companies that are called to demonstrate the necessary good will to prioritize sustainability over their financial benefit, since usually the most ecological way to operate is rarely the most profitable one.

---

235. "The UN on Social Media," https://www.un.org/en/social-media/index.html.
236. "The World Bank Group—Follow Us," https://www.worldbank.org/.
237. Client Earth https://www.clientearth.org/, "WWF—Follow Us" https://www.wwf.org.uk/, The Climate Group https://www.theclimategroup.org/.
238. EU TTIP web page, http://ec.europa.eu/trade/policy/in-focus/ttip/have-your-say/.
239. Olivier De Schutter, *Trade in the Service of Sustainable Development*, Hart, 2015, 152.
240. Organization for Economic Cooperation and Development (OECD), "Processes and Production Methods (PPMs): Conceptual framework and considerations on use of PPM-Based trade measures," OCDE/GD(97)137 (Paris, 1997), 7.
241. "EU Ecolabel," http://ec.europa.eu/environment/ecolabel/; "Energy Star," https://www.energystar.gov/ and "Nordic Swan Ecolabel" https://www.nordic-ecolabel.org/.

The aforementioned suggestions to increase citizen participation in the shaping of sustainable trade have a twofold goal: first, to contribute to bringing sustainability to a sector often accused of undermining it; and second, to augment the feeling of participation in a democratic procedure regarding processes directly affecting citizens' lives. Since the first goal is more likely to be measurable in the long run, it remains to be examined how the citizen participation could be measured.

One idea could be to utilize the same online platform that could serve as catalogue of the citizen groups functioning as stakeholders in order to conduct surveys measuring the groups' satisfaction from the information they are receiving and their participation in the procedures. To reach a broader sample of citizens, trustworthy survey-conducting authorities, such as the Eurobarometer within the EU,[242] could be chosen for the same purpose.

VII CONCLUSION

Regarding how globalization has affected international trade and climate change impacts on major economies, the economic means of the major economies of the world and their efforts should be maximized to facilitate and help the world overall, not just within their own domain. Evidently, this is not the precedence of most nations. Due to political disagreements and differences, there has recently been a shift to a more hostile and sovereign attitude. Those who want the benefits of globalization should share and resolve the problems that it comes with.

We conclude that it is possible to reduce energy dependence and grow macro-economically and that nature protection does not necessarily have to translate into being the opposite of economic production. But there needs to be a greater effort from the major economies and the world's biggest GHG emitters not to simply reach set targets. Undoubtedly, they must combine financial efforts with knowledge and technology for the benefit of the rest of the world. It is not enough for nations to continue to operate their climate preservation within their borders anymore[243] or stay within the comfortability of belonging to an organization for a higher political merit. However, this is easier said than put into practice. It is clear from many events, both recent and past, that there has always been a cloud of cautiousness over almost every nation to contribute collectively with finances, political power, technology until it is absolutely necessary. Therefore, such a global effort is unlikely to happen, unless perhaps a disastrous event occurs.

On one hand, globalization and the expansion of international trade have, undoubtedly, damaged our environment and increased global warming. On the other hand, globalization and free trade have played a major role in the enrichment of life conditions and social conditions of billions of people across the world. The point here is not to blame any part of the system, but to highlight responsibilities and to explore

242. "European Commission—Public Opinion," http://ec.europa.eu/commfrontoffice/publicopinion/index.cfm.
243. F. Branger and P. Quirion, "Climate Policy and the 'Carbon Haven' Effect," WIREs Climate Change 2014, 5:53-71. DOI: 10.1002/wcc.245, p. 54.

the possibilities offered in order to fight global warming and to adopt a sustainable development. Most likely, youth will vote in liberal democracies to change politics toward more sustainability since there is ever more recognition of the issue.

On the role of the EU and the WTO in climate change, if the EU and the WTO have, in our opinion, damaged environment through the years, they have at their disposal formidable tools in order reverse the trend. The harmonization of European rules in matter of environment, the adoption of ambitious standards in international treaties and the creation of climate clubs seem to offer some hope.

Some would argue that international based solutions, requiring consensus, are not the best solutions as they will always be less ambitious than unilateral ones. We have argued that the only sustainable solution that would allow the creation of a new logic of trade and production taking into account environment is required at an international scale. The bottom-up approach of the Paris Agreement offers wide possibilities for parties to act unilaterally, bilaterally or multilaterally in order to tackle global warming. In a world where supply chains are highly vulnerable, protectionism, economic nationalism, and mistrust are not the solution.

To tackle global climate change, international agreements were created; however, they were not always successful. The Paris Agreement is one such agreement. Even if it has shown the unity of the world in the climate change war, it still just a promise. There is no concrete and suitable plan shown in the Paris Agreement. The NDCs model is just the way that can drag every nation to agree on the accord. With an alternative type of cooperation, the climate club seems to be more functional than the world accord. With the incentive-based system, the club can directly tackle environmental issues while promoting international trade and investment. However, to be a success the climate club needs the strong founder countries that can make visible incentive and punishment.

By being called the leader in the field of environmental protection, the EU has already shown the effort by making the policies and the strategies that seem to be effective. Therefore, it is not difficult to prove that with the sustainable-development mindset, the EU has become the world leader for fighting global climate change.

As for BITs, having looked at the notions of climate action and sustainable energy, the investment relationship between the EU and Vietnam, and having done a textual analysis of the IPA and the EVTA, it can be concluded that EU IPAs have, till date, not sufficiently addressed the issues of sustainable development and in particular those related to climate action and sustainable energy. In particular, the IPA contained several nonbinding obligations, derogations from NT and stringent conditions to apply sustainable development rules, the chapter on sustainable development contained weak and not sufficiently enforceable obligations, and the liberalization of areas related to sustainable development in the broader context are limited.

While recognizing the latter, we have proposed three ways in which this IIA could, however, be used as a vehicle to contribute to climate action and sustainable energy as IIAs have, although arguably, been contributing to market-orientated reforms and domestic standards, such as transparency.

By revising weak and ambiguous standards, including an enforcement mechanism used for other controversial aspects of investment law, and, upholding internationally recognized standards, EU IIAs could contribute significantly to climate action, sustainable energy and sustainable development.

Technology is a great enabler for education, climate change mitigation, access to energy for all, and for getting the benefits of international trade and investment. It democratizes the system to make it inclusive, irrespective of the economic background. One can learn from anywhere and therefore have a prosperous future thanks to digital transformation. We need to shift from the future we fear to the future we want.

Regarding the role of citizens in sustainable trade, in summation, it is indeed possible that citizens could aid in providing sustainable trade with a structure closer to bottom-up, to an extent, similar to how it has happened in the climate change and energy regime, although not identically, due to the particular nature of the WTO. Substantial weight should be given to the use of online resources, namely the internet and the social media, as well as the labelling of products and services.

Observing the steps that already have been taken toward this direction, it is clear that the countries of the Western, developed world are the ones dominating the field at the moment, represented mainly by the European Union and the USA. However, for truly achieving democratization, it should be stressed that no country or group of countries should be left behind, and therefore the engagement of all states should be a priority.

Finally, citizen empowerment for the shaping of sustainable trade will not happen unilaterally. It demands the coordinated efforts of national and international political and legal authorities on the one hand, since they are the ones assigned with the creation, evolution and enforcement of law and policy, and of the civil society on the other, since it an essential factor in the shaping of the above. In the hope of a more democratic and sustainable future, it would be ideal that they meet halfway. To that, we should add optimism resulting from the inter-generational perception of climate change:[244] just over half of Americans aged 55 and over believe climate change is

244. For analyses of climate change perceptions, *see* Yale Program on Climate Change Communication, "Politics and Global Warming." Executive Summary (December 2018), available at http://climatecommunication.yale.edu/publications/politics-global-warming-december-2018/2/; William Nordhaus, *The Climate Casino*. "Climate Science and Its Critics," Chapter 24, pp. 293-326, Yale University Press, 2013; Renee Lertzman, "In Climate Change, Psychology Often Gets Lost in Translation," *Pacific Standard*, 2015, available at https://psmag.com/environment/in-climate-change-psychology-often-gets-lost-in-translation; Cameron Brick and Sander van der Linden, "Yawning at the Apocalypse," The Psychologist: Climate Change (September 2018), pp. 30-34, available at https://thepsychologist.bps.org.uk/volume-31/september-2018/yawning-apocalypse; Hiroko Tabuchi, "In America's Heartland, Discussing Climate Change Without Saying 'Climate Change'," *The New York Times*, 2017, available at https://www.nytimes.com/2017/01/28/business/energy-environment/navigating-climate-change-in-americas-heartland.html; Shannon Hall, "Exxon Knew about Climate Change Almost 40 Years Ago," *Scientific American* (October 26, 2015); Campbell and Kay, "Solution Aversion: On the Relation Between Ideology and Motivated Disbelief," *Journal of Personality and Social Psychology* 107, 809 (2014); Marx et al. "Experiential and Analytic Processing of Uncertain Climate Information," 17 *Global Environmental Change* 47-58 (2007); Suzanne Goldberg, "Work of Prominent Climate Change Denier Was Funded by Energy Industry," *The Guardian*, 2015, available at https://www.theguardian.com/environment/2015/feb/21/climate-change-denier-willie-soon-funded-energy-industry; Julia Corbett, "Media Power and Climate Change," *Nature Climate Change* (2015); Maxwell Boykoff and Gesa Luedecke, "Elite News Coverage of Climate Change," *Climate Change Communication* (2016); Dan Kahan, "Climate-Science

anthropogenic and worry about it, whereas in the case of Americans aged 18-34, it is 75%.[245] That is great news.

Communication and the Measurement Problem," 36 *Advances in Political Psychology* 1-43 (2015), available at https://ssrn.com/abstract=2459057; Connie Roser-Renouf et al., "Mes sage Strategies for Global Warming's Six Americas," *Handbook of Environment and Commun ication*, 2014, Yale Program on Climate Change Communications, available at http://climatec ommunication.yale.edu/publications/global-warmings-six-americas-book-chapter-preview/.

245. The Economist, "Climate Change: States' Rights," June 29, 2019, pp. 13-14, at 14.

Conclusion

Few people in the rich world may want to volunteer to reduce their living standards. At the same time, it is difficult to ask people from parts of the world that are still developing to sacrifice their chance to become rich. A middle-ground compromise could be technology that brings economies away from using fossil fuels as their principal source of energy. Technology could be instrumental to understand the links between international trade, sustainable development, and environmental protection, which have become increasingly important to understand how we can reach a sustainable future. We know by now that international trade can help decarbonize the economy and create jobs. This duality can be reached through the liberalization of trade in green goods, the circular economy, border carbon adjustment mechanisms (BCAM), or the blue economy, to mention but a few.

The WTO Agreement clearly states in its preamble that the international community should pursue free trade in the context of sustainable development. As a result, the future of the world economy cannot be separated from the future of the environment. This means that we need to reimagine the rules of international trade (whether multilateral, regional, bilateral) for sustainable development. It also means that free trade needs to be consistent with environmental protection. This book concludes that potential trade concerns should not be an obstacle for the formation of a club of carbon markets in a climate club.

One wonders whether existing WTO rules are sufficient for the envisaged transformation towards a low-carbon society in the context of the EU's Green Deal or is it necessary to have a new agreement/interpretation of existing rules (e.g., on CBAM or subsidies)? If this is a necessary step, should it be part of an overall WTO multilateral reform package discussion or a separate discussion with a possible plurilateral agreement? If it is a plurilateral agreement outside the WTO, would that pose a risk for the most-favored-nation principle, which is a core principle in WTO law?

Going forward, the international community should aim at concluding agreements to eliminate barriers and tariffs on green goods; it should eliminate barriers to trade in environmental services; it should put an end to fisheries subsidies; and both the WTO and the World Bank should work together to phase down and gradually out fossil fuel subsidies, which are diametrically opposed to climate change mitigation.

Moreover, the WTO rules (and other rules of international trade) should be drafted through the prism of sustainable development to serve the needs of the 21st century. This may only be achieved via plurilateral agreements, not multilaterally, since the WTO has proven time and time again that it is not feasible. For instance, a group of like-minded countries could take this initiative. In addition, carbon border adjustment mechanisms throughout the world may help put a price on carbon via a carbon tax. With climate a major global priority, it remains to be seen whether the EU will disrupt the global trading system as it will inevitably implement carbon border adjustment taxes.

In recent years, and certainly since the collapse of multilateral trade negotiations in 2008, we have seen the rise of FTAs. Recent FTAs have environmental chapters that promote environmental protection. A case in point is USMCA, although its deficiency is that anything related to climate change is omitted in the agreement. The good news is that many countries are promoting climate change-related technology and many governments throughout the world would like to see a pro-climate agenda in their trade policies (largely because it is in their interest—both in terms of health for their citizens and economic sustainability—to do so). That may mean trade restrictions as part of climate change mitigation measures.

The future trade agenda is full of mega-trends, side-effects of geopolitical conflicts (like that between the US and China, which will most likely be the most important bilateral relationship in the world for years to come in fields such as energy security, international trade, climate change, or finance, to name but a few),[1] it is about e-commerce (especially when it comes to data flows), as well as open, sustainable, and assertive trade. More cooperation in areas such as climate change and public health is urgently required. The recovery from covid-19 will imply the interaction between international trade policy and domestic policies and that trade policy is an enabler of other domestic policies. Countries are prioritizing the implementation and enforcement of their FTAs, especially the sustainability commitments in FTAs. A case in point where sustainability is at the heart of trade policy is the EU-Mercosur FTA, as without sustainability clauses, there would be no political support in the EU for the ratification of this FTA as of early 2021. In fact, the notion of sustainability is present in all labor and environmental protection chapters in recent EU FTAs.

Equally, an increasing number of countries are aiming at carbon neutrality by 2050 or 2060 (which is perceived as benign unilateralism) as well as greater integration of trade policy with other domestic policies (such as sustainability—in its three dimensions, namely development, environmental, and social—and the digital economy). Similarly, governments could enact policies that greener consumption. Putting a price on carbon is a good way forward, making sure that companies and consumers pay for their emissions.

1. President Xi of China speaks of "ecological civilization" to show its assertiveness in becoming a climate leader. Areas for potential cooperation with the US are, among others, carbon capture and storage, hydrogen power, and the development of green financial instruments to fund such cooperation. That said, Chinese leaders seem cautious in their bilateral relations with the West, which they see as a region of the world in economic decline and political instability, as opposed to the economic rise and social stability of China.

Lastly, we should not underestimate the phenomenal positive impact green technology can have on climate change mitigation. In recent times, there have been scientists stating negative views on the future of humanity and that humans should forget about solving the climate crisis and, instead, invest their efforts and money in migrating from Earth to other planets. However, technology evolves very rapidly. The current debate is that green hydrogen will guarantee a sustainable future for our planet. For instance, hydrogen sustains three times as much energy as kerosene and is lighter. In addition, public spending in research and technology is growing in most OECD countries, and more and better subsidies for R&D may take place.[2]

Therefore, there are many reasons to believe that tomorrow's technology will be able to tackle climate change effectively, especially if we continue to invest in green technology. In addition, many countries increasingly have green policies. Technological advancement is a clear example of human progress and, as a result, governments and companies should aim at the promotion of green technology to fight climate change. This can be done with the creation and proliferation of climate clubs, whether for countries or companies.

2. The Economist, "The Roaring 20s?," January 16, 2021, p. 7.

Index

D

E

K

L

M

W

ENERGY AND ENVIRONMENTAL LAW & POLICY SERIES

1. Stephen J. Turner, *A Substantive Environmental Right: An Examination of the Legal Obligations of Decision-makers towards the Environment*, 2009 (ISBN 978-90-411-2815-7).
2. Helle Tegner Anker, Birgitte Egelund Olsen & Anita Rønne (eds), *Legal Systems and Wind Energy: A Comparative Perspective*, 2009 (ISBN 978-90-411-2831-7).
3. David Langlet, *Prior Informed Consent and Hazardous Trade: Regulating Trade in Hazardous Goods at the Intersection of Sovereignty, Free Trade and Environmental Protection*, 2009 (ISBN 978-90-411-2821-8).
4. Louis J. Kotzé and Alexander R. Paterson (eds), *The Role of the Judiciary in Environmental Governance: Comparative Perspectives*, 2009 (ISBN 978-90-411-2708-2).
5. Tuula Honkonen, *The Common but Differentiated Responsibility Principle in Multilateral Environmental Agreement's: Regulatory and Policy Aspects*, 2009 (ISBN 978-90-411-3153-9).
6. Barbara Pozzo (ed.), *The Implementation of the Seveso Directives in an Enlarged Europe: A Look into the Past and a challenge for the Future*, 2009 (ISBN 978-90-411-2854-6).
7. Henrik M. Inadomi, *Independent Power Projects in Developing Countries: Legal Investment Protection and Consequences for Development*, 2010 (ISBN 978-90-411-3178-2).
8. Nahid Islam, *The Law of Non-Navigational Uses of International Watercourses: Options for Regional Regime-Building in Asia*, 2010 (ISBN 978-90-411-3196-6).
9. Yasuhiro Shigeta, *International Judicial Control of Environmental Protection: Standard Setting, Compliance Control and the Development of International Environmental Law by the International Judiciary*, 2010 (ISBN 978-90-411-3151-5).
10. Katleen Janssen, *The Availability of Spatial and Environmental Data in the European Union: At the Crossroads between Public and Economic Interests*, 2010 (ISBN 978-90-411-3287-1).
11. Henrik Bjørnebye, *Investing in EU Energy Security: Exploring the Regulatory Approach to Tomorrow's Electricity Production*, 2010 (ISBN 978-90-411-3118-8).
12. Véronique Bruggeman, *Compensating catastrophe victims: A Comparative Law and Economics Approach*, 2010 (ISBN 978-90-411-3263-5).
13. Michael G. Faure, Han Lixin & Shan Hongjun, *Maritime Pollution Liability and Policy: China, Europe and the US*, 2010 (ISBN 978-90-411-2869-0).

14. Anton Ming-Zhi Gao, *Regulating Gas Liberalization: A Comparative Study on Unbundling and Open Access Regimes in the US, Europe, Japan, South Korea and Taiwan*, 2010 (ISBN 978-90-411-3347-2).
15. Mustafa Erkan, *International Energy Investment Law: Stability through Contractual Clauses*, 2011 (ISBN 978-90-411-3411-0).
16. Levente Borzsa´k, *The Impact of Environmental Concerns on the Public Enforcement Mechanism under EU law: Environmental protection in the 25th hour*, 2011 (ISBN 978-90-411-3408-0).
17. Tarcísio Hardman Reis, *Compensation for Environmental Damages under International Law: The Role of the International Judge*, 2011 (ISBN 978-90-411-3437-0).
18. Kim Talus, *Vertical Natural Gas Transportation Capacity, Upstream Commodity Contracts and EU Competition Law*, 2011 (ISBN 978-90-411-3407-3).
19. WangHui, *Civil Liability for Marine Oil Pollution Damage: A Comparative and Economic Study of the International, US and Chinese Compensation Regime*, 2011 (ISBN 978-90-411-3672-5).
20. Chowdhury Ishrak Ahmed Siddiky, *Cross-Border Pipeline Arrangements: What Would a Single Regulatory Framework Look Like?*, 2012 (ISBN 978-90-411-3844-6).
21. Rozeta Karova, *Liberalization of Electricity Markets and Public Service Obligations in the Energy Community*, 2012 (ISBN 978-90-411-3849-1).
22. Sandra Cassotta, *Environmental Damage and Liability Problems in a Multilevel Context: The Case of the Environmental Liability Directive*, 2012 (ISBN 978-90-411-3830-9).
23. Mark Wilde, *Civil Liability for Environmental Damage: Comparative Analysis of Law and Policy in Europe and US*, 2013 (ISBN 978-90-411-3233-8).
24. Bernard Taverne, *Petroleum, Industry and Governments: A Study of the Involvement of Industry and Governments in Exploring for and Producing Petroleum*, 2013 (ISBN 978-90-411-4563-5).
25. Anton Ming-Zhi Gao & Chien Te Fan (eds), *Legal Issues of Renewable Energy in the Asia Region: Recent Developments in a Post-Fukushima and Post-Kyoto Protocol Era*, 2014 (ISBN 978-90-411-4856-8).
26. Sabina Manea, *The Instrumentalization of Property: Legal Interests in the EU Emissions Trading System*, 2014 (ISBN 978-90-411-5420-0).
27. Joseph A. Tolorunse, *Protection of Property Rights in Discovered Petroleum Reservoirs*, 2014 (ISBN 978-90-411-5604-4).
28. Katelijn Van Hende, *Offshore Wind in the European Union: Towards Integrated Management of Our Marine Waters*, 2015 (ISBN 978-90-411-5613-6).
29. Ken'ichi Matsumoto & Anton Ming-Zhi Gao (eds), *Economic Instruments to Combat Climate Change in Asian Countries*, 2015 (ISBN 978-90-411-5408-8).

30. Eduardo G. Pereira & Damilola S. Olawuyi (eds), *Practical Considerations to Negotiate an Enforceable Joint Operating Agreement under Civil Law Jurisdictions*, Second Edition, 2020 (ISBN 978-94-035-0664-7).
31. Ying Shen, *China's Way to Carbon Emissions Reduction: The Choice of Regulatory Instruments and Its Legal Challenges*, 2015 (ISBN 978-90-411-6049-2).
32. Roy Andrew Partain, *Environmental Hazards from Offshore Methane Hydrate Operations: Civil Liability and Regulations for Efficient Governance*, 2017 (ISBN 978-90-411-8730-7).
33. Anton Ming-Zhi Gao & Chien-Te Fan (eds), *The Development of a Comprehensive Legal Framework for the Promotion of Offshore Wind Power: The Lessons from Europe and Pacific Asia*, 2017 (ISBN 978-90-411-8397-2).
34. Tade Oyewunmi, *Regulating Gas Supply to Power Markets: Transnational Approaches to Competitiveness and Security of Supply*, 2018 (ISBN 978-90-411-9869-3).
35. Huseyin Cagri Corlu, *Application of Anti-manipulation Law to EU Wholesale Energy Markets and Its Interplay with EU Competition Law*, 2018 (ISBN 978-90-411-9603-3).
36. Kaisa Huhta, *Capacity Mechanisms in EU Energy Law: Ensuring Security of Supply in the Energy Transition*, 2019 (ISBN 978-94-035-1451-2).
37. Maria João C. Pereira Rolim, *Reconciling Energy, the Environment and Sustainable Development: The Role of Law and Regulation*, 2019 (ISBN 978-94-035-1461-1).
38. Eduardo G. Pereira, Alexandra Wawryk, Heike Trischmann, Catherine Banet & Keith B. Hall (eds), *The Regulation of Decommissioning, Abandonment and Reuse Initiatives in the Oil and Gas Industry: From Obligation to Opportunities*, 2020 (ISBN 978-94-035-0693-7).
39. Rasa Engstedt, *Handbook on European Nuclear Law: Competences of the Euratom Community under the Euratom Treaty*, 2021 (ISBN 978-94-035-2830-4).
40. Cinzia Caporale, Ilja Richard Pavone & Maria Pia Ragionieri (eds), *International Food Law: How Food Law can Balance Health, Environment and Animal Welfare*, 2021 (978-94-035-1761-2).
41. Rafael Leal-Arcas, *Climate Clubs for a Sustainable Future: The Role of International Trade and Investment Law*, 2021 (ISBN 978-94-035-3715-3).